ACS SYMPOSIUM SERIES **622**

Nanotechnology

Molecularly Designed Materials

Gan-Moog Chow, EDITOR
Naval Research Laboratory

Kenneth E. Gonsalves, EDITOR
University of Connecticut

Developed from a symposium sponsored
by the Division of Polymeric Materials:
Science and Engineering, Inc.
at the 210th National Meeting
of the American Chemical Society,
Chicago, Illinois,
August 20–24, 1995

American Chemical Society, Washington, DC 1996

Library of Congress Cataloging-in-Publication Data

Nanotechnology: molecularly designed materials / Gan-Moog Chow, editor, Kenneth E. Gonsalves, editor.

    p.   cm.—(ACS symposium series, ISSN 0097–6156; 622)

"Developed from a symposium sponsored by the Division of Polymeric Materials: Science and Engineering, Inc., at the 210th National Meeting of the American Chemical Society, Chicago, Illinois, August 20–24, 1995."

Includes bibliographical references and indexes.

ISBN 0–8412–3392–6

1. Nanostructure materials.

I. Chow, Gan-Moog, 1957– . II. Gonsalves, Kenneth E. III. American Chemical Society. Division of Polymeric Materials: Science and Engineering, Inc. IV. American Chemical Society. Meeting (210th: 1995: Chicago, Ill.) V. Series.

TA418.9.N35N357   1996
620.1′1—dc20
                                      96–6023
                                           CIP

This book is printed on acid-free, recycled paper.

Foreword

THE ACS SYMPOSIUM SERIES was first published in 1974 to provide a mechanism for publishing symposia quickly in book form. The purpose of this series is to publish comprehensive books developed from symposia, which are usually "snapshots in time" of the current research being done on a topic, plus some review material on the topic. For this reason, it is necessary that the papers be published as quickly as possible.

Before a symposium-based book is put under contract, the proposed table of contents is reviewed for appropriateness to the topic and for comprehensiveness of the collection. Some papers are excluded at this point, and others are added to round out the scope of the volume. In addition, a draft of each paper is peer-reviewed prior to final acceptance or rejection. This anonymous review process is supervised by the organizer(s) of the symposium, who become the editor(s) of the book. The authors then revise their papers according to the recommendations of both the reviewers and the editors, prepare camera-ready copy, and submit the final papers to the editors, who check that all necessary revisions have been made.

As a rule, only original research papers and original review papers are included in the volumes. Verbatim reproductions of previously published papers are not accepted.

ACS BOOKS DEPARTMENT

Contents

SEMICONDUCTORS, METALS, AND NANOCOMPOSITES

CERAMICS AND SOL–GELS

INDEXES

Preface

MOLECULARLY DESIGNED NANOSTRUCTURED MATERIALS have been investigated for more than a decade with a wide range of experimental methods. Disciplines including chemistry, biology, physics, materials science, engineering, and medicine have contributed to recent advances in the control of the unique properties of nanostructured materials by rational design in synthesis and processing. Specific research topics covered in this book include vapor-phase synthesis; metal colloids in polymers and membranes; semiconductors, metals, and nancomposites; and ceramics and sol–gel-derived materials. This volume will be a valuable resource for scientists, engineers, and technologists interested in learning how to apply their disciplines to this emerging field.

We acknowledge the encouragement and generous support for the symposium upon which this book is based provided by Charles Y-C Lee of the Materials and Chemistry Division of the Air Force Office of Scientific Research, Hoechst Celanese Corporation, and the ACS Division of Polymeric Materials: Science and Engineering, Inc.

GAN-MOOG CHOW
Naval Research Laboratory
Code 6930
Center for Bio/Molecular Science
   and Engineering
Washington, DC 20375

KENNETH E. GONSALVES
Institute of Materials Science
University of Connecticut
Storrs, CT 06269

November 29, 1995

Chapter 1

Nanoscience and Nanotechnology: A Perspective with Chemistry Examples

W. M. Tolles

Naval Research Laboratory, Code 1007, 4555 Overlook Avenue, S.W., Washington, DC 20375–5321

Selected programs and opportunities in nanoscience are reported from a recent tour of 42 laboratories in Europe. Some additional perspective gained from the Naval Research Laboratory as well as other U.S. programs is included. Those aspects of nanoscience offering opportunities for fabrication through lithography and molecular self-assembly are emphasized. Behavior such as molecular switching is reviewed. Goals for information storage and retrieval should be recognized in terms relative to existing capabilities today. Potential applications in areas of electronics, optics, and materials are offered as examples of technological interest.

The definition of the word "nanoscience" or "nanotechnology" is not always clear. The terms are best reserved for phenomena associated with structures roughly in the 1-100 nm size range where the properties are of interest due to the size of the structure, and are typically different than those of a molecule or a comparable bulk material. Advances in fabrication of small features with lithography and, alternatively, in preparing structures with comparable feature sizes by methods of self-assembly, provide a number of opportunities for chemistry. This author spent six months working out of the Office of Naval Research in Europe on a focused technology assessment in the area of nanoscience and nanotechnology (1). During this period 42 laboratories in eight different countries were visited. Research programs related to nanostructures in various academic disciplines were viewed with the goal of determining advances and emerging directions. Europe has maintained substantial research in solid state chemistry and physics, and has enjoyed three Nobel Prizes within the last dozen years in research related to nanostructures. The first of these was that for the discovery of the Quantum Hall Effect, awarded to Klaus von Klitzing at the Max Planck Institute, Stuttgart, Germany. The second is for the discovery of the scanning tunneling microscope (STM) awarded to Heine Rohr and Gerd Binnig at IBM, Zurich, Switzerland. The third was shared by Jean-Marie Lehn at the Collège de France in Paris, France, along with two American chemists, (D. J. Cram, UCLA, and C. J. Pedersen, DuPont). Stimulation and growth from these seminal research efforts have expanded in many directions, and account for a number of quality research programs in the field of nanostructures today.

Fabrication and Lithography

Lithographic fabrication of features for chips remains an exceptionally important frontier with substantial barriers beyond 0.18 micron. Massively parallel methods for irradiating and aligning such dimensions are critical to writing chips containing 10^{12} bits in a reasonable period of time. The question of future lithographic processes that will be used is a critical question filled with uncertainty (Broers, A. N., *to be published*).

Resist Processes. Research in resist processes continues to provide additional necessary refinements (2) by understanding the chemistry responsible for feature definition. Etching a semiconductor introduces damage sites below the surface. Some model experiments have determined the depth of this damage by measuring the conductance as a function of the width of a structure (3). Without damage, the conductance should extrapolate to nearly zero at zero width. The experimental plot, however, produces a functional dependence parallel to that of an undamaged structure. By noting the extrapolated width at zero conductance, the depth of the damaged surface is determined. Selected etching conditions can have less than 2 nm damage, or as much as 20 nm (4-5). Electron cyclotron resonance (ECR) tends to damage surfaces excessively (it is Cl^- that does the damage). Magnetically confined plasma etching produces relatively smooth surfaces. At power levels of less than 15 watts, low semiconductor damage occurs since the principal constituent is the chlorine molecule rather than the chloride ion. It appears that methane/hydrogen gas is a useful "universal etch" for II-VI materials (6).

Very Thin Resists. Research involving resist processes continues to reveal new and innovative approaches to extending the limits of lithography. Several relatively new and innovative methods of obtaining higher resolution include the use of very thin or monomolecular layers. Silylation of thin resist layers has the advantage that it is a straightforward extension of commonly used resists with an additional silylation step. Very thin or monomolecular layers are able to define smaller features due vertical wall irregularity, characteristic of commonly-used resists. Future fabrication techniques are likely to consider these methods of defining feature sizes less than 0.1 micron; it is not clear which approach will best demonstrate the advantages required for the demands of that future era.

One technique is the use of a self-organized monolayer of a chlorosilane (7). By exposing this monolayer to radiation (photons, electrons, or ions), adhesion of a catalyst used for electroless metal deposition allows patterning of metallic features (8-9) with 20 nm feature sizes (10) using electrons emitted from a scanning tunneling microscope tip. This process has been extensively studied, and is also capable of fabricating a wide variety of surfaces having chemical and physical properties in a defined pattern.

Chemical changes induced on passivated silicon have also recently been found to introduce a convenient method for fabricating high resolution features (11-12). Electrons remove hydrogen atoms from the passivated silicon surface, allowing oxidation and the formation of a silicon oxide layer able to withstand subsequent etching. Exposing the surface to an atomic force microscope with a voltage applied, along with a subsequent ion etch, leads to linewidths of approximately 10 nm (13). This approach has been used to fabricate silicon field-effect transistors (FETs) (14) with feature sizes as small as 30 nm. This method of writing on passivated silicon with tunneling tips has been used at Philips Laboratory in Eindhoven, The Netherlands, where such patterns have been successfully written on amorphous silicon layers. This process may be used to deposit silicon on metals with a subsequent transfer of the patterns to the underlying metals (15).

Molecular Self-Assembly

The subject of molecular structure is relatively well-studied and understood, however structures of molecules larger than a nanometer have been generally considered of biological interest. Larger molecules originating from cluster studies, such as the derivatives of fullerenes, approach one nanometer in dimensions, and are the subject of many research studies due to their novelty. Molecular self-assembly uses the interactive forces of solid state lattice structures, chemical bonds, and van der Waals forces to form larger aggregates of atomic or molecular units with specific geometries. Molecular recognition leads to molecular aggregates demonstrating an approach to the design of a wide variety of nanostructures. This "bottoms up" approach can be used to make nanostructures that are identical to one another (a truly monodisperse sample size). Most fabrication techniques such as lithography result in a degree of polydispersivity due to the statistical nature of the chemical changes brought about by the exposure/development process.

It is conceivable that the "bottoms up" approach may lead to some functionality currently being pursued by the "top down" approach that has been so successful in the past several decades. The challenge remains to fabricate nanostructures having desirable properties from purely chemical forces. These are of interest mainly due to either chemical properties (dominated by reactive groups at the surface) or physical properties that are different from those of molecular or bulk materials.

Amphiphillic materials. The formation of nanostructures from the interaction of amphiphillic molecules has been well documented (16). At Shell Research B.V., Amsterdam, it is found that the formation of structures involving surfactants and oil in water may be modeled surprisingly well using relatively simple assumptions about the electrostatic interaction (17). Amphiphillic molecules represent a class of surfactants allowing water and oil to mix through the formation of emulsified clusters. The manner in which these amphiphillic molecules provide the stimulus for emulsification has been modeled, with some interesting prediction of nanostructures through self-assembly as well as predictions of macroscopic behavior of commercial importance.

A simplified model was assumed in which two "blocks" were used: 1) a block attracted to water, and 2) a block attracted to oil. A Leonard-Jones potential was assumed for the oil-oil and water-water interactions, but only the repulsive potential function was assumed for the oil-water interaction. The degree of prediction available from this simple model is impressive (18-25). The formation of a bilayer through the attraction of the "hydrophobic hydrocarbon tails" was clear with a surfactant molecule involving one hydrophilic head group. However, if the number of hydrophilic head groups was increased to several "blocks," different shaped clusters appeared. In some cases the geometry of the shapes of this surfactant material became elongated (cigar shaped). Further, it was predicted that tubular or cylinder structures should be particularly effective in emulsification processes. Experimental verification of this behavior was obtained, leading to a factor of approximately ten increase in the emulsification rates.

Organic Thin Films. Molecular thin films (nanodimensional in one dimension) are of interest for several reasons in addition to use as alternatives to conventional resists. Langmuir Blodgett (LB) films have lateral and vertical order. The need for vertical and/or lateral order should be assessed based on the requirements envisioned for a particular system. For applications such as a resist or insulator, lateral order does not seem to add advantages unless sub-nanometer resolution is desired. Vertical order has been useful in determining the interaction length between neighboring molecules through physical measurements as a function of distance, but is not necessary for many of the thin layer approaches being studied today.

Substituted Alkanes on a Surface. Chemisorbed monolayer films of thiols on gold have lateral order as they form epitaxially on the substrate (26), and a number of variations have followed these initial observations. Alkyl derivatives such as thiols have been examined with scanning probes on a well-formed graphite substrate at temperatures just above the melting point of the thiol. At these temperatures a molecular layer is immobilized on the substrate with well-defined patterns that may be observed with STM (27-35). The systems are prepared either in the neat melt form or in a solution with a solvent present.

Patterns observed at the Universität Mainz show clear crystalline arrangements with the head groups aligned in various geometries. The alignment variations correspond to different two-dimensional phases of the molecules in the graphite environment Some molecular arrangements show Moire patterns when the natural two-dimensional crystalline spacings of the molecule do not quite match that of the underlying base structure. Molecular dynamics simulations assuming the Leonard-Jones potential give a fairly good representation of the patterns observed. The model even predicts the tilting of adjacent molecules, as observed experimentally.

Philips Laboratory in Eindhoven has examined self-assembled monolayers of aliphatic thiols on gold. Microscopic observations revealed apparent "holes" in the coatings fabricated. Under higher resolution with the STM probe, the holes are really domains with a different 2-dimensional structure (36).

Organic Thin Film Transistors. The idea of using conducting organic polymers for the components of a transistor occurred to a number of researchers. These have been studied extensively in Thiais, France (37-43). The types of organic materials introduced in these "organic transistors" included polyacetylenes, polypyrroles (44) and polythiophenes (45-51), with many variations involving chain length, side-groups, and methods of depositing film (evaporation, electrolytic growth, spin coating, etc.). The design of these transistors resembled the standard MISFET, with a gate voltage applied through an insulator, typically silica (initially). Current-voltage curves were obtained that characterized these devices, and which soon demonstrated where the utility may be advantageous and where certain limitations were to be found. The carrier mobility for many of these films typically was on the order of 10^{-7} to 10^{-2} cm^2/volt-second (as compared with approximately 1000 for silicon), and this limited high frequency performance. By careful selection of fabrication techniques and of gate insulating materials, mobilities of somewhat greater than unity were obtained - still shy of semiconductor performance. The advantages for these devices appear to be due to 1) the ease of preparation (no high vacuum or expensive equipment was needed), 2) the mechanical flexibility of the transistors once they were fabricated, and 3) the potential use of these materials as sensors. By introducing ion-specific groups (such as crown ethers) this transistor-like behavior demonstrates sensitivities to ions or materials for which it has specifically been designed. They become equivalent to Ion Selective Field Effect Transistors (ISFETs), where the sensor is simultaneously the gate of the transistor.

Organic Monolayers for Electrical Insulation. Thin layers of organic molecules chemisorbed on silicon have been shown to have excellent insulating and dielectric breakdown strengths. Monolayer films of siloxanes have formed high dielectric strength films for insulation in microelectronic devices (52). Insulating layers of silica in 0.1 micron semiconductors approach 5 nm, close to the limit expected for good performance of silica as an insulator. Recently they have demonstrated breakdown voltages of up to 12 MV/cm, indicating that a 1.9 nm monolayer may perform on a par with 5 nm silica insulating layers (53). Also, recent results indicate that the barrier for electron transfer across these organic monolayers is 4.5 eV, considered to be a very high for hydrocarbon monolayers (Boulas, C.; Davidovits, J. V.; Rondelez, F. and Vuillaume, D. *to be published*).

Clusters. Forming a monodisperse condensed-phase sample having large molecular weights can represent a challenge. Molecular recognition may be used to form units having high molecular weights (54). Chemical methods of forming monodisperse assemblies of clusters should lead to interesting material properties, not the least of which is the unusual optical behavior recently recognized.

Optical Properties of Clusters. Recently optical luminescence of InP clusters has revealed surprisingly narrow (less than one millielectron volt) spectral lines (55). Photoluminescence and cathodoluminescence has been used at the University of Lund, Sweden to excite InP imbedded between layers of GaInP (Carlsson, N.; Seifert, W.; Petersson, A.; Castrillo, P.; Pistol, M. E. and Samuelson, L., *to be published*). STM has been used to excite some of these materials (56). The strained InP layer with approximately 10 monolayers of thickness spontaneously reforms into 100 nm quantum dot structures. These quantum dots luminesce at 1.6 to 1.85 eV and have demonstrated line widths of less than 0.1 meV at 77 K.

Quantum dots have been fabricated at the Technische Universität Berlin by growing one Angstrom layers of InAs/GaAs using MBE methods. The product is a pattern of well-shaped quantum dots 12 nm in dimension with a size dispersion of 20%. Luminescence of these structures at 1.1 eV has a width of 0.06 eV. The wavelength of maximum luminescence shifts with changing particle size (obtained through a change in processing conditions). The overall width is limited by the polydispersivity of the sample. A most interesting spectral pattern is obtained by looking at a very small portion of the sample at the "high energy side" of the sample with 42 nm spatial resolution (in effect, probing a very small number of particles) (57). The spectrum reveals a number of resolved peaks, each due to a single quantum dot having a 0.17 nm line width (the spectral resolution of the instrument) at a wave length of 880 nm.

The above behavior suggests that the spectral purity of a truly monodisperse sample of quantum dots could exhibit a most interesting strong and very narrow absorption peak.

Routes to Monodispersivity: Large Molecules or Clusters. True monodispersivity may be obtained with mass spectrometric separation. Although only microscopic samples are likely to be prepared by this approach, a program at Cavendish Laboratory in Cambridge, England is examining this method to prepare clusters with a monodispersivity of one part in 500.

One approach to obtaining a monodisperse sample of high molecular weight is that of preparing dendrimers (58), of which one example is illustrated in Figure 1. In this preparation each reaction leading to the final product adds a number of molecular fragments to the existing molecular framework with an overall molecular weight of M^n, where M is the molecular weight of the fragment and n is the number of synthesis steps. Large molecules have been synthesized at Eidgenoessische Technische Hochschule (ETH) Zurich around a basic porphyrin unit and are soluble even with a molecular weight of 19,000.

Routes to Monodispersivity: Molecular Recognition. Molecular recognition provides a basis large complex nanostructures based on non-bonding attractive forces between molecules. Many examples of this have appeared in the literature (59), largely with organic molecules or complexes containing a few inorganic ions. A relatively recent structure combines the chelating ability of linearly-arranged organic binding sites with inorganic ions to form complexes having a relatively large number of metal ions. This is illustrated in Figure 2. A complex consisting of six units of 6,6'-bis[2-(6-methylpyridyl)]-3,3'-bipyridazine and nine silver ions indicates this

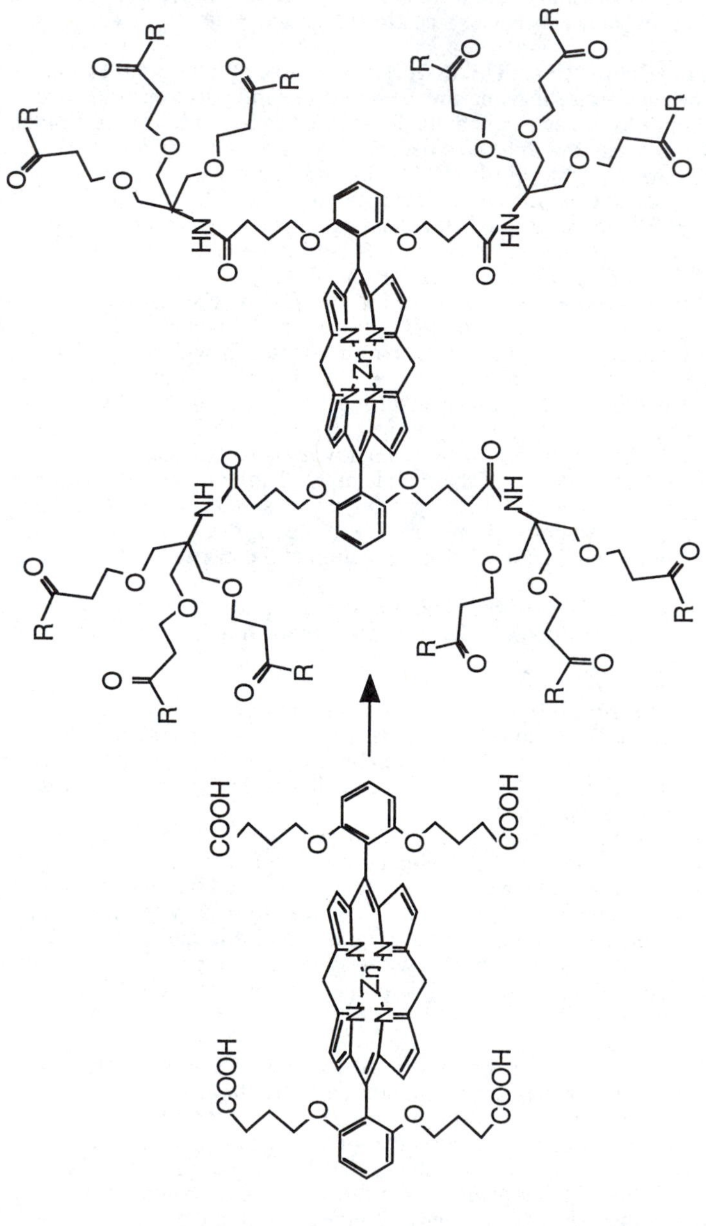

Figure 1. Illustration of first step in building a dendrimer.

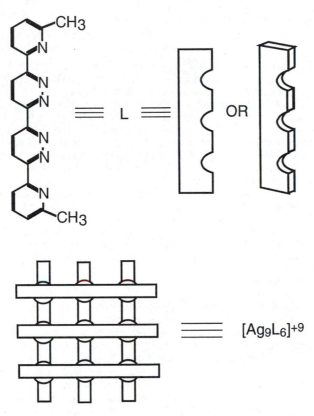

Figure 2. Illustration of cluster formation with high percentage of metallic ingredient.

unusual structure (60). The interesting observation about these structures is the relatively large nucleus of inorganic atoms/ions in a compact structure, possibly suitable for further processing towards the objective of a monodisperse sample of clusters.

The base pairs of DNA have demonstrated how amide linkages may provide larger structures through molecular recognition. The full spectrum of base pairs may be synthetically fabricated. Using the complementary base pairs (including those not found in nature with DNA) unusual geometrical shapes (such as cubes) have been fabricated by reacting base pairs that bind selectively (61-62). This represents an alternative means of building nanostructures using the "bottoms up" approach.

Routes to Monodispersivity: Small Crystallites. Spin-coating methods containing precursor salts can produce very small crystallites on the surface of a silicon wafer (63-64). Compounds of copper nitrate, for example, have been readily produced by this method with particle sizes ranging from 4 μ down to 4 nm. Further chemical treatments (such as heating and hydrogenation) are used to produce catalytically-active materials (such as copper clusters, in this case).

A method of fabricating surprisingly uniform clusters using an electrostatic spray technique is under investigation at Oxford University. A solution is forced through a thin capillary electrode at several thousand volts. The field causes the liquid to form a "Rayleigh cone" that emits small droplets that flash evaporate. Under the proper conditions a spread in sizes of 0.1% is obtained, which is quite monodisperse. Rates of production correspond to microliters of fluid per second. Samples that have been prepared by this method include GaAs (65), CdS (66), PbS (67), nitrides, arsenides, and metals. Some of these materials are deposited in conjunction with polymer precursors to form a dispersion and to avoid sintering of the aggregated clusters. A most surprising observation has been made when silver nitrate (concentration about 0.001M in methanol) is injected using this process. The high electric field (and associated charge balance) reduces the silver ion to silver metal clusters 5-10 nm in diameter (Dobson, P. J.; private communication).

Molecular Switching

Switching behavior of molecule systems can lead to changes in spectral characteristics, electromotive potentials, conductivity, polarization and optical rotatory dispersion. In all cases the physical properties exhibited are characterized by measurements of macroscopic properties. Such properties have been studied and understood for years. What is new is the introduction of additional materials and a possible better understanding of the microscopic factors that lead to the macroscopic properties. The goals of this research at the present time are scientific in nature.

A molecular configuration demonstrating flexibility for molecular switching has been examined at the University of Gröningen. This structure has a large 3-ring structure (see Figure 3) bonded to a substituted anthracene at the 10 position. Substituted units X, and Y include -CH$_2$-, S, and O. Position Z has been investigated for the addition of methyl and OH groups. The chiral activity of this class of materials is very high, and the spectral changes of each conformation are significantly different, demonstrating reversible photochromism. The barrier from one conformation to another may be varied with the substituents X and Y (68).

The group of materials known as "calixarenes" have demonstrated considerable flexibility for specific interactions. These molecules have a large cage consisting of four bridged benzene rings with substituent groups attached to each benzene ring (see Figure 4). The overall structure resembles a bowl designed for molecules having specific dimensions. By starting with the calixarene structure and modifying the substituents to include another (inverted) bowl structure, a cavity may be synthesized (69). Introduction of a molecule in the center of this cage gives a structure that can

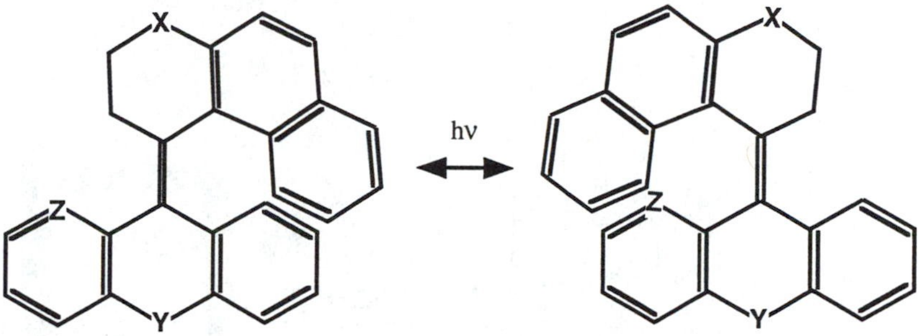

Figure 3. Molecular structures exhibiting large change in optical rotation with change in conformation.

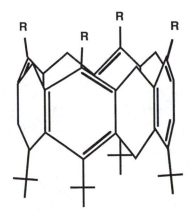

Figure 4. Example of a calixarene

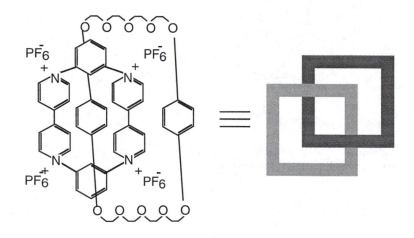

Figure 5a. Example of a catenane

Figure 5b. Schematic
representation of olympiadene

Figure 5c. Schematic of a
rotaxane

have multiple orientations of the trapped molecule relative to the stationary cage. It is anticipated that one or the other configuration will be induced by an externally-applied electric field, hence the concept of a "ferroelectric molecule." Technically speaking, true ferroelectric behavior represents a cooperative phenomenon with ordered phases, however. Such an entity should be "switched" by devices such as a scanning tunneling probe.

The search for improved molecular switches is a major focus for efforts in the molecular electronics program (70) in France. One approach involves intravalent charge transfer, where two ions in different charged states are linked chemically. By studying the charge transfer spectra of these complexes insight may be gained into the strength of the interactions between the two oxidizable/reducible groups. This overall approach has been termed "intra-molecular electronics (71)." The unit studied often is the ruthenium ion in the +2 and +3 states. Two of these ions are linked together by a variety of chelating systems, including various pyridines, cyanobenzenes and other arrangements (72-75). Under certain circumstances the angle between two ring structures will twist in an excited state, reducing the interaction between the two ruthenium ions of an excited complex, increasing the lifetime of the excited state (76). Other "mixed-valence systems involving organic molecules have been studied (77). A related series exhibiting molecular switching is that of a twisted internal charge transfer complex (78). The twisted bond between donor and acceptor of a charge-transfer complex enhances the excited state lifetime and modifies the spectral characteristics significantly.

The catenanes represent a series of molecules that are characterized by interlocking rings, where one ring molecule is formed such that it penetrates a second ring molecule (79). One of the smaller structures is represented in Figure 5a. The fabrication and characterization of these structures represent a fascinating new field of endeavor. Synthesis of these molecules makes use of a "self-assembly" process reminiscent of the self-assembly of viruses, where covalent and non-covalent associations are alternated to obtain the final product. The substituents on each ring give various electrophilic or lyophilic properties. Questions such as which positions of each ring are most often in contact, and in what proportion, are being revealed by this research. This type of information is most often revealed by NMR spectra, mass spectrometry, secondary ion mass spectrometry (SIMS), and X-Ray diffraction. NMR spectra also reveal the rate of exchange between different conformations in the molecule. The photochemistry and electrochemistry of these molecular systems demonstrate "switching" of the preferred conformation with excitation or oxidation/reduction (80). Interlocking five of these rings has recently been demonstrated with the preparation of olympiadane (81), illustrated in Figure 5b, suggesting a series of structures that could be extended to polymer length.

The molecular structure resembling a ring circling the central portion of a dumbbell is known as a rotaxane (82) (Figure 5c). The ring molecules resemble those used in the catenanes. The end "stopper" groups are substituents such as tri-isopropyl-silyl. The ring on this molecule can shuttle back and forth between the two ends, with variations in the rate of transit as well as the amount of time spent on each. As with the catenanes, the association of the ring molecule with the end groups (or with substituents A and B in the middle rod) depends on the electrophilic and lyophilic nature of A and B with the ring. This changes with oxidation and the energy level of excited states. With oxidation and reduction, the basic unit thus becomes a "memory unit," depending on the previous history of the molecule. This is also envisioned as a possible source of mechanical energy at the molecular level. Such motions within biological molecules are found to be responsible for the contraction of muscle cells within living species. Surprisingly good agreement is found between the experimentally determined molecular geometries and those predicted by the models (83).

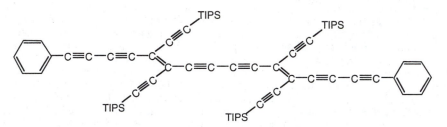

Figure 6. An example of a poly-triacetylinic molecular system

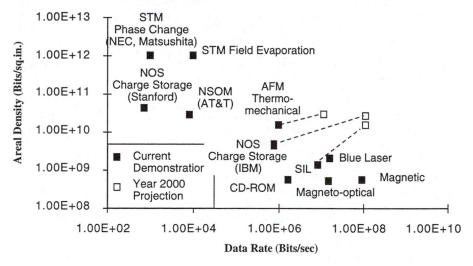

Figure 7. Approaches to High Density Recording

Molecular "Conductivity." A wide variety of poly-triacetylenic materials have been synthesized (84) (see example in Figure 6). The basic material properties resemble those of polyacetylene. It is interesting to observe oligomer units of this material in conjunction with UO_2^{++}. The UO_2^{++} forms a complex with the oligomer unit at positions where the R groups branch from the main polymer chain (electrical neutrality is assured with additional ions present). The chain thus becomes a very heavy unit, with a sequence of UO_2^{++} units positioned along the chain. It is possible to observe the individual polymer chains with Transmission Electron Microscopy (TEM) when these uranium dioxide ions are complexed with the chain.

Information Density

It is anticipated that understanding the molecular properties leading to this behavior may be useful in information storage and retrieval. With the advent of local probes such as scanning tunneling microscopy the ability to examine this behavior at a local level is becoming a reality.

It is true that molecules and matter may now be modified by external probes at dimensions approaching several Angstroms. The major question about the feasibility of a method for storing and retrieving this information will determine the most promising research areas. It is useful to consider the performance of various methods of information storage and retrieval under consideration today (Figure 7) (Mamin, H. J.; Terris, B. D., Fan, L. S., Hoen, S., Barrett, R. C. and Rugar, D. *IBM J. Res. Dev., to be published*[*]). In this figure, NOS refers to nitride-oxide-semiconductor structures and SIL refers to a solid immersion lens. The unmistakable trend shows that the rate of information transfer decreases with decreasing size of the storage bit. Factors such as energy density related to bit rates per unit area may ultimately represent limitations on therate of storage and retrieval of information. Research into the limits of these processes represents an important and fascinating objective to consider. Of course, once these limitations are understood, the most promising research directions should be chosen.

Sensors and Devices

Expansion of research in any particular field arises from either rapid scientific progress in understanding a field or the potential for technological use. The ability to fabricate and introduce molecular sensors on a silicon or GaAs chip in close proximity to the electronics on the same chip represents a source of inexpensive sensing elements able to provide extensive information in response to environmental changes. This objective has been recognized and is a goal for a number of research programs involving thin films and nanostructures.

Chemical sensors are under intensive investigation at the University of Tübingen, Germany, where developments are leading to molecular design for desired interactions. Phenomena being incorporated into a wide variety of recognition events include 1) changes in mass (as sensed with a quartz balance or a surface acoustic wave device); 2) changes in transport properties (especially doped oxides and semiconductors); 3) heats of reaction (as sensed on miniature thermocouples); 4) changes in work function (as sensed with current changes); 5) changes in capacitance (between interdigitated electrodes); electrochemical potential; and 6) optical spectral changes.

With the wealth of data available from a multiple sensor suite it is possible to envision a great range of information specific to each of many different chemicals. For

---

[*]   Reproduced with permission, IBM Corporation, Research Division, Almaden Research Center.

gas mixtures, pattern recognition techniques serve to unravel interrelationships and identify the constituents. A book has been written on this subject specifically for detailed sensor identification (85). An electronic nose is on the market today, using a sequence of twelve conducting polymers, each of dimensions 10 μ x 1 mm. Sensitivities of this device are on the order of ppm for a large number of gases

The calixarenes, mentioned earlier, represent a useful class for designed recognition (86-87) (also Schierbaum, K.-D.; Gerlach, A.; Göpel, W.; Müller, W. M.; Vögtle, F.; Dominik, A. and Roth, H. J. *Fresenius J. Anal. Chem.* Vol. 348 in press.). A program in Enschede, The Netherlands, prepared a large calixarene cage with 140Å^2 area (external) attached to thiol groups that would bind to a gold surface. Attempts were finally successful when additional "spacer" chains were introduced such that the thiol linkage had the same area as that of the calixarene cage. This illustrated a general principle: in order to obtain good monolayer definition with long-chain molecules, the group attached to the surface must have a similar area as the group extended away from the surface. The material demonstrated a high selectivity for the tetrachloroethylene molecule. Closely related to the sensing function is that of selective membrane transport. Cage molecules designed for specific dimensions may be placed in membrane materials such as polysiloxanes; placed on a gate of an FET produces an ion-sensitive FET (ISFET). Research in Enschede. A number of specific ion sensors have been demonstrated, including one for sodium dihydrogen phosphate with a selectivity ratio of better than 100:1. Sensing this latter chemical is of interest due to its predominance in the fertilizer industry.

The surface acoustic wave (SAW) device has also been used for sensing changes in mass of a thin film, and is inherently more sensitive than a vibrating quartz balance because the mass of the backing material is much less; also the sensitivity is linearly dependent on frequency. At a liquid interface, however, the ordinary Rayleigh wave generates waves in the liquid that dissipate energy and reduce the sensitivity. Alternative modes such as the Love wave is under study at the Paul Drude Institute in Berlin (88). This transverse shear wave becomes a surface wave if a second layer of solid material having a slower propagation speed is added to the surface, and does not dissipate energy through radiation. Recent publications (89-90) have modeled the conditions for optimum sensitivity for these waves and indicate optimistic performance.

Calorimetric information for very small sample sizes may be observed experimentally by using a bimetallic strip. With dimensions of 2 μ x 20 μ it is possible to observe less than one femtojoule of energy absorbed by measuring the optical deflection off the metal strip (91). The light is modulated at 1000 Hz; detection uses phase sensitive methods. By shining light on a metallic surface containing a molecular monolayer, the spectrum of an adsorbed species may be observed. The optical power on the strip is about 2-20 nW; the strip used experimentally thus far has a thermal relaxation time of about 0.5 msec. It is anticipated that the metabolism of a single cell may be measured with this apparatus.

Summary. The scientific frontiers involving nanostructures provide new approaches for fabricating and characterizing materials. Lithographic methods continue to contain sources of innovation and scientific insight. Additionally, chemical insight could yield alternative methods of fabricating well defined nanostructures in quantity. Such structures can have beneficial properties for neighboring disciplines such as materials science. Self-assembly, considered the domain of the chemist and biologist for centuries, is attracting the attention of other disciplines due to the potential opportunities involving desired physical properties of materials. Simple questions such as the transport behavior of a single molecule have not been answered, and contain the opportunity for fascinating discoveries. Research into the behavior of molecular switching will bring additional insight into the physical properties of interesting materials that undergo change with a variety of stimuli. With the powerful fabrication and characterization tools emerging today, such measurements are likely to yield new insight into the behavior of molecules as nanostructures.

Technological interest is considered in light of these research products. Most of these prospects are directed new materials due to their physical properties. The continuing frontier of lithography contains many examples for chemical innovation. Organic thin films offer opportunities for novel resist behavior or for the dielectric strength advantages they offer. Using nanostructures for the storage and retrieval of information seems a bit further on the horizon; a firm understanding of the principles leading to viable systems must be recognized to define the most useful directions for such activity. Organic transistors and other information devices of a macroscopic nature offer unique properties, but the niche technological opportunities have not yet been recognized. Molecular switching appears to hold some possibilities for new materials involving photochromism, ferroelectric materials and display devices. The optical properties of cluster materials appear to offer advantages if monodisperse samples can be prepared; it is likely that such a path will largely be through a chemical synthesis rather than based on statistically dominant processes such as crystallization or lithographic definition. New materials fabricated with nanostructure grain sizes have advantageous mechanical strength, resiliency, and wear properties. The great variety of materials possible with these new nanostructures should offer surprising new properties for years to come. Perhaps the most widespread near-term application of many of these ventures is that of chemical sensors. A great variety of phenomena combine with the opportunities to fabricate sensors contiguous to electronics and associated logic units. These will provide inexpensive and sensitive smart sensors will open opportunities for environmental and biological/medical applications for years to come.

With a field so diverse and yet unexplored we can expect many opportunities for unexpected physical properties, or for new routes to desired nanostructures through chemistry.

Acknowledgments. This effort was supported by the Office of Naval Research and the Naval Research Laboratory. Researchers at the Naval Research Laboratory are recognized for their contributions to their work mentioned. A large number of individuals contributed to the final itinerary in Europe. Thanks are extended to the Deputy Chief of Naval Research, Fred Saalfeld; the Director of Research at NRL, Dr. Timothy Coffey; the Commanding officer at ONR-Europe, CDR Dale Milton; and the Chief Scientist at ONR-Europe, Dr. John Silva. Thanks are offered to many scientists in the U.S. for suggesting contacts in Europe and for information about their research programs: Ari Aviram, Nick Bottka, Jeff Calvert, Rich Colton, Elizabeth Dobisz, Dave Ferry, Hal Guard, Wiley Kirk, Christie Marrian, Tom McGill, Jim Murday, Harvey Nathan, Marty Peckerar, Mark Reed, Joel Schnur, Gerry Sollner, Jim Tour, and George Whitesides.

Literature cited

1. Tolles, W. M. *NRL Tech. Rept.* **Dec. 30 1994** NRL/FR/1003-94-9755.
2. Tsutsui, K.; Hu, E. L. and Wilkinson, C. D. W. *J. Vac. Sci. Technol. B*, **1993**, *11*, 2233.
3. Murad, S. K.; Wilkinson, C. D. W.; Wang, P. D.; Parkes, W.; Sotomayor-Torres, C. M. and Cameron, N. *J. Vac. Sci. Technol. B* **1993**, 11, 2237.
4. Wang, P. D.; Foad, M. A.; Sotomayor-Torres, C. M.; Thoms, S.; Watt, M.; Cheung, R.; Wilkinson, C. D. W. and Beaumont, S. P. *J. Appl. Phys.* **1992**, *71*, 3754.
5. Rahman, M.; Foad, M. A.; Hicks, S.; Holland, M. C. and Wilkinson, C. D. W. *Mat. Res. Soc. Symp. Proc.* **1993**, *279*, 775.
6. Foad, M. A.; Wilkinson, C. D. W.; Dunscomb, C. and Williams, R. H. *Appl. Phys. Lett.* **1992**, *60*, 2531.

7. Calvert, J. M. *J. Vac. Sci. Technol. B* **1993**, *11*, 2155.
8. Calvert, J. M., In *Thin Films: Organic Thin Films and Surfaces*; Ulman, A., Ed.; Academic Press, Boston, MA, 1995, Vol. 20; pp 109-141.
9. Perkins, F. K.; Dobisz, E. A.; Brandow, S. L.; Koloski, T. S.; Calvert, J. M.; Rhee, K. W.; Kosakowski, J. E. and Marrian, C. R. K., *J. Vac. Sci. Technol. B*, **1994** *12*, 3725.
10. Marrian, C. R. K.; Perkins, F. K.; Brandow, S. L.; Koloski, T. S.; Dobisz, E. A. and Calvert, J. M., *Appl. Phys. Lett.* **1994** *64*, 390.
11. Snow, E. S.; Juan, W. H.; Pang, S. W. and Campbell, P. M., *Appl. Phys. Lett.* **1995**, *66*, 1729.
12. Campbell, P.M.; Snow, E. S. and McMarr, P. J., *Appl. Phys. Lett.* **1995**, *66*, 1388.
13. Snow, E. S.; Juan, W. H.; Pang, S. W. and Campbell, P. M. *Appl. Phys. Lett.* 1995, *66*, 1729.
14. Campbell, P. M.; Snow, E. S. and McMarr, P. J. *Appl. Phys. Lett.* 1995, *66*, 1388.
15. Kramer, N.; Jorritsma, J.; Birk, H.; Schonenberger, C. *Microelectron. Eng. (Netherlands)*, **1995**, *27*, 47.
16. Schnur, J. M., *Science*, **1993**, *262*, 1669.
17. Karaborni, S.; van Os, N. M.; Esselink, K. and Hilbers, P. A. J. *Langmuir* **1993** *,9*, 1175.
18. Smit, B.; Hilbers, P. A. J.; Esselink, K.; Rupert, L. A. M.; van Os, N. M. and Schlijper, A. G. *Nature* **1990**, *348*, 624.
19. Smit, B.; Esselink, K.; Hilbers, P. A. J.; van Os, N. M.; Rupert, L. A. M. and Szleifer. *Langmuir* 1993 9, 9.
20. Smit, B.; Hilbers, P. A. J.; Esselink, K.; Rupert, L. A. M.; van Os, N. M. and Schlijper, A. G. *J. Phys. Chem.* **1991**, *95*, 6361.
21. Smit, B.; Schlijper, A. G.; Rupert, L. A. M. and van Os, N. M. *J. Phys. Chem.* **1990**, *94*, 6933.
22. Karaborni, S. *Langmuir* **1993**, *9*, 1334.
23. Karaborni, S.; van Os, N. M.; Esselink, K. and Hilbers, P. A. J. *Langmuir* **1993**, *9*, 1175.
24. Buontempo, J. T.; Rice, S. A.; Karaborni, S. and Siepmann, J. I. *Langmuir* **1993**, *9*, 1604.
25. Karaborni, S. and Toxvaerd, S. *J. Chem. Phys.* **1992**, *97*, 5876.
26. Strong, L. and Whitesides, G. M., *Langmuir*, **1988**, *4*, 546.
27. Rabe, J. P. and Buchholz, S. *Phys. Rev. Lett.* **1991**, *66*, 2096.
28. Rabe, J. P.; Buchholz, S. and Askadskaya, L. *Physica Scripta* **1993**, *T49*, 260.
29. Rabe, J. P. *Atomic and Nanometer-Scale Modification of Materials: Fundamentals and Applications;* Kluwer Academic Pub., The Netherlands, 1993; 263.
30. Rabe, J. P.; Buchholz, S. and Askadskaya, L. *Synthetic Metals* 1993, *54*, 339.
31. Cincotti, S. and Rabe, J. P. *Appl. Phys. Lett.* **1993**, *62*, 3531.
32. Hentschke, R.; Schürmann, B. L. and Rabe, J. P. *J. Chem. Phys.* **1992**, *96*, 6213.
33. Hentschke, R.; Askadskaya, L. and Rabe, J. P. *J. Chem. Phys.* **1992**, *97*, 6901.
34. Askadaskaya, L. and Rabe, J. P. *Phys. Rev. Lett.* **1992**, *69*, 1395.
35. Rabe, J. P. and Buchholz, S. *Science* **1991**, *253*, 424.
36. Schönenberger, C.; Sondag-Huethorst, J. A. M.; Jorritsma, J. and Fokkink, L. G. J. *Langmuir* **1994**, *10*, 611.
37. Garnier, F. *Angew. Chem.* **1989**, *101*, 529.
38. Peng, X.; Horowitz, G.; Fichou, D. and Garnier, F. *Appl. Phys. Lett.* **1990**, *57*, 2013.
39. Horowitz, G.; Peng, X.; Fichou, D. and Garnier, F. *J. Appl. Phys.* **1990**, *67*, 528.

40. Garnier, F.; Horowitz, G.; Peng, X. and Fichou, D. *Adv. Mater.* **1990**, 2, 592.
41. Horowitz, G. and Delannoy, P. *J. Appl. Phys.* **1992**, 70, 469.
42. Horowitz, G.; Hajlaoui, R.; Deloffre, F. and Garnier, F. *Macromolecules* **1993**
    323.
43. Horowitz, G.; Deloffre, F.; Garnier, F.; Hajlaoui, R.; Hmyene, M. and Yassar,
    A. *Synthetic Metals* **1993**, *54*, 435.
44. Youssoufi, H. D.; Hmyene, M.; Gernier, F. and Delabouglise, D. *J. Chem. Soc.,
    Chem. Commun.* **1993**, 1550.
45. Charra, F.; Lavie, M-P.; Lorin, A. and Fichou, D. *Synthetic Metals*, **1994**, *65*,,
    13.
46. Fichou, D.; Nunzi, J-M.; Charra, F. and Pfeffer, N. *Adv. Mater.* **1994**, *6*, 64.
47. Knobloch, H.; Fichou, D.; Knoll, W. and Sasabe, H. *Adv. Mater.* **1993**, *5*, 570.
48. Nunzi, J-M.; Pfeffer, N.; Charra, F. and Fichou, D. *Chem. Phys. Lett.* **1993**,
    215, 114.
49. Fichou, D.; Horowitz, G.; Xu, B. and Garnier, F. *Synthetic Metals* **1992**, *48*,
    167.
50. Birnbaum, D.; Fichou, D. and Kohler, E. *J. Chem. Phys.* **1992**, *96*, 165.
51. Cheng, X.; Ichimura, K.; Fichou, D. and Kobayashi, T. *Chem. Phys. Lett.*
    1991, *185*, 286.
52. Fontaine, P.; Goguenheim, D.; Deresmes, D.; Vuillaume, D.; Garet, M. and
    Rondelez, F.*Appl. Phys. Lett.*, **1993**, *62*, 2256.
53. Boulas, C.; Davidovits, J. V.; Rondelez, F. and Vuillaume, D; *Microelect. Eng.*,
    1995, *28*, 217.
54. Drain, C. M. and Lehn, J.-M. *J. Chem. Soc. - Chem. Commun.* **1994**, 2313.
55. Samuelson, L.; Lindahl, J.; Montelius, L.; and Pistol, M.E. *Inst. Phys. Conf.
    Ser.*, **1994**, 51.
56. Samuelson, L.; Gustafsson, A.; Lindahl, J.; Montelius, L.; Pistol, M-E.; Malm,
    J-O.; Vermeire, G. and Demeester, P. *J. Vac. Sci. Technol. B*; **1994**, *12*, 2521.
57. Grundmann, M.; Christen, J.; Ledentsov, N. N.; Böhrer, J.; Bimberg, D.;
    Ruvimov, S. S.; Werner, P.; Richter, U.; Gösele, U.; Heydenreich, J.; Ustinov,
    V. M.; Egorov, A. Y.; Kop'ev, P. S and Alferov, Z. I. *Phys. Rev. Lett.*, **1995**,
    74,, 4043.
58. Dandliker, P.J.; Diederich, F.; Gross, M.; Knobler, C.B.; Louati, A. and
    Sanford, E.M. *Angew. Chem. Int. Ed. Engl.*, **1994**, *33*, 1739.
59. Lehn, J.-M., *Pure & Appl. Chem.*, **1994**, *66*, 1961.
60. Baxter, P. N. W.; Lehn, J.-M.; Fischer, J. and Youinou, M.-T., *Angew. Chem.
    Int. Ed. Engl.*, **1994**, *33*, 2284.
61. Johnsson, K.; Allemann, R. K.; Widmer, H. and Benner, S. A. *Nature* **1993**,
    365, 530.
62. Bain, J. D.; Switzer, C.; Chamberlin, A. R. and Benner, S. A. *Nature* **9 April
    1992**, 356.
63. Kuipers, E. W.; Laszlo, C. and Wieldraaijer, W. *Catalysis Lett.* **1993**, *17*, 71.
64. Kuipers, E. W.; Doornkamp, C.; Wieldraaijer, W. and van den Berg, R. E.
    Materials **1993**, *5*, 1367.
65. Salata, O. V.; Dobson, P. J.; Hull, P. J. and Hutchison, J. L .*Appl. Phys. Lett.*
    1994, *65*,, 189.
66. Salata, O. V.; Dobson, P. J.; Hull, P. J. and Hutchison, J. L .*Thin Solid Films*
    1994, *251*, 1.
67. Salata, O. V.; Dobson, P. J.; Hull, P. J. and Hutchison, J. L .*Advanced Materials*
    1994, *6*, 772.
68. Feringa, B. L.; Fager, W. F. and de Lange, B. *Tetrahedron* **1993**, *49*(37), 8267-
    8310.
69. Verboom, W.; Rudkevich, D. M. and Reinhoudt, D. N. *Pure & Appl. Chem.*
    1994, *66*, 679.
70. Warman, J. M.; Schuddeboom, W.; Jonker, S. A.; de Haas, M. P.; Paddon-

Row, M. N.; Zachariasse, K. A. and Launay, J-P. *Chem. Phys. Lett.* **1993**, *210*, 397.
71. Joachim, C. and Launay, J. P. *J. Mol. Elect.* **1990**, *6*, 37.
72. Collin, J-P.; Lainé, P.; Launay, J-P.; Sauvage, J-P. and Sour, A. *J. Chem. Soc. Chem. Commun.* **1993**, 434.
73. Marvaud, V. and Launay, J-P. *Inorg. Chem.* **1993**, *32*, 1376.
74. Marvaud, V.; Launay, J-P. and Joachim, C. *Chem. Phys.* **1993**, *177*, 23.
75. Joachim, C.; Launay, J-P. and Woitellier, S. *Chem. Phys.* **1990**, *147*, 131.
76. Gourdon, A.; Launay, J-P.; Bujoli-Doeuff, M.; Heisel, F.; Miehé, J. A.; Amouyal, E. and Boillot, M-L. *J. Photochem. Photobiol. A: Chem* **1993**, *71*, 13.
77. Bonvoisin, J.; Launay, J-P.; Van der Auweraer, M. and De Schryver, F. C. *J. Phys. Chem.* **1994**, *98*, 5052.
78. Gourdon, A.; Launay, J.-P.; Bujoli-Doeuff, M.; Heisel, F.; Miehé, J. A.; Amouyal, E. and Boillot, M.-L. *J. Photochem. Photobiol. A: Chem.*, **1993**, *71*, 13.
79. Amabilino, D. B.; Ashton, P. R.; Reder, A. S.; Spencer, N. and Stoddart, J. F. *Angew. Chem. Int. Ed. Engl.* **1994**, *33*, 433.
80. Bissell, R. A.; Córdova, E.; Kaifer, A. E. and Stoddart, J. F. *Nature* **1994**, *369*, 133.
81. Amabilino, D. B.; Ashton, P. R.; Reder, A. S.; Spencer, N. and Stoddart J. F. *Angew. Chem., Int. Ed.* **1994**, *33*, 1286.
82. Ballardini, R.; Balzani, V.; Gandolfi, M. T.; Prodi, L.; Venturi, M.; Philp, D.; Ricketts, H. G. and Stoddart, J. F. *Angew. Chem. Int. Ed. Engl.* **1993**, *32*, 1301.
83. Ricketts, H. G.; Stoddart, J. F. and Hann, M. M. In *Computational Approaches in Supramolecular Chemistry;* Wipff, G. Ed.; Kluwer Academic Pub., The Netherlands, 1994.
84. Anthony, J.; Boudon, C.; Diederich, F.; Gisselbrecht, J-P.; Gramlich, V.; Gross, M.; Hobi, M. and Seiler, P. *Angew. Chem. Int. Ed. Engl.* **1994**, *33*, 763.
85. *The Electronic Nose*; Gardner, J. W. and Bartlett, P. N., Eds.; NATO ASI Series, Series E: Applied Sciences 5-8 August 1991, Kluwer Academic Publishers, Boston/London, ,Vol. 212 1991.
86. Schierbaum, K. D.; Hierlemann, A. and Göpel, W. *Sensors and Actuators B* **1994**, *19*, 448.
87. Schierbaum, K. D. *Sensors and Actuators B*, **1994**, *18*, 71.
88. Drobé, H.; Leidl, A.; Rost, M. and Ruge, I. *Sensors and Actuators A* **1993**, *37-38*, 141.
89. Enderlein, J.; Chilla, E. and Fröhlich, H-J. *Sensors and Actuators A* **1994**, *41-42*, 472.
90. Kovacs, G.; Lubking, G. W.; Vellekoop, M. J. and Venema, A. *1992 Ultrasonics Symp., IEEE* **1992**, 281.
91. Barnes, J. R.; Stephenson, R. J.; Woodburn, C. N.; O'Shea, S. J.; Welland, M. E.; Rayment, T.; Gimzewski, J. K. and Gerber, C .*Rev. Sci. Instrum.*, **1994**, *65*, 3793.

RECEIVED November 9, 1995

VAPOR PHASE

Chapter 2

On-line Sizing of Colloidal Nanoparticles via Electrospray and Aerosol Techniques

L. de Juan and J. Fernández de la Mora

Mechanical Engineering Department, Yale University, 9 Hillhouse Avenue, New Haven, CT 06520

The size distributions of colloidal suspensions of nanoparticles 74 nm to 14 nm in diameter are analyzed on-line. The sols are first diluted in water seeded with enough TFA to attain electrical conductivities in the range of 0.01 S/m. The solution is then finely dispersed into an atmosphere of CO_2 via a Taylor cone-jet. The resulting electrospray of ultrafine droplets dries, transferring the solution particles virtually uncontaminated into the gas. There they are sized by means of a differential mobility analyzer and an inertial impactor of unusually high resolution. The technique is first tested successfully with previously calibrated monodisperse polystyrene latex (PSL) spheres 74 to 21 nm in diameter. It is then used to size a solution of colloidal silica with particle diameters nominally between 10 and 14 nm.

As a result of the materials applications of particles with dimensions in the nanometer range, a variety of schemes are emerging for their large scale production. However, except for the special case of certain monodisperse sols which may be sized spectroscopically (*1*), the characterization of the products from such synthetic processes tends to be based on tedious post-mortem procedures such as electron microscopy. Evidently, the availability of a rapid and general procedure to determine the size distribution of nanoparticles as they are being produced would allow a much faster progress in this field. Accordingly, the objective of this research is to develop a methodology to determine on-line the size distributions of colloidal nanoparticles or large dissolved molecular clusters. The procedure we use involves (i) transferring the suspended particles from the liquid into air, and (ii) sizing them in the gas phase. The first step involves an extension of the *electrospray* technique introduced by Fenn et al.(*2*) for the mass spectrometric (MS) analysis of large polymers. The liquid is electrostatically atomized into very fine and highly charged droplets, each holding at most one nanoparticle, and negligible amounts of involatile impurities. Solvent evaporation then leaves the bare particle as a charged entity suspended in the gas.

0097–6156/96/0622–0020$15.50/0

Transferring Uncontaminated Nanoparticles from Solution to a Gas.

Bringing a particle from a liquid into a gas involves in principle just the atomization of the liquid into droplets, followed by evaporation of the solvent. In practice, however, the droplet must be sufficiently small to (i) carry one suspended particle at most, and (ii) contain negligible amounts of involatile impurities. Otherwise, after evaporation, these impurities will form a crust over the particle (if the droplet contains one), or will alternatively leave a solid aerosol residue which would complicate the task of relating the measured aerosol size distribution to that in the original colloid. It is thus essential that the quantity of involatile impurity per initial liquid droplet be small compared to the volume of the smallest suspended nanoparticle of interest. For the case of colloidal particles smaller than 100 nm, these requirements are not met by pneumatic nebulizers, which produce initial drop sizes in the range from 1 to 10 microns. For instance, a drop 10 microns in diameter containing only 0.1 parts per million by volume of dissolved involatile residues would still leave after complete evaporation a solid residue 46 nm in diameter. Likewise, if the initial solution contained a suspension of silica spheres 10 nm in diameter at a volume fraction of 10%, a drop 10 microns in diameter would carry 10^8 such spheres. It is thus evident that the successful transfer of unagglomerated and uncontaminated nanoparticles from the liquid into the gas phase requires dividing the liquid into initial droplets only a few times larger than the nanoparticles themselves.

Electrospray Atomization
Taylor Cone-Jets and Electrospray Mass Spectrometry (ESMS).
Electrospray (ES) atomization is the only known procedure capable of fragmenting a liquid into pieces the largest of which has dimensions in the nanometer range. This method has recently become quite familiar to the analytical chemist thanks to so-called electrospray ionization (ESI) (2). In it, small charged droplets are first produced from a microjet emitted from the tip of a Taylor cone (3). Two such objects are illustrated in Figure 1. They appear naturally when the interface between a conducting liquid (of sufficiently small surface tension) and a gas is charged to an electrical potential of several kV above that of a neighboring electrode. After some evaporation without charge loss, the droplets become unstable and explode into smaller fragments. Following several such fissions, the surface of the smallest resulting drops develops electric fields large enough to field-emit small ions into the gas. In the case of very large solutes such as chain polymers with molecular weights in the range of millions (4), a more likely road to ion formation is Dole's charged residue mechanism (5). In this scenario, the solvent in a drop small enough to contain initially only one polymer molecule and negligible quantities of involatile residue would evaporate completely leaving behind the charged polymer molecule by itself.

As in pneumatic nebulizers, in addition to producing molecular ions, electrosprays also create solid residues from involatile materials originally in solution. Their diameters vary typically from a few nm up to tens of nm, which complicates the task of relating the size spectrum of the ions and particles appearing in the gas phase to that originally in solution. But this difficulty has not affected earlier electrospray mass spectrometry (ESMS) work because the mass over charge ratio m/z is generally above 30,000 Dalton for compact solid residues, which is much larger than the m/z values typical of solution ions studied so far via ESMS. Indeed, the only available mass spectrometric evidence for such dense residues is in the work of Smith and his colleagues with a low frequency quadrupole (6). Consequently, ESI has developed quite successfully for the analysis of small ions as well as large organic chain polymers (whose m/z values are typically smaller than 3000) on the basis of initial drops with diameters generally in the range of several microns. The small m/z values observed for large chain polymer are due to their small electrostatic energy relative to that for a similarly charged compact solid residue. Accordingly, interference with the

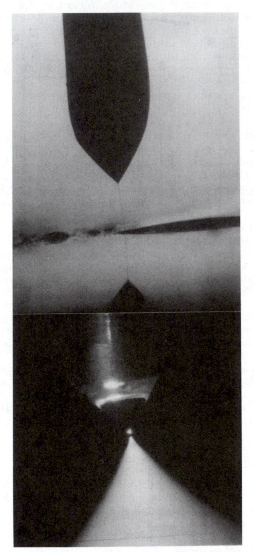

Figure 1. Photographs of two cone-jets. The one with the least conducting liquid (right) emits a long jet. The jet on the more conducting liquid (left) is barely discernible, but the spray of drops into which it breaks up is clearly visible. (Reproduced with permission from reference 3b. Copyright 1995 Academic Press, Inc.)

aerosol residue from involatile solutes will not be so easily avoided for the case when the analyte of interest is itself relatively compact.

Previous Studies Involving Electrosprays and Nanoparticles. The fine atomization features of electrosprays have been exploited already to some degree in relation to the generation of nanomaterials. In one instance, the product of interest was the residue from the salts initially dissolved in the sprayed liquid (7). Although the corresponding size distributions were relatively broad, their mean diameter could be made quite small. Two recent investigations have electrosprayed monodisperse quantum dots originally in solution and incorporated them into a solid matrix material co-deposited by CVD (1, 8). However, the presence of agglomerated dots on the final composite indicates that the initial droplets were too large and must have contained several particles.

The studies just discussed illustrate the convenience of "designing" rationally the droplet size best suited for a particular application. The required information has been available for some time, and will be briefly reviewed in the following section . It has been used previously (9) to generate nanoparticles (as small as 3 nm in diameter; see Figure 3 in reference 10) of salt residues from electrosprays of highly conducting solutions. It has also allowed studies on the kinetics of ion evaporation from liquid surfaces (10). Wilm and Mann (11) have used also high liquid conductivities and small flow rates to greatly enhance the analytical sensitivity of electrospray ionization. Fuerstenau and Benner (12) have exploited the smallness of the droplets formed by electrosprays of relatively conducting liquids to generate aerosols from colloidal suspensions of spheres of polystyrene latex (PSL) 100, 50 and 26 nm in diameter. Although they have not measured the diameter of their aerosol particles, they have convincingly demonstrated monodispersity as well as lack of aggregation in their aerosol, creating an important precedent to our work here.

Scaling Laws for the Droplets Emitted from Cone-Jets. Recent studies on the laws governing the size of electrosprayed drops have made it possible in some cases to control their initial diameter d at will. Experiments show that relatively inviscid liquids (water, methanol, formamide, etc.) exhibit electrospray regimes where they tend to produce fairly uniform drops. In contrast, highly viscous liquids such as glycerol tend to break up into more complex sprays which have been studied only to a limited degree. We shall therefore restrict the discussion to solutions with relatively small viscosities. Likewise, we will deal exclusively with polar liquids, which are best suited to form electrosprays of nanodroplets. A limited discussion on alternative situations is available for nonpolar liquids,(13) and on viscous effects (14, 15).

For moderately viscous solutions with a large dielectric constant ($\varepsilon > 9$), one finds empirically that stable cone-jets of a given substance can only be formed within a finite domain for the flow rate Q of liquid pushed through the cone-jet: $Q_{min} < Q < Q_{max}$ (see the appendix for an exception). Furthermore, the size distribution of the resulting droplets is narrowly centered around a unique diameter d only in the smaller band $Q_{min} < Q < Q_{bif}$, where Q_{bif} tends to be substantially smaller than Q_{max}. The lower range $Q < Q_{bif}$ offers the conditions of greatest interest for our present purposes (15), and will be the only one to be considered subsequently. The geometry and voltage V of the electrodes determines the range of values of V where a cone-jet is stable. But (given K and Q) V and the geometry are relatively irrelevant in fixing the resulting droplet size distribution, whose mean diameter d and monodispersity depend principally on Q and on the electrical conductivity K of the solution. The main parameters are the dimensionless liquid flow rate η and the characteristic length d_{min} (15, 16, 17, 18):

$$\eta^2 = \rho Q/(\sigma t_r); \quad t_r = \varepsilon\varepsilon_0/K; \quad d_{min} = (\sigma t_r^2/\rho)^{1/3}; \quad d \sim d_{min}\eta^{2/3} = (Qt_r)^{1/3} \quad (1\text{-}4)$$

where ρ, σ and ε are the density, surface tension coefficient and dielectric constant of the liquid. ε_0 is the electrical permittivity of vacuum, and t_r is generally called the electrical relaxation time.

The key criterion to produce smaller and smaller droplets by increasing the electrical conductivity K of the solution and decreasing the flow rate Q of liquid had been advanced qualitatively in (19). Its quantitative form $d \sim (Qt_r)^{1/3}$ is most useful, but is still only part of the description. One must also know how Q_{min} depends on the fluid properties. This issue has been investigated only superficially, but all the data available for polar liquids show that Q_{min} corresponds to conditions of equation 1 where:

$$\eta_{min} \sim 1, \tag{5a}$$

implying that the smallest drop diameter that may be electrosprayed is of the order of the characteristic length d_{min} defined in equation 3:

$$\text{smallest drop diameter} \sim d_{min} = (\sigma t_r^2/\rho)^{1/3}. \tag{5b}$$

Equation (5a) applies to all the polar liquids which we have investigated, from rather viscous substances such as glycerol, or tetra and tri-ethylene glycol, to fairly inviscid fluids such as water, formamide, benzaldehyde, benzyl alcohol, and the simplest alcohols from methanol to at least 1-octanol. It holds also at all the salt concentrations we have investigated (up to 1 molar), and for all the types of solutes tried, from alkali halides to larger organic salts such as tetrabutyl ammonium halides.

Although measured values of η_{min} have roughly spanned the relatively broad range between 1/2 and 2, for the case of water and formamide, we have discovered that Q_{min} can actually be predicted much more precisely than through equation 5a. Its value corresponds well with the condition at which the positive and negative electrolyte ions carried by the liquid are fully separated from each other. In other words, for a positively charged Taylor cone near the minimum flow rate, all the incoming negative ions would be neutralized on the positive electrode, while the emitted drops would contain exclusively the positive ions without any counter-ion. This condition corresponds to the maximum possible emission of charge per unit volume:

$$[I/Q]_{max} = ne, \tag{6}$$

where I is the electrospray current (current ejected by the cone-jet), e is the charge on the positive ions and n is the number density (molecules/cm^3) of the incoming dissolved salt, taken to be fully ionized. Because I/Q varies with a negative power of Q, $[I/Q]_{max}$ actually corresponds to the smallest I and Q attainable. From a large set of data on polar liquids, the following empirical approximation for the spray current has been found (16):

$$I = f(\varepsilon) \, (\sigma KQ/\varepsilon)^{1/2}, \tag{7}$$

where the function $f(\varepsilon)$ asymptotes towards the value 18 for $\varepsilon > 40$. Extrapolating equation 7 down to $Q = Q_{min}$ and making use of equation 6 leads to

$$\eta_{min} = (Z^+ + Z^-) \, (\rho/\varepsilon_0)^{1/2} \, [f(\varepsilon)/\varepsilon], \tag{8}$$

where $(Z^+ + Z^-)$ is the sum of the electrical mobilities of the positive and negative ions. For a fully ionized salt, it is related to the electrical conductivity of the solution via

$$K = ne \, (Z^+ + Z^-). \tag{9}$$

Although it seems intuitively strange that complete charge separation might be achieved in practice, the quantities I_{min}, Q_{min} and n are readily measured to confirm the approximate validity of equation 6 for the cases of water and formamide. For LiCl solutions of these two substances, the right hand sides of equation 8 were measured to be 0.53 and 0.41, respectively, while the corresponding values of η_{min} (as defined in equation 1) were in both cases in the vicinity of 0.5 (16).

The conclusion is that one can predict the approximate value of the diameter of electrosprayed drops as functions of K and Q, and one can also estimate fairly well its smallest value, which is comparable to d_{min} in equation 3. This determines straightforwardly the lowest conductivity a given solvent must have in order to be sprayable into drops of a given diameter d. Accordingly, we have control of one fundamental parameter of the spraying process: the ratio between the diameter d_p of the colloidal particle to be sized, and that for the electrospray droplet used to disperse it into the gas, for which we will use the quantity $(Qt_r)^{1/3}$ of equation 4:

$$\lambda = (Qt_r)^{1/3}/d_p \qquad (10)$$

To avoid agglomerated colloids in the aerosol, one additional issue is to use suspensions sufficiently dilute to lead to no more than one particle per droplet. As a result, many of the electrosprayed droplets will contain no colloidal particles, and will give rise to a solid residue even in the purest solvent. However, at sufficiently small values of $(Qt_r)^{1/3}/d_p$, the diameter d_r of the residues will be considerably smaller than the diameter d_p of the dissolved particle. Furthermore, d_r depends on Q according to equation 4, while d_p does not, providing a reliable criterion to discriminate between particles originally in solution, and residues from solution droplets.

A practical issue that may be anticipated here is that typical electrical conductivities required for the generation of drops a few tens of nanometers are in the range of 1 S/m. Attaining such high K values requires the addition of substantial concentrations of salts (in the range of 0.1 molar for water), which may destabilize the sol one wishes to size. Although this problem remains to be solved in general, a recent finding by Tang and Gomez (*20*) offers a possible strategy to deal with it (see the appendix).

Size Analysis

Once the colloid has been transferred into the gas, its size distribution may be found by aerosol techniques.However, the high particle charge and the accompanying aerosol of involatile residues call for a number of modifications to available sizing procedures.

A Size-Spectrometer based on an Inertial Impactor. Inertial impactors are devices where an aerosol is accelerated through a nozzle, forming a jet which impacts perpendicularly against a collector plate (*21*) This simple scheme has the remarkable property of acting as a low-pass filter, in which particles smaller than a critical diameter remain in suspension, while those supercritical impact on the surface and are captured. The parameter governing the subcritical-supercritical transition is generally called the Stokes number S:

$$S = t_p U/d_n; \qquad U = 4c^2 m'/(\pi \gamma d_n^2 p), \qquad (11, 12)$$

where d_n is the diameter of the nozzle forming the jet, and U is the average jet velocity based on its mass flow rate m' (3.6 x 10^{-3} g/s for all the results to be reported) and the gas pressure p in the plenum chamber just upstream the nozzle. γ is the specific heat ratio of the gas (1.4 for air), and t_p is the particle relaxation time, related to their diffusion coefficient D and mass m_p via Einstein's law:

$$D = kTt_p/m_p. \qquad (13)$$

k is Boltzmann's constant and T is the absolute temperature. c is the speed of sound of the gas in the plenum chamber, $c^2 = \gamma kT/m$, where m is its molecular mass.

The critical value S* of S, above which impaction arises depends on the geometrical and fluid dynamic parameters of the impactor, and takes typically values near 0.1. S* can be determined by calibration, so that the experimental conditions at which the transition is observed for a well defined cluster do fix t_p in equation 11. Accordingly, impactors measure the product $m_p D$, which is also proportional to the ratio between mass and drag. For objects much smaller than the gas mean-free-path, t_p is proportional to the ratio of m_p over the cluster's cross sectional area. In the

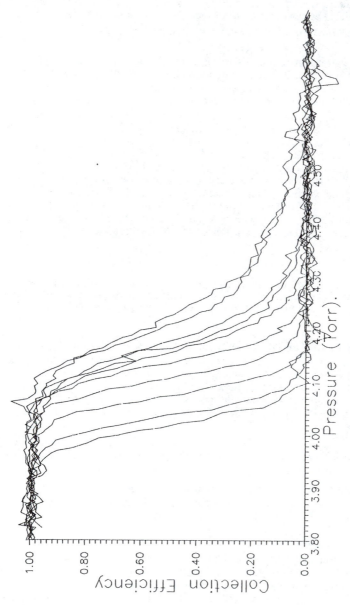

Figure 2.   Aerodynamic spectra I(p) from a monodisperse aerosol of singly charged oil droplets.   Each curve corresponds to a different value of the repulsive voltage V_R, which sharpens the steps and shifts their position to the left (see reference 23 for more detail).

particular case of a sufficiently small spherical particle of density ρ_p and diameter d_p, one may write after using standard expressions for the free-molecule drag of a sphere:

$$S = 0.178 \, \rho_p d_p m' c^3/(p^2 d_n^3). \tag{14}$$

Thus, what the impactor measures is the product $\rho_p d_p$, which suitably divided by $1 g/cm^3$ is sometimes called the aerodynamic diameter.

The impactor used in this work is a variant of that described by Rao et al. (*22*). The aerosol is passed through a critical orifice to fix the mass flow rate m', and is accelerated through a nozzle by a pump. Varying the pumping speed changes U (thus S for a given particle), which is monitored continuously through the pressure p upstream from the nozzle ($p \sim 1/U$). The current I associated to the captured fraction of the particles is measured with an electrometer connected to the collector plate. The impactor thus yields aerodynamic spectra $I(p)$ which are closely connected to the aerosol size distribution. For instance, as illustrated in each of the curves of Figure 2, the spectrum $I(p)$ associated to monodisperse particles would exhibit a single step dropping from a maximum current at $p < p^*$ down to zero current for $p > p^*$. A measurement of p^* yields the aerodynamic particle diameter. Recent improvements in this instrument (*23*) have increased its resolution to values near 50. This performance is attained by incorporation of two novel features. First, the particles are concentrated on the axis of the impactor nozzle by aerodynamic focusing with a lens system type introduced in (*24*). In addition, there is now freedom to impose a repulsive voltage difference V_R between the nozzle and the collector. By avoiding the deposition of subcritical particles by diffusion and electrophoresis, this "electrostatic blowing" is able to sharpen further the rightmost step shown in Figure 2 (for $V_R = 0$). Notice, however, that V_R also shifts to the left the position of these steps, which amounts to increasing slightly the value of S^*. In other words, S^* is a function of V_R:

$$S^* = S^*(V_R). \tag{15}$$

The particular instrument used in this work is nearly identical to a previous model (*23*), except for some changes in the focusing lenses. Also, the impactor nozzle is now a sharp-edge orifice of diameter $d_n = 3.61$ mm, drilled in a plate 0.8 mm thick. It has been calibrated by standard procedures using singly charged monodisperse oil droplets. S^* is defined based on the pressure at the middle of the step (where the collected current is half-way between its values just above and below the beginning and the end of the step, respectively). As shown in Figure 3, S^* depends slightly on the gas speed as a result of compressibility phenomena. This effect is quantified in terms of the Mach number, defined here as the ratio between the average velocity U of equation 12 and the speed of sound c of air in the plenum chamber upstream the nozzle:

$$M = U/c. \tag{16}$$

Notice a certain irreproducibility in these data from day to day. We attribute it to variations in the mass flow rate m' through the critical orifice, which are probably due to partial clogging of the orifice with dust. This long range variation problem would be of serious concern for high precision sizing of particles, but is relatively irrelevant for our preliminary purposes here. Consequently, we did not monitor m' for each measurement.

Effect of the High Charge Level on Impactor Response. The aerodynamic spectra shown in Figure 2 correspond to singly charged oil droplets. But the charge levels to be expected from aerosol particles originating from electrosprayed colloids are orders of magnitude higher, which leads to important differences in impactor response. Figure 4 illustrates this effect through the aerodynamic spectra of an aerosol of approximately monodisperse PSL spheres 74 nm in diameter. It has been generated by electrospraying a dilute water suspension of the PSL, seeded with trifluoroacetic acid (TFA) to achieve an electrical conductivity $K = 4.7 \times 10^{-3}$ S/m. The aerosol is then passed through a differential mobility analyzer (DMA, reference *25*), which

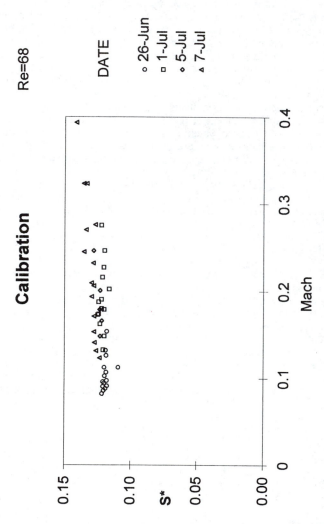

Figure 3.   Impactor calibration curve S*(M) for singly charged drops.

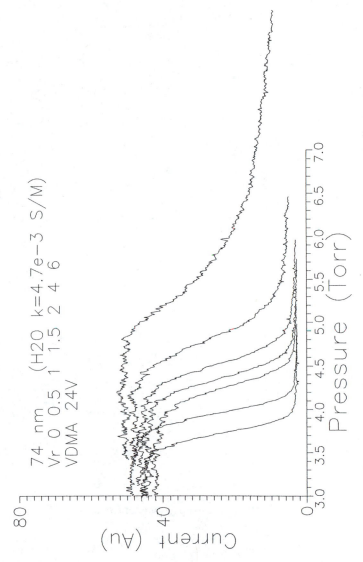

Figure 4.   Aerodynamic spectra I(p) at various repulsion voltages V_R for highly charged PSL spheres 74 nm in diameter.

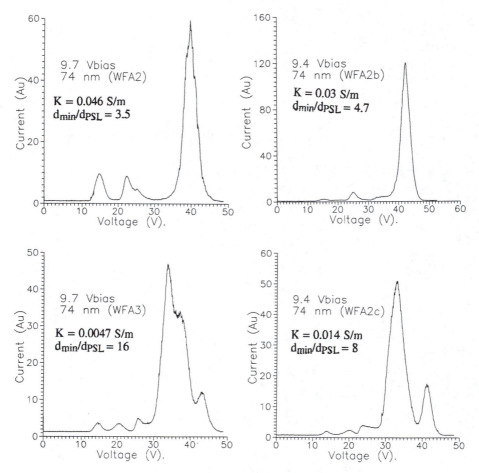

Figure 5.   Mobility distribution for electrosprays of aqueous solutions of 74 nm PSL. Notice the effect of the drop diameter quantified through d_{min}/d_{PSL}.

selects particles having a relatively narrow distribution of charge around $q = 544$ e in this case. The various curves shown correspond to several values of the repulsion voltage V_R. Two main differences with the spectra of Figure 2 are: (i) the curves for $V_R = 0$ and $V_R = 1$ Volt do not exhibit a sharp fall down to zero current, but decay rather slowly to a finite current level. (ii) Such long tails are eliminated for $V_R > 1.5$ V, but at the expense of a considerable shift in the position of the step. The second phenomenon is just a magnified version of that observed in Figure 2. The first results from particle deposition below the critical value of S due to image attraction to the plate. This phenomenon is here $544^2 \sim 0.3 \times 10^6$ times larger than for a singly charged particle of the same diameter.

Notice that, even with $V_R > 2$ V, the steps in Figure 4 are relatively broad, with $\Delta p/p \sim 7\%$. This result, however, does not reflect the resolution limit of our impactor, but rather the size spread of this particular PSL hydrosol. The actual resolution of the impactor would lead to $\Delta p/p$ values between 1 and 2% with really monodisperse sprays, as confirmed with singly and multiply charged PSL particles 0.3 microns in diameter.

We have seen that, in order to attain a reasonable resolution, one must suppress the image attraction tails via a sufficiently intense electrostatic blowing. But because the associated shift in S* is not negligible for these highly charged particles, one must be able to quantify this effect precisely. At values of V_R large enough to eliminate the tails in the aerodynamic spectra, the effect (equation 15) of V_R in S* for an aerosol particle with charge q is accounted for by the dimensionless parameter £:

$$S^* = S^*(\pounds); \quad \pounds = qV_RD/(kTUd_n). \qquad (17, 18)$$

£ is just a ratio between the electrical velocity drift of the particle in the field V_R/d_n and the fluid velocity U. The fact that this parameter is proportional to q introduces a new complication in the interpretation of the impactor spectra, because q is not known precisely, nor is it just a simple function of d_r. In fact, the aerosol formed by electrospraying a monodisperse colloid exhibits a relatively broad charge distribution. This is evident from Figure 5, which shows the distribution of electrical mobilities of the aerosol resulting from electrospraying various PSL hydrosols seeded with TFA. In this graph, the electrical mobility Z is related to the experimental variable V shown in the horizontal axis via $Z(V - V_{bias}) = 6.134$ cm^2/s, where V_{bias} is either 9.7 V or 9.4 V. For singly charged spherical particles, Z is a known function of their diameter d_p: $Z_1 = Z_1(d_p)$, so that $Z = Z_1(d_p)$ q/e. Consequently, for given d_p, the mobility distribution is a faithful indicator of the charge distribution. Among the various peaks shown in the mobility spectra of Figure 5, all corresponding to V above 30 V are due to PSL, while those below are due in part to involatile residues. For the most conducting solutions in Figure 5, the mobility distribution of PSL shows a single peak, though with q varying by nearly 50% from its minimum to its maximum value. For the least conducting solutions, the PSL has at least three mobility peaks spread over a comparably broad range.

Selecting the Charge State of a Given Particle Size via a DMA. We have seen that the shifts in S* resulting from the electrostatic elimination of the tails in the size spectra are not negligible and depend substantially on q. But, because q has a broad distribution, the aerodynamic spectra cannot be turned into size spectra except with a considerable inversion indeterminacy. A reasonable way out of this difficulty is to narrow down drastically the charge distribution. This can be done by passing the aerosol through a differential mobility analyzer similar to that used to obtain the spectra of Figure 5. This instrument samples out of the electrospray only a narrow range of mobilities selected at will around a certain value Z_0 (for instance, the highest peak of the distributions of Figure 5). Given an unknown aerosol sampled from an electrospray of a polydisperse colloid, passage through the DMA would yield neither a fixed charge nor a known particle diameter, but a distribution of particles whose

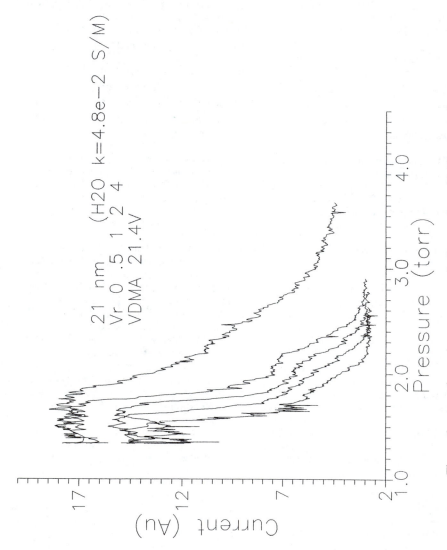

Figure 6. Aerodynamic spectrum similar to those in Figure 4, for a suspension of PSL particles 21 nm in diameter. The first step is from solution residues.

charge q and diameter d_p are uniquely related via $Z_0 = Z_1(d_p) q/e$. Subsequent passage through the impactor would provide an independent piece of information relating q and $\rho_p d_p$. The inversion of the aerodynamic spectra could then be carried out unambiguously provided ρ_p and the relation $S^*(\pounds)$ of equation 9a are known. In fact, the aerodynamic spectra of Figure 4 have been obtained by first sampling the aerosol through the DMA at the peak of the mobility distribution.

Results with Colloidal Suspensions of Monodisperse PSL Spheres.

We have carried out a series of measurements with water suspensions of PSL spheres seeded with TFA. The spheres were supplied by Duke Scientific as calibrated size standards with nominal diameters as shown in Table I, and with an unknown standard deviation. The table shows also the electrical conductivity of the corresponding solutions as well as the ratio between d_{min} from equation 3 and the nominal sphere diameter. All the solutions were electrosprayed into CO_2 because the high surface tension of water requires large voltages for the cone-jet to set in, which tends to produce electrical breakdown when the gas is air (*26, 19*). The aerosol is sampled from the spray region and passed through the DMA. This instrument operates in such a way that the exiting aerosol is carried in dry air, and its selected mobility corresponds nearly to that in air rather than in CO_2.

Table I: Characteristics of PSL Solutions Electrosprayed

d_{PSL} (nm)	K (S/m)	$\dfrac{d_{min}}{d_{PSL}}$	η
74	4.70E-3	16	0.65-0.9
74	1.38E-2	8	1.03-1.29
74	3.00E-2	5	1.34
74	4.60E-2	3.5	1.16
63	1.38E-2	9.5	0.93
54	1.38E-2	11	1.04
40	1.38E-2	15	0.84-1.26
30	4.8E-2	8	0.87-1.34
21	4.8E-2	12	0.95-1.5

The effect of varying the electrical conductivity of the liquid has been explored in greatest detail for the solutions with $d_{PSL} = 74$ nm. The following qualitative trends were found. As d_{min}/d_{PSL} decreases, the current through the impactor decreases, while the electrospray becomes less stable. At the lowest ratio at which we could run ($d_{min}/d_{PSL} = 3.5$) there was some agglomeration of the particles at the tip of the Taylor cone, making it unstable. Because the jet diameter is typically 1/2 of the drop diameter, it is tempting to speculate that the jet is becoming too narrow to transmit the particles efficiently. On the other hand, as d_{min}/d_{PSL} increases, the size of the residues from involatile solutes can become comparable with the size of the particles one wishes to measure. The conclusion is that a value of d_{min}/d_{PSL} between 5 and 12 works fine. In fact, this statement holds for $d_{PSL} = 74$ nm as well as for all the other particle sizes tried.

A variety of aerodynamic spectra similar to those of Figure 4 have been obtained for $d_{PSL} = 63, 54, 40, 30$ and 21 nm. All the data show similar trends, and allow a clear cut definition of the position of a step. By selecting a sufficiently large mobility in the DMA, it was possible in some cases to sample into the impactor simultaneously the PSL particles as well as the residue aerosol from the droplets not containing PSL. The result can be seen in Figure 6 for $d_{PSL} = 21$ nm, where two distinct steps are apparent. Which corresponds to the PSL and which to the residues may be determined

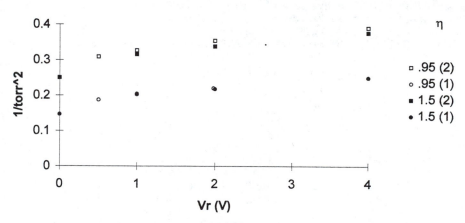

Figure 7. Effect of the repulsion voltage V_R on the position of the step in the aerodynamic spectra for PSL spheres 21 nm in diameter. The liquid flow rate shifts the steps for the residues, but not for the PSL particles.

by taking the spectra with different initial drop diameters, whereupon the residues become larger while the PSL remains nearly unchanged. This behavior is illustrated in Figure 7, which plots $(p_{1/2})^{-2}$ at the middle of each step (proportional to S^*) as a function of V_R for $d_{PSL} = 21$ nm. Data are shown for two different liquid flow rates, and only those corresponding to the (PSL) step at the largest $p_{1/2}$ fall into one curve for the two flow rates.

The contention (equation 17) that S^* is a function mostly of £ has been checked through changes in q (by varying the DMA voltage V_{DMA} for a given d_{PSL}), in V_R and in d_{PSL}. The effect of V_{DMA} can be seen in Figures 8a and 8b for $d_p = 74$ nm. Figure 8a shows raw data, while Figure 8b demonstrates how they collapse approximately on top of each other when represented in terms of £. Figure 9 provides a confirmation of the same point for all the particle sizes and various DMA voltages. Notice that the $S^*(£)$ curve is approximately linear, except at the smallest values of £, at which image attraction phenomena are clearly betrayed in the aerodynamic spectra by the large subcritical tails. Excluding these small £ data, the unknown size of a monodisperse aerosol can be obtained by extrapolating to zero V_R the curve $1/p^{*2}$ versus V_R. We find that this extrapolated value of S^* agrees well with the one obtained from singly charged particles of the same diameter.

In conclusion, electrospray atomization followed by a DMA and an aerosol size spectrometer provides a scheme suitable for on-line sizing of monodisperse nanoparticles originally in solution. However, one must carefully avoid interferences from residue particles formed in the electrospray.

Analysis of a Colloidal Suspension of SiO$_2$

The sols analyzed so far have served the purpose of illustrating the fact that very fine colloids can be properly dispersed and analyzed via electrospray and aerosol instrumentation. However, a somewhat more stringent test of this technique can be given by attempting the analysis of more complex colloidal suspensions. Our first attempts with colloidal gold (kindly supplied by Dr. Bret Halpern of Jet Process Corporation, New Haven) failed because the sol precipitates when the required quantity of acid is added. We have been more successful with an aqueous solution of silica (21% weight), kindly given to us by NISSAN, chemical branch (commercially available under the name SNOWTEX C). This solution is nominally stable even at relatively low PH. It was diluted greatly in deionized water before adding enough TFA to achieve an electrical conductivity K = 0.0931 S/m. Like in the case of water-PSL suspensions, the solution remained stable. The corresponding mobility spectrum is shown in Figure 10 where now Z (V-9.4 Volt) = 6.134 cm^2/s. The peak to the left corresponds to ions. Notice that there is only one additional peak (rather than one for the SiO$_2$ and another for the residue particles). This can be attributed to the fact that all the particles in the aerosol are ion emitters, which fixes the electric field on their surface independently of their size, thus determining also approximately their electrical mobility ($Z \sim q/d^2$ in the free-molecule range *27, 10*).

Figure 11 shows the aerodynamic spectra corresponding to the voltage at the peak of the mobility spectrum of Figure 10. These steps are shown in Figure 12 to arise at pressures which are independent of liquid flow rate, which establishes them as corresponding to the dissolved particles rather than to residues from involatile salts. Figure 9 represents also the $S^*(£)$ data corresponding to the SiO$_2$ colloid, which have been reduced based on the diameter $d_p = 13.7$ nm. This value was obtained by extrapolating to $V_R = 0$ the curves $1/p^{*2}$ versus V_R. It corresponds well with the nominal diameter of these particles (10-14 nm), while the associated $S^*(£)$ points fall right through the data for the PSL.

In conclusion, although there is plenty of room for improvements in the measurement technique proposed, it does clearly offer promise for on-line sizing of colloidal nanoparticles.

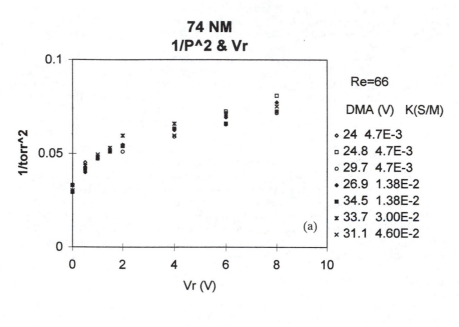

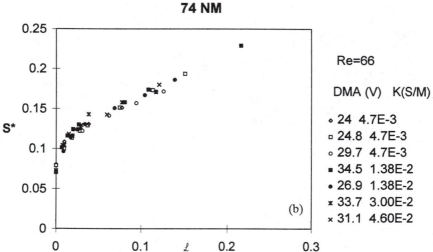

Figure 8.   Effect on S* of the charge q on PSL particles 74 nm in diameter.
q is controlled through the DMA. (a) raw data; (b) collapsed S*(£) data.

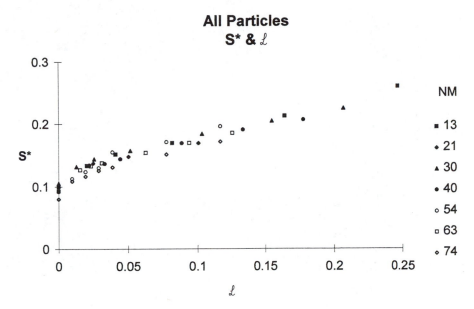

Figure 9. S*(£) curves for all data obtained with different values of the DMA voltage, the liquid flow rate and the diameter of the PSL particles.

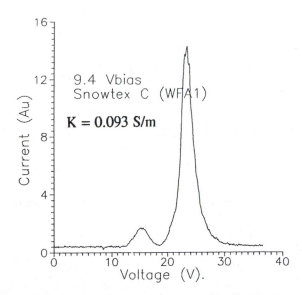

Figure 10. Mobility spectrum for an electrospray of colloidal silica (Snowtex C; 10-14 nm)

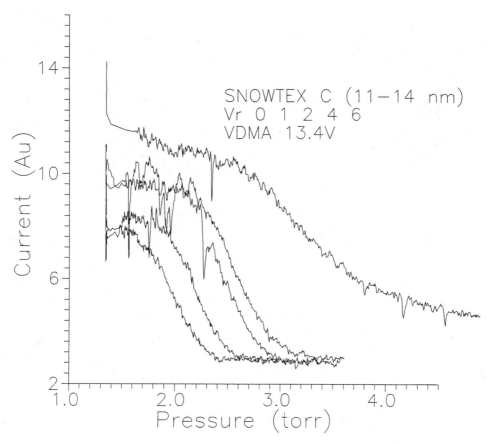

Figure 11. Aerodynamic spectra for the colloidal silica solution of figure 10

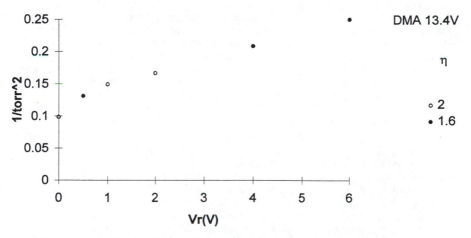

Figure 12. Effect of the repulsion voltage V_R on the position of the step in the aerodynamic spectra for the colloidal silica particles at two liquid flow rates.

Acknowledgment This work has would have been impossible without the contributions of J. Rosell, K. Serageldin, S. Brown, J. Lazcano and N. Davis to the development of the impactor. It has been sponsored by NSF grant CTS-9319051 and NIH Grant 5-R01-GM3 1660-07-08 in collaboration with Professor J. B. Fenn (Virginia Commonwealth University).

Appendix: Overcoming the High-Conductivity Requirement for Fine Electrosprays?

The scaling laws discussed for cone-jets require fairly large conductivities in order to generate very fine drops. But this constraint may perhaps be overcome with the help of an electrospray regime recently discovered by Tang and Gomez. (*20*) Notice, however, that much work remains to be done before this possibility is demonstrated under conditions relevant for the analysis of suspended nanoparticles.

When $Q = Q_{min}$, equation 6 implies that all the negative ions brought by the liquid into the cone are neutralized at the capillary needle, and cannot therefore reach the cone-jet. But the positive ions are transmitted in full, so that the liquid cone cannot remain neutral and must become unstable. However, this restriction holds only when the gas surrounding the cone-jet is an insulator. Otherwise, charge neutrality can be assured by negative ions entering the cone tip from the gas and moving towards its interior. This would, incidentally, also supply a current larger than the minimum of equation 6. What Tang and Gomez (*20*) have discovered is a regime in which water can be electrosprayed at flow rates Q some ten times smaller than allowed by the limit (8). When faced with this paradox, they went on to show that: (i) their spray current was indeed larger than allowed by equation 6, and (ii) mass spectrometric analysis of the ions in the spray region revealed the presence of gas ions. Their regime thus undoubtedly involves a mildly conducting gas. Another most interesting finding for the Tang-Gomez electrospray regime is that d scales with the 2/3 power of Q,

$$d \sim (\rho Q^2/\sigma)^{1/3}, \qquad \qquad (A1)$$

rather than as the 1/3 power of equation 4. Interestingly enough, the empirically found dependence of equation A1 is exactly as one would expect for an equipotential cone-jet, while equation 4 is due to the finite conductivity of the liquid (*16*). It thus appears that conduction through the gas has two important effects relaxing the need to increase the conductivity, and to reduce the flow rate. On the one hand, it removes the limitation on a minimum flow rate or a minimum d for a given K. On the other, it eliminates charge relaxation effects leading to the much faster 2/3 power law (equation A1), which allows reaching smaller drop diameters with less drastic reductions in the flow rate.

Literature Cited
1   Danek, M.; Jensen, K.F.; Murray, C.B.; Bawendi, M. G., *J. Crystal Growth*, **1994**, *145*, 714-729, 1994; also *Appl. Phys. Lett.*, **1994**,*65*, 2795-2797.
2   Fenn, J. B.; Mann, M. ; Meng, C. K.; Wong, S. K.; Whitehouse, C., *Science*, **1989**, *246*, 64-71.
3   Cloupeau, M.; Prunet-Foch, B., *J. Electrostatics*, **1989**, *22*, 135; Aguirre de Carcer, I.; Fernández de la Mora, J., *J. Coll. Int. Sci.* **1995**,*171*, 512-517.
4   Nohmi, T.; Fenn, J. B. *J. Am. Chem. Soc.*, **1992**,*114*, 3241-3246.
5   Dole, M., Mach, L. L., Hines, R. L., Mobley, R. C., Ferguson, L. P., Alice, M. B., *J. Chem. Phys.*, **1968**,*49*, 2240.
6   Winger, B. E.; K. J. Light-Wahl. R. R. O. Loo, H. R. Usdeth and R. D. Smith, *J. Am. Mass Spectrom.*, **1993**,*4*, 536-545.
7   Rulison, A.; Flagan, R. C., *J. Am. Ceram. Soc.*, **1994**, *77*, 3244-3250.
8   Salata, O. V.; Dobson P. J.; Hull, P. J.; Hutchison, J. L., *Thin Solid Films*, **1994**, *251*, 1-3.

9   Loscertales I. G.; Fernández de la Mora, J. In *Synthesis and Characterization of Ultrafine Particles*, Marijnissen, J ; Pratsinis, S., Eds., Delft University Press, 1993, pp. 115-118.
10 Loscertales, I. G.; Fernández de la Mora, J.,**1995**, *J. Chem. Phys,193,* 5041-5060.
11 Wilm, M. S.; Mann, M., *Int. J. Mass Spectrom. Ion Processes* , **1994**,*136,* 167.
12 Fuerstenau, S.; Benner, W. H., Proceedings of the Annual Meeting of the American Association for Aerosol Research, Pittsburgh, PA, 10-October-1995.
13 Fernández de la Mora, J., On the outcome of the Coulombic fission of a charged isolated drop, J. Coll. Interface Sci., accepted, 1995.
14 Rosell-Llompart, J., *Ph.D. Thesis,* Yale University, New Haven, CT, 1994.
15 Rosell-Llompart, J.; Fernández de la Mora, J., *J. Aerosol Sci.,* **1994**, *25,* 1093.
16 Fernández de la Mora, J.; Loscertales, I. G., *J. Fluid Mech.,* **1994**, *260,* 155.
17 Barrero, A.; Gañán, A., In *Proceedings of the 4th International Aerosol Conference,* Flagan, R.C., Ed., Los Angeles, California, 29 August-1994.
18 Chen, D.R.; Pui, D. Y. H.; K.; Kaufman, S., *J. Aerosol Sci.,* **1995**, *26,* 963-977
19 Smith, D. P. H.; *IEEE Trans. Ind. Appl.,* **1986**, IA-*22,* 527-535.
20 Tang, K.; Gomez, A., *J. Coll. Int. Sci.,* in press, 1995.
21 Marple, V. A., *Ph.D. Thesis,* U. Minnesota, 1970.
22 Rao, N.; Fernández de la Mora, J.; McMurry, P., *J. Aerosol Science,* **1992**, *23,* 11.
23 Fernández de la Mora, J., *Chem. Eng. Communications,* to appear, 1995.
24 Liu, P.; Ziemann, P. J.; Kittelson, D. J.; McMurry, P., *Aerosol Sci. & Tech.,* **1995**,*22,* 293.
25 Liu, B. Y. H.; Pui, D. Y. H. , *J. Coll. Interface Sci.,* **1974**,*47,* 155-171.
26 Zeleny, J., *Proc. Phil. Cam. Soc.,* **1915**, *18,* 71-83.
27 Katta, V.; Rockwood, A. L.; Vestal, M. L.,*Int. J. Mass Spectrom. Ion Proc.,* **1991**, *103,* 129-148.

RECEIVED January 11, 1996

Chapter 3

In Situ Characterization and Modeling of the Vapor-Phase Formation of a Magnetic Nanocomposite

M. R. Zachariah[1], R. D. Shull[2], B. K. McMillin[1,4], and P. Biswas[3]

[1]Chemical Science and Technology Laboratory and [2]Materials Science and Engineering Laboratory, National Institute of Standards and Technology, MS 221/B312, Gaithersburg, MD 20899–0001
[3]Department of Environmental Engineering, University of Cincinnati, Cincinnati, OH 45221–0071

Gas phase combustion synthesis offers the potential to produce bulk quantities of particles with controllable morphologies and chemistries. In this paper we demonstrate the use of an atmospheric pressure flame to produce a nanocomposite with unique magnetic properties and investigate the salient features of their formation through the application of in-situ diagnostics and modeling.
    Magnetically isolated nanometer sized magnetic particles can show magnetic behavior different than those found in the bulk and represent an interesting example of size dependent properties. We have investigated the application of flame processing to the synthesis of this class of materials, in which the goal has been to encapsulate a magnetic cluster within a non-magnetic host particle. A premixed methane/oxygen flame diluted with nitrogen has been used as the reacting environment in which iron carbonyl and hexamethyldisiloxane was added as the magnetic and non-magnetic precursor materials. Nanometer sized composite particles were formed, containing 5-10 nm γ-Fe_2O_3 encased in a silica particle whose diameter ranged from 30-100 nm depending on loading and flame temperature and are shown to have superparamagnetic behavior.
    Planar laser-based imaging measurements of fluorescence and particle scattering have been obtained during flame synthesis of the iron-oxide/silica superparamagnetic nanocomposites. The results indicate that the vapor phase FeO concentration is very sensitive to the amount of precursor added, indicating a nucleation controlled growth. The FeO vapor concentration in the main nucleation zone was insensitive to the amount of silicon precursor injected, implying that nucleation of each component occurred independently from the other.

[4]Current address: LAM Research, Fremont, CA 94538

Modeling of the gas phase chemistry and particle growth shows that homogeneous nucleation occurs early on in the process followed by heterogeneous condensation processes. Molecular dynamics computation have indicated that the iron oxide clusters will phase segregate and migrate toward the inside edge of the silica cluster consistent with experimental observation.

Gas phase combustion is well proven method for the bulk production of fine powders. On the other hand it has scarcely been investigated from the fundamental perspective, of the chemistry and physics of particle formation and growth. Our interest is in the development of a more thorough understanding of how to control particle formation in order to enable the economical production of materials with unique and useful properties. With the increasing interest in composite powders, control of both morphology and chemistry take on greater importance.

One example of the use of nanostructured materials has been the recent theoretical investigations into magnetic refrigeration technology which have suggested a novel approach to obtaining a higher operating temperature than that achievable through conventional paramagnetic materials (1,2). The principles of magnetic refrigeration invoke a magnetic entropy cycle in which application of a magnetic field results in alignment of magnetic moments and a resulting decrease in the magnetic entropy. This implies that the temperature increases (also known as the magnetocaloric effect) and results in a rejection of heat to the surroundings. Upon removal of the field the system will absorb heat and complete the cycle. The basic features relevant to the magnetic properties of importance for this work are illustrated in Fig. 1 for para, ferro and superparamagnetic materials. The magnetocaloric effect for the purposes of our discussion is what controls the extent of cooling, and is proportional to the temperature dependence of the magnetization at constant field $(\delta M/ \delta T)_H$, and inversely proportional to the heat capacity of the material. As seen in Fig. 1, the magnetization of a paramagnet decreases rapidly to a small value at higher temperatures and would be of limited value for cooling purposes. For a ferromagnet, the magnetization is high below the Curie point where the magnetic spins are aligned, and only weakly interact above the Curie point, which means that the magnetochloric effect peaks sharply only at the Curie point. Never the less at high temperatures, the magnetocaloric effect for a ferromagnet is larger than that for a paramagnet. If properly constructed ferromagnetic clusters embedded in a non-magnetic host can act as individual domains where the spins within a cluster are aligned, but randomly oriented with respect to neighboring clusters when the clusters are spaced sufficiently far apart. The result is a magnetization-temperature curve shifted from the paramagnetic case to higher temperatures. The implications are that for the same applied field the nanocomposite should give an enhanced temperature change or conversely the same change at lower applied field. Under such conditions one obtains a *"superparamagnetic nanocomposite"* with a much larger magnetocaloric effect. i.e. $dT_{nano} > dT_{para}$.

Strategy for Flame Synthesis

One of the issues in the practical application of nanostructured materials is the ability to synthesize sufficient quantities with appropriate and controllable properties at low cost. Gas phase combustion processes provide one of the best methods for the bulk production of nanostructured materials due to the relative ease of scale-up and the relative simplicity of the process. The magnetic system chosen for study was γ-Fe_2O_3/SiO_2 because a composite of this class has been demonstrated to yield superparamagnetic properties (3) via a liquid phase synthesis route.

The reactor configuration chosen for this work was a premixed methane/air flame as the reaction environment to which iron and silicon bearing precursors were added to obtain a nanocomposite composed of the magnetic and non-magnetic host

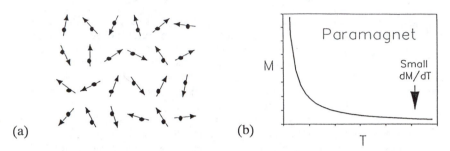

Schematic of the (a) magnetic spins in a paramagnetic material and (b) the temperature dependence of its magnetization at constant applied field.

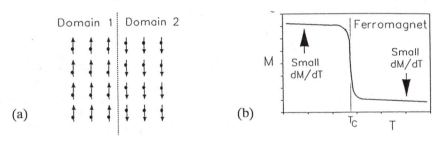

Schematic for a ferromagnetic material of the (a) magnetic spins showing a boundary (dashed) between domains and (b) the temperature dependence of its magnetization at constant H.

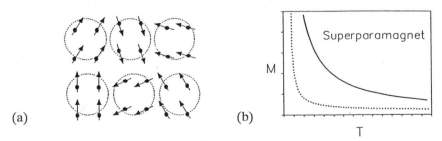

Schematic of the (a) magnetic spins in a superparamagnet with the magnetic clusters indicated by dashed circles and (b) the temperature dependence of M for a superparamagnetic nanocomposite (solid line). Also shown for reference in (b) is the M vs T curve (dashed) for a paramagnet possessing only 1/3 the magnetic moment of the superparamagnet.

Figure 1. a) Strucure of magnetic spins for para, ferro and superparamagnets.
b) Magnetization vs. temperature at constant magnetic field.

respectively. Figure 2 shows schematically the flame reactor and the particle quenching/collection cold finger. The flame was sustained with a premixed composition of methane/oxygen and nitrogen in various proportions (always lean in fuel) to obtain temperatures that varied between 1800 and 2500 K which were measured by two color optical pyrometery. The precursor species, iron carbonyl and hexamethyldisiloxane was added as the magnetic and non-magnetic precursor materials, and were delivered to the reactor as saturated vapors in argon. The logic behind this choice of reactor geometry was that such a configuration enjoys the advantage of a very rapid rise in temperature (< 1 ms) and the ability to sustain a high temperature for a relatively long period of time (> 100 ms) if desired. This was an important processing consideration in that based on phase equilibria considerations what we hoped to do was to form nanodroplets of two immiscible liquids (Temperature > 2000 K) of iron oxide and silicon oxide, which could be very rapidly quenched to preserve this structure and minimize any chemistry between the iron and silicon which might lead to Fe_2SiO_4, the favored product below 1500 K. The overall growth mechanism envisioned for this process is illustrated in the cartoon depicted in Fig. 3 and serves to illustrate the variety of processes that might be taking place.

The cold finger had a dual purpose. Quenching the chemistry and subsequent particle growth processes, and as an efficent method of collecting bulk quantaties of powder in which the residence time could be reasonably well defined and controlled. In general, most of the bulk samples studied were collected with residence times less than 5 ms. Samples for TEM analysis were obtained by the rapid insertion of grids directly into the flame at specified heights above the exit nozzle of the reactor.

Morphology and Chemical Composition

X-ray diffraction studies on bulk samples of the powders have been conducted on both the Fe/Si/O nanocomposite and Fe/O particles and are shown in Fig. 4. The XRD pattern from the nanocomposite shows a high background and a broad peak at 22° indicative of amorphous silica. The more defined lines correspond closely, both in location and relative magnitude to Fe_3O_4/γ-Fe_2O_3 , with line widths suggestive of very small crystallites. A definitive determination of the phase as either Fe_3O_4 or γ-Fe_2O_3 is difficult due to the fact that the peaks are broad and the overlap of magnetite and maghemite are severe. However, raman spectra did indicate that the material is mainly γ-Fe_2O_3. By contrast, XRD results when only the iron precursor are added indicate the iron particles are more crystalline than in the nanocomposite. The fact that the XRD patterns of the nanocomposite show distinct patterns for iron oxide implies that we are making a nonhomogeneous particle in which there are iron rich regions. Interestingly, the chemical state of the iron in the nanocomposite and the isolated iron-oxide were essentially the same. The clearest difference seems to be that the iron-oxide particles show reflection from at least two phases hematite and maghemite/magnetite while hematite is not observed in the composite.

Figure 5 shows micrographs of the nanocomposite particles captured above the exit nozzle of the reactor (total residence time < 3 ms) by rapid insertion of the TEM grid (4). The TEM images clearly indicate that the particles are highly spherical and unagglomerated and composed of two distinct regions. The darker encapsulated regions are iron rich and are encased with a silicon rich matrix. The composite particles range in size from 25 - 100 nm in diameter with the inclusions typically less than 10 nm in diameter. EELS analysis confirmed that the dark inclusions within the silica matrix were iron rich with an Fe/O atom ratio between 0.7-0.9 and that iron inclusion within

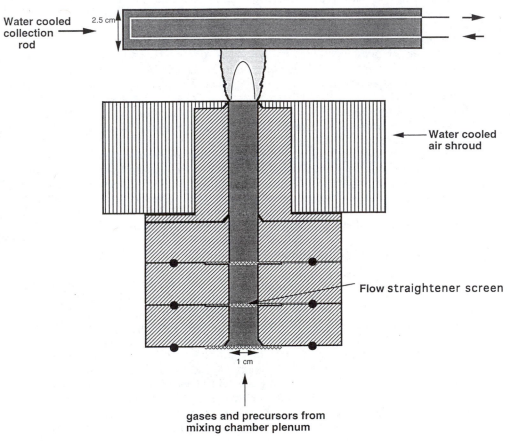

Figure 2.   Schematic of premixed burner and cold finger.

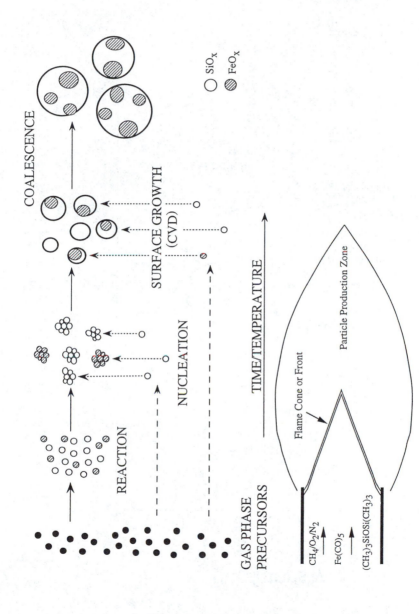

Figure 3.   Illustration of the various processes important in the formation of the nanocomposite.

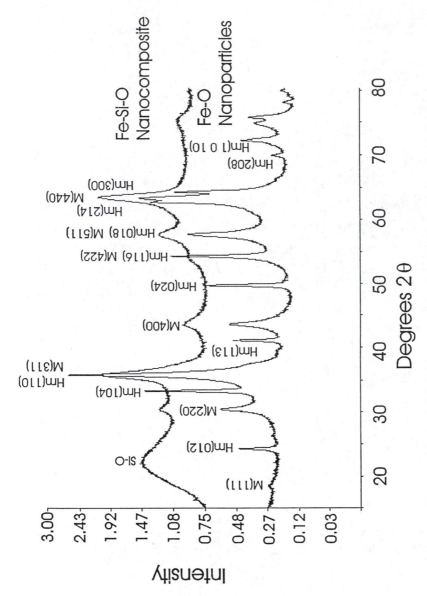

Figure 4. X-ray diffraction patterns of iron-oxide and nanocomposite. M(hkl) = maghemite/magnetite; Hm(hkl) = hematite.

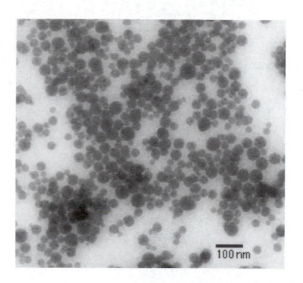

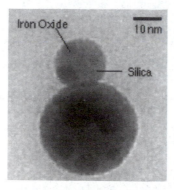

Figure 5. TEM's of as-formed iron oxide/silica nanocomposite.

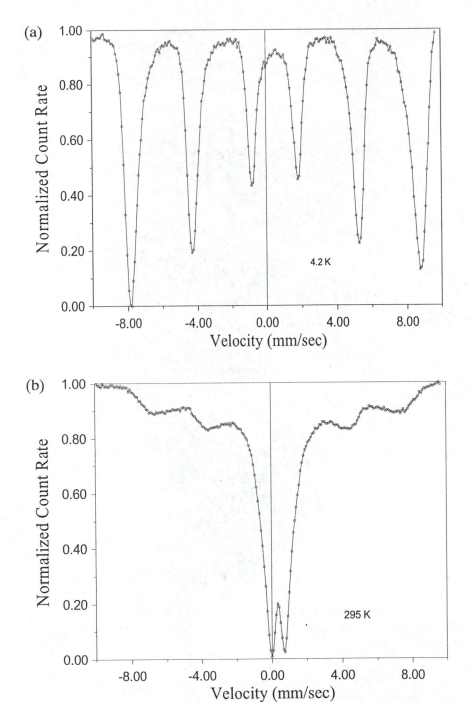

Figure 6. Mössbauer patterns measured with Si/Fe=1.3 at (a) 295 K and (b) 4.2 K.

the silicon rich matrix went as low as Fe/O=0.04 indicating that some iron is finely dispersed throughout the matrix, either as very small clusters or in chemically bound form to silicon. There were, however regions where no detectable traces of iron were found.

By varying the Si/Fe ratio the separation within each of the magnetic clusters could be varied, while varying the Fe precursor concentration resulted in changes in the average size of the iron containing inclusions. However, the fact that the particle sizes and chemical composition of the iron rich regions was similar in both the composite and pure iron particle is suggestive that gas phase chemistry and subsequent homogeneous nucleation is unaffected even at these high temperatures by the presence of the other nucleation component. More direct evidence of this point will be discussed later in this paper when we investigate some of the gas phase chemistry.

Magnetic Properties

Both magnetic susceptibility and Mössbauer spectra were measured on compacts of these powders. At room temperature, the Mössbauer spectrum for the sample produced with a Si/Fe ratio of 1.3 is shown in Fig. 6a. The spectrum is a superposition of at least two spectra: (1) a less intense very broad multiple-line spectrum extending to high velocities indicative of varying-sized regions of magnetically-ordered iron oxide and (2) a much larger intensity central doublet (possessing an isomer shift of 0.37 mm/sec and quadrupole splitting of 0.74 mm/sec) indicating superparamagnetism similar to that reported by Shull et al. (5) for submicron particles of Fe_3O_4. The form of the superparamagnetic particles present in this composite may be deduced from the Mössbauer spectrum measured at temperatures lower than the blocking temperatures of these submicron particles. Upon cooling from room temperature, the magnetic susceptibility data do not indicate the presence of any magnetic phase transition, but the Mössbauer spectrum measured at 4.2 K showed a dramatic change. At 4.2 K, Fig. 6b shows that only a single six-line spectrum, although broadened, is observed. The loss of the central Mössbauer doublet on cooling to 4.2 K indicates this temperature is below the blocking temperatures for the submicron iron-oxide particles. In addition, the single spectrum observed at low temperature also implies that the two components which comprised the room temperature spectrum are just different size distributions of the same form of iron oxide; the magnetically-ordered portion being simply those large particles with blocking temperatures above room temperature. The measured magnetic hyperfine field of 41.2 MA/m (518 kOe) and 0.51 mm/sec isomer shift are only slightly different from the values (41.7 MA/m and 0.47 mm/sec respectively) expected for γ-Fe_2O_3. The only other possible identification, consistent with the earlier described x-ray results and the high magnetization possessed by these materials, would have been Fe_3O_4 (magnetite), but then a low temperature Mössbauer spectrum comprised of at least eight readily separated absorption peaks would have been observed. Consequently, the form of the iron oxide embedded in the silica is deduced to be the high temperature form of Fe_2O_3, maghemite.

Magnetization measurements as a function of magnetic field and temperature were performed in a SQUID magnetometer, and the data above 155 K are presented in Fig. 7 as a function of the reduced parameter H/T. Separately measured magnetization loops while the magnetic field was cycled to positive and negative values showed no hysteresis in this temperature range. The lack of hysteresis and the superposition of all the data in Fig. 7 show that this material is indeed superparamagnetic above 155 K with a particle size which is not a function of temperature, consistent with the Mössbauer

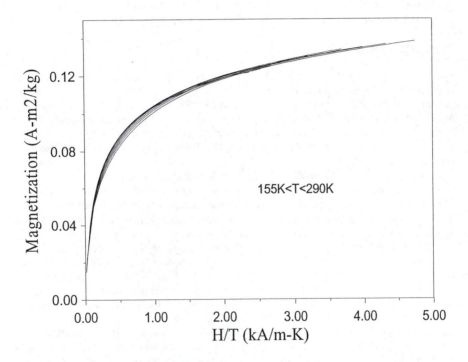

Figure 7. Magnetization (M) vs. temperature normalized magnetic field (H/T)
taken between 155 K and 290 K.

data described above. Below 155 K magnetic hysteresis was observed, again consistent with the Mössbauer results.

In-Situ Measurements

In situ measurements aimed at investigating the gas phase species concentration, temperature field, and particle size offer the best potential for a better understanding of the underlying chemical/transport phenomena that occur during particle synthesis. Laser-based diagnostics offer nonintrusive, sensitive methods for measuring particle size/number density, gas temperature, and species concentration in these reactors (6-9).

Advances in CCD camera technology have enabled the application of standard spectroscopic and light scattering methods to two dimensions (10). In these two-dimensional measurements, a thin sheet of laser light is used to illuminate the flow and the resulting signal is collected with a lens and imaged onto a CCD camera. In addition to providing efficient data collection, these planar imaging measurements are useful because they provide a means to simultaneously measure flow properties and visualize flow structures in two-dimensions, with excellent spatial resolution.

Experimental Details: As shown schematically in Fig. 8 the laser excitation source used for the optical measurements was a XeCl excimer-pumped dye laser, operating at 10 Hz with a ~30 ns pulse duration, ~5 mJ pulse energy, and ~0.2 cm^{-1} bandwidth. Using a cylindrical and spherical lens combination, the beam was expanded vertically, formed into a thin sheet, and directed through the center of the flame. Within the imaged region of the flame, the laser sheet measured ~300 μm x 35 μm. The laser energy and spatial distribution were monitored during the experiments by directing a 5% reflection of the laser sheet onto a static dye cell, and recording the resulting fluorescence with a video CCD camera and frame-grabber computer board. For the laser induced fluorescence (LIF) measurements, the laser was tuned to ~559.5 nm to excite the P(17)+R(40) transition in the $^5\Delta_4 \leftarrow X^5\Delta_4$ (0,0) orange system of FeO. For the particle scattering measurements, the laser was tuned to 530 nm, where no gas phase transitions exist.

For both the fluorescence and particle scattering measurements, the signal was collected at a right angle to the illumination plane and imaged onto an intensified, cooled CCD camera (576 X 384 pixels, each 23 μm square) using an f/4.5 lens. To suppress flame emission within the images, the intensifier was synchronized to, and gated on for a ~75 ns during each laser pulse. For the fluorescence measurements, long and short-pass filter combinations were used to block laser scattering and prevent flame luminosity, outside of the LIF wavelength region, from reaching the detector. The images obtained were typically averaged over 250-300 laser shots to improve the signal-to-noise ratio, and were spatially averaged 2 X 2 pixels which is the effective resolution of the intensifier.

All of the images were corrected for camera dark background, flatfield uniform response of the camera and collection lens, and laser energy and spatial distribution. The fluorescence images were also corrected for laser-induced particle incandescence and scattering by subtracting images obtained with the laser detuned from the absorption transition. The video CCD images of laser-induced fluorescence from a static cell of dilute laser dye solution were used to normalize the laser energy and vertical spatial distribution in the corrected images. The laser profile images were re-mapped from the video CCD to the intensified CCD coordinates based on images obtained with the laser sheet masked.

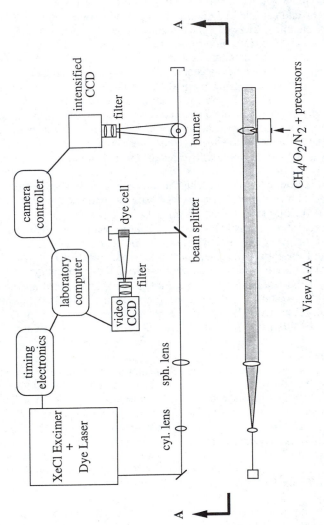

Figure 8.   Schematic of planar laser induced/scattering experimental apparatus.

Results and Discussion: Two-dimensional images of FeO(g) LIF and particle scattering were obtained for a variety of cases in which the iron precursor, and iron and silicon precursors were added.

An example of the characteristic images obtained are shown in Fig's 9a-c, for flame emission, FeO fluorescence and Mie scattering for the iron only precursor feed, while Fig. 9d shows an additional scattering image for a flame seeded with both iron and silica precursors.

The FeO(g) LIF image of the flame region shown in Fig. 9b, indicates that FeO(g) appears only outside the flame front, implying that little if any precursor decomposition occurs prior to the front. As the flow continues away from the flame front, there is an onset of particle formation and the FeO(g) concentration decreases downstream. The corresponding scattering image (Fig. 9c) reveals a central region wherein particle concentrations are low or the particles are too small to yield detectable signals (white region). Further downstream, an increase in the scattering signal is observed, consistent with the decreasing FeO(g) concentration. The scattering signal is highest near the radial boundary of the flame, presumably because particles are much larger there owing to faster nucleation rates resulting from the lower temperature, and the longer times available for growth by coagulation.

Fig. 9d shows a particle scattering image from a flame seeded with both iron and silicon precursors. In comparing the two scattering images in Fig. 9, we find that the case with both silicon and iron precursors present is markedly different in two ways: (1) the scattering signals are generally much larger for the cases with the silicon precursor present, compared to those with only the iron precursor; and (2) significant particle scattering is observed within the central region of the flame just above the cone tip, for cases with silicon precursor present.

While the two-dimensional images provide a great deal of qualitative insight into the chemistry-flowfield interactions, examining the centerline profiles simplifies the analysis, since the flow is ideally one-dimensional along that streamline. Figure 10 shows the concentration of FeO(g) as a function of time (or axial position) for different inlet concentrations   The results are presented as a function of time from the flame front, since the precursor decomposition essentially begins at the conical flame front where the temperature rapidly increases.

As illustrated in Fig. 10, the concentration of FeO(g) increases rapidly at the flame front, with a similar rate of increase for each precursor feed rate. This is followed by a decrease in the FeO(g) concentration due to the conversion of the vapor to the particle phase. The two processes of precursor oxidation (FeO(g) formation) and that of particle formation (FeO(g) consumption) take place simultaneously; however, the profiles indicate that the precursor oxidation is the faster of the two initially. This behavior is expected since one might approximate the FeO(g) formation rate to be (pseudo) first order with respect to the precursor concentration due to the excess oxygen, while the consumption (at least at early times) will be proportional to the square of the FeO(g) concentration resulting from FeO(g) dimerization. An analysis of this type has been performed by expressing the particle formation or nucleation rate as a kinetic process (6, 11-13) details of which are presented in this paper in the section on modeling.

Using an iron only precursor feed rate as a baseline condition, experiments were conducted with iron and two different silicon precursor feed conditions. Representative centerline profiles for the FeO(g) relative concentration are shown in Fig. 11. In comparing the FeO(g) concentrations for cases with just iron and with both iron and silicon feeds, there is no significant difference in the initial rise and peak in

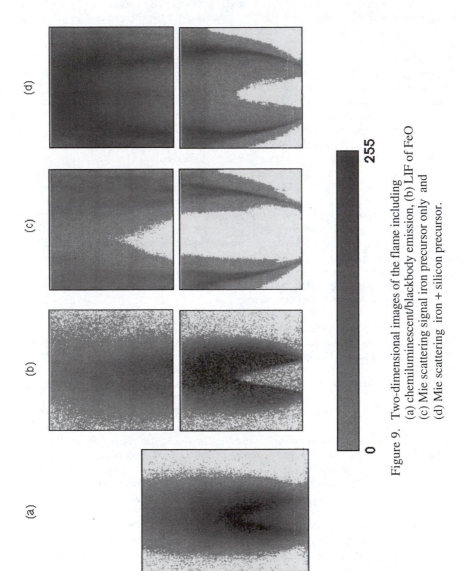

Figure 9. Two-dimensional images of the flame including (a) chemiluminescent/blackbody emission, (b) LIF of FeO (c) Mie scattering signal iron precursor only and (d) Mie scattering iron + silicon precursor.

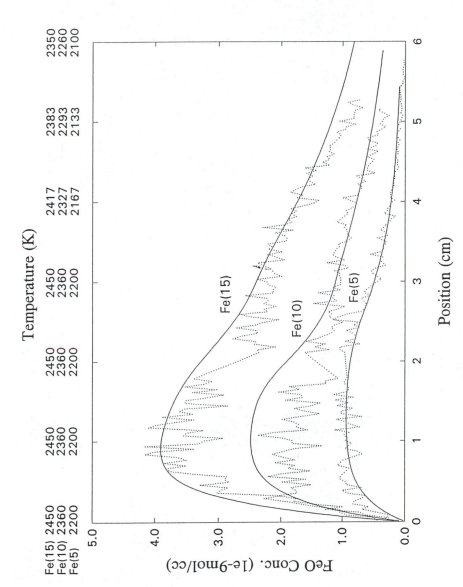

Figure 10. Centerline profile of FeO(g) concentration as a function of time from the flame front. Mole fractions: Fe(5) = 1.6E⁻⁴; Fe(10) = 5E⁻⁴; Fe(15) = 8E⁻⁴.

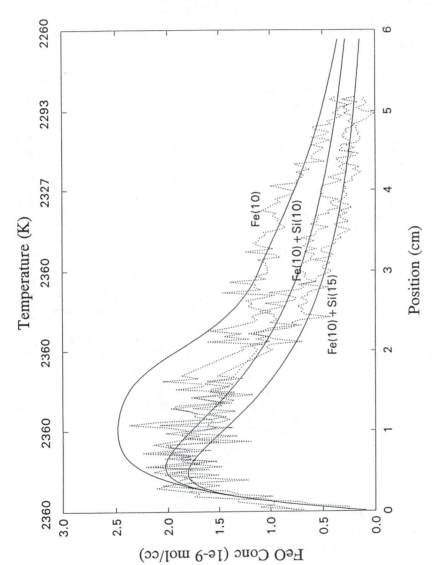

Figure 11. Centerline profile of FeO(g) concentration as a function of time from the flame front with just iron Fe(10)=5E^{-4} and both iron oxide and silica precursors. Si(10)=5.5E^{-4}; Si(15)=9.2E^{-4}.

FeO concentration. This indicates that the chemistry of FeO is independent of the added silicon. Recall that at these temperatures iron-silicate is not stable in the solid, the results indicate that iron-silicon gas phase species are not either. Some differences are observed at distances downstream, with the FeO(g) concentration being slightly lower for the "iron and silicon" feed conditions. For the iron only feed condition, the FeO vapor is converted to particles by homogenous growth. When both precursors are used, there is an additional pathway for transformation of FeO vapor; that is, the silica particles provide an additional surface. This results in a slightly faster decrease in the FeO(g) concentrations as observed in Fig. 11.

Phenomenologic Aerosol Modeling

Phenomenologic modeling of the aerosol growth has been applied to the formation of this nanocomposite through the application of the aerosol general dynamic equation (GDE) (14,15). We apply the GDE to describe the transition between chemistry which leads to a condensable species, and the subsequent processes of gas phase polymerization, via cluster-cluster collisions and cluster-monomer interactions. Essentially we are implying that gas to particle conversion is controlled by kinetic processes and not thermodynamic (12). In general this is a good assumption for ceramics, since the chemistry in most practical systems will be sufficiently fast as to leave the monomers in a highly supersaturated state such that issues of critical cluster size employed in classical nucleation theory can be ignored. This being the case, if we know the monomer generation rate then subsequently processes can be accounted for as follows:

Monomer Balance

$$\frac{dN_1}{dt} = -N_1 \sum_{i=2}^{\infty} \beta_{1,i} N_i + R_m \tag{1}$$

Cluster/aerosol balance

$$\frac{dN_i}{dt} = \frac{1}{2} \sum_{j=2}^{i-1} \beta_{j,i-j} N_{i-j} - N_i \sum_{i=2}^{\infty} \beta_{i,j} N_j \tag{2}$$

where N_1 represents the monomer concentration, R_m is the monomer generation rate (obtained from gas chemical kinetics), N_i is the concentration of molecule, cluster or particle of i'mers. The first term in cluster/aerosol balance equation represents the formation rate of species i from smaller structures, while the second term accounts for its removal rate through polymerization. The rate coefficients β_{12} accounts for hard sphere collisions with no activation energy and are obtained from kinetic theory. Because, the size range of interest extends from angstroms to hundreds of nanometers, the set of equations as expressed in there general form encompass several million ordinary differantial equations. In order to solve this system, the size domain of interest is discretized into size bins numbering typically less than 50. The exact solution procedure used here has been described extensively by others (14,15) and in more detail for this work in an upcoming publication (11). As written, equations 1 and 2 are for a single component system, but there is in principle no limitation on the number of monomeric species and there interparticle reactions. For the system under study here it was assumed that the primary monomeric species N_1 was Fe_2O_2 and $H(OH)Si_2O_2$ for the iron and silicon species respectively which were assumed to be formed as follows:

$Fe(CO)_5 ===> FeO$ $k_1 = 1.5 \times 10^{12} \exp(-10060/T)$

$FeO + FeO => Fe_2O_2$ = monomer $k_2 = 10^{(5440/T + 8.9)}$

$(CH_3)_3Si-O-Si(CH_3)_3 ===> HSi(O)OH$ $k_3 = 4.9 \times 10^{17} \exp(-48288/T)$

$HSi(O)OH + HSi(O)OH => H(OH)Si_2O_2$ = monomer $k_4 = 2.0 \times 10^{23} T^{-3.0} \exp(-5700/T)$

where k is in [cm-mole-sec] units; Values are obtained from reference 11.

These rates are used to calculate R_m as a function of time, which in turn are the source rates for the aerosol dynamic equations. Notice that k_2 and k_4 have a negative temperature dependance. This is characteristic for recombination reactions and in the language of classical nucleation theory imply that as the temperature increases the nucleation rate should drop. A more thorough discussion of this point can be found in reference 12.

Results from the model calculations are presented as the solid lines in Fig's 10 and 11. It should be pointed out here that the results were obtained by first finding a reasonable fit to the Fe(10) case in Fig. 10, which primarily involved defining the production rate of FeO (k_1) to be 1/40th the reported literature value for the thermal decomposition of iron carbonyl. This is a reasonable assumption since the formation of FeO should be considerably slower than the known rate constant for the initial decomposition of the iron carbonyl. Rates constant for k_2 was obtained by scaling to the temperature dependance of the rate constant for SiO dimerization (12) and a best fit to experiment. k_4 was obtained from quantum chemical calculation combined with reaction rate theory (12,13). As seen the model does a reasonable job of predicting the precursor concentration sensitivity to the measured FeO concentration. The model indicates that the FeO dimer concentration is the rate limiting process during the early stages of particle formation and that during the later stages surface growth (heterogeneous condensation) on existing particles becomes the dominant mechanism. Model comparison with added silicon shown in Fig. 11 predict a relative independence in the FeO concentration to the formation of silica. However, the model does show some enhancement in nucleation, due to the presence of larger particle surface area when the amount of the silicon precursor is increased.

Atomistic Modeling

Atomistic classical molecular dynamics modeling as well as quantum chemical calculation for small molecular species can give both insight and fundamental data on nucleation and cluster growth processes.

We have applied ab-initio molecular orbital methods in conjunction with reaction rate theory in a systematic procedure for calculating rates for chemical nucleation/reaction processes (12,13). An example of such an approach has been the calculation of the nucleation of SiO, a monomer to silica particle formation. The results of the computations indicate the rate constants are both pressure and temperature sensitive. This sensitivity results from the rate of energy transfer with the surrounding bath gas (M). Higher temperatures and lower pressures both lower the bath gas encounter rate with the growing cluster and therefore the nucleation rate. The individual kinetic steps in a nucleation event can be expressed as follows:

$A_{n-1} + A$	=>	$A_n\#$	; Cluster growth/heating
$A_n\#$	=>	$A_x + A_y$	; Cluster fragmentation
$A_n\# + M$	=>	$A_n + M$	; Cluster stabilization/cooling where M is
			the third body stabilizer.

Data calculated in this manner are used in the phenomonological model described above. As the cluster becomes larger the fragmentation rate becomes smaller until it is no longer contributes. Under these conditions we turn to molecular dynamics computation to understand the cluster-cluster growth process.

Molecular dynamics methods afford a unique view of the dynamical feature of non-equilibrium systems as well as the ability to obtain a wide variety of data of physical/chemical properties. We have used this approach to model the equilibrium and kinetic properties of silicon nanoparticles containing up to 1000 atoms (16,17). However for such an analysis to yield even semi-quantitative results requires a knowledge of the interaction potential. For a system as complicated as the one under consideration here, interatomic potentials are beyond the scope of current state of the art. However, one can still, under certain circumstances obtain qualitative information.

In the TEM imaging of the iron oxide/silica nanocomposite for example, it was observed that the iron oxide clusters were arranged in such a way as to suggest that the iron oxide inclusions were positioned to maximize the distance between them. Rather than buried deep within the silica particle the iron oxide seemed to reside within the inside edge of the silica. The question arises as to whether this results from some type of magnetic interaction or phase segregration that might cause the particles to migrate away from each other. A simple Lennard-Jones (L-J) dynamics study was constructed in which two types of molecular entities were randomly arranged. These two types of species represent the iron and silicon phases in which we scale the homonuclear interaction potential (ε_{ii}) of each component to the known melting points of iron oxide and silica. For silicon-iron interaction, (which presumably would be ironsilicate), we scale to the melting point of ironsilicate. Under these asumptions we obtain $\varepsilon_{Si-Si} > \varepsilon_{Fe-Fe} > \varepsilon_{Fe-Si}$ ($\varepsilon_{Si-Si} = 1.5$; $\varepsilon_{Fe-Fe} = 1.3$; $\varepsilon_{Fe-Si} = 1.0$).

The computation was initiated with approximately 25 % of the atoms comprising the iron type and the remainder silica. The clusters were prepared in the liquid state with constant temperature molecular dynamics. The results are represented pictorially in Fig. 12 and clearly show that with time the minor component has associated together and migrated toward the edge of the silica. Given enough computation time the iron components would all be associated as one cluster and one would have a composite particle made up of essentially the fusion of two pure particles. The implication for experimental results are that the nanocomposite is a metastable structure and that give sufficient residence time in the reactor they would show complete segregation. This should be ubiquitous to any composite particle formed from rapid condensation at high temperatures, where the diffusion lengths are relatively short.

Summary and Conclusions

In this paper we have demonstrated the application of gas-phase combustion processes to produce nanocomposites with superparamagnetic properties. The combustion process was designed in such a way as to allow for the rapid formation of nano-liquid droplets which would phase segregate, combined with rapid quenching to preserve the microstructure and chemical composition. The resulting material showed superparamagnetic behavior between 155 K and room temperature.

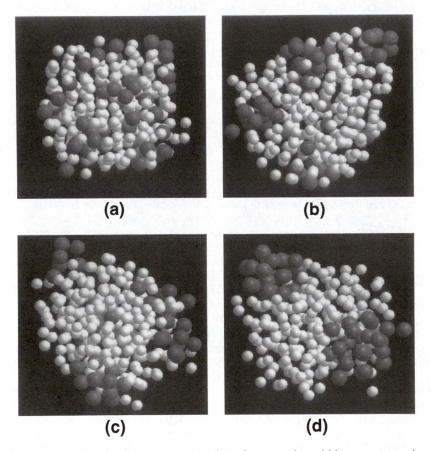

Figure 12. Molecular dynamcs computation of segregration within nanocomposite.

Planar laser-induced fluorescence and Rayleigh/Mie imaging measurements have been used to examine the mechanisms of particle formation from gas phase species in a flame reactor. During synthesis of iron-oxide/silica nanocomposites, the relative gas phase precursor (FeO(g)) concentration was measured for different feed conditions and compared with particle scattering measurements. The results indicate that the iron oxide particle formation rate is slower than the gas phase precursor decomposition rate, and essentially proceeds after all of the FeO(g) has formed. Measurements in flames seeded with both silicon and iron precursors show no significant changes in the rate of formation of FeO(g), indicating that the iron and silicon components nucleate out independently. Light scattering measurements indicate that sintering effects are faster for the nanocomposite than for silica in keeping with the known sintering rates for silica and iron oxide.

Both aerosol modeling and more fundamental atomistic and molecular level models have been applied to this problem. Aerosol dynamics modeling has lead to a better understanding of the individual steps that comprise the formation of particles, all the way from nucleation to subsequent growth. Both molecualar orbital and reaction rate theory was used as sources of fundamental data for input to the aerosol dynamics model. A simplistic molecular dynamics computation has been used to explain the particle morphology observed.

References

1. Shull, R.D., Swartzendruder and Bennett, L.H., *Proceedings of the Sixth International Cryocoolers Conference*; Ed. G. Green and M. Knox, David Taylor Research Center Publ. #DTRC-91/002, Annapolis, Md, p. 231, (1991)
2. Shull, R.D., Swartzendruder and Bennett, L.H., *Proceedings of the International Workshop on Studies of Magnetic Properties of Fine Particles and their Relevance to Materials Science*, Rome, Italy (Nov,1991)
3. Shull, R.D., Ritter, J.J., Shapiro, A.J., Swartzendruber and Bennett, L.H., (1990) *J. Appl. Phys.* **67**, 4490
4. Zachariah, M.R., Chin,D., Semerjian, H.G., and Katz, J,L. (1989) *Comb. Flame*, **78**, 287
5. Shull, R.D., Atzmony, U., Shapiro, A.J., Swartzendruber, L.J., Bennett, L.H., Green, W.J., and Moorjani, K. (1988) *J. Appl. Phys.* **63** (8), 4261
6. Zachariah, M.R., and Semerjian, H.G. (1989) *AIChE. J.*, **35**, 2003
7. Zachariah, M.R., and Joklik, R.G. (1990) *J. Appl. Phys.* **68**, 311
8. Zachariah, M. R., Chin, D., Katz, J. L., and Semerjian, H. G. (1989) *Applied Optics* **28**, 530.
9. Zachariah, M. R. and Burgess, Jr., D. R. F. (1994) *J. Aerosol Sci.* **25**, 487.
10. McMillin, B.K., and Zachariah, M.R. (1995) *J. Appl. Phys.*, **77**, 5538
11. Biswas, P., Wu, C-Y., Zachariah, M.R., and McMillin, B.K., submitted
12. Zachariah, M.R. and Tsang W. (1993) *Aeros. Sci. Tech.* **19**, 499.
13. Zachariah, M.R. and Tsang W. (1995) *J. Phys. Chem.* **99**, 5308
14. Gelbard, F., and Seinfeld, J.H. (1980) *J. Colloid Int. Sci.*, **78**, 485
15 Landgrebe, J.D., and Pratsinis, S.E. (1990) *J. Colloid Int. Sci.*, **139**, 63
16. Zachariah, M. R.. Carrier, M.J., and Blaisten-Barojas, E. in *Gas-Phase and Surface Chemistry in Electronic Materials Processing,* MRS Proceedings, Eds, T.J. Mountziaris, (1994) **334**, 75-80
17. Zachariah, M. R.. Carrier, M.J., and Blaisten-Barojas, E., *Molecularly Designed Ultrafine/Nanostructured Materials* , MRS Proceedings (1994) **351**, 343-348

RECEIVED January 25, 1996

Chapter 4

Flame Synthesis of Nanosize Powders

Effect of Flame Configuration and Oxidant Composition

Wenhua Zhu and Sotiris E. Pratsinis[1]

Department of Chemical Engineering, University of Cincinnati, Cincinnati, OH 45221−0171

Flame reactors have high potential for synthesis of inexpensive nanosize powders since submicron size particles are made this way on an industrial scale at a rate of 100 ton/day. Thus, the formation of nanosize titania and silica powders by oxidation of $TiCl_4$ and $SiCl_4$ in a diffusion flame reactor were investigated using air or O_2 as oxidant and methane as fuel. The product titania and silica particle size as well as phase composition of titania particles were precisely controlled by the reactant mixing and oxygen partial pressure in the flame. Faceted and/or spherical particles were obtained in the range of 18 to 250 nm for TiO_2 and 13 to 110 nm for SiO_2. High oxygen concentrations in the flame atmosphere retarded the formation of rutile phase titania powders though the flame residence time was only in the order of several ms. The precursor conversions of both $TiCl_4$ and $SiCl_4$ were incomplete due to the short residence time and low flame temperature.

Nanosize particles with diameters in the range of 1 ~ 100 nm are of interest for fabrication of advanced ceramics and semiconductors. Typical ways of synthesizing nanosized powders are wet chemical processes and aerosol processes. The later are advantageous for manufacture of particulates since they do not involve the tedious steps, high liquid volumes and surfactants of wet chemistry processes (*1*). As a result, they can produce materials of high purity at high yields and provide a molecular level mixing for composite powders.

Various aerosol processes have been developed for the generation of ultrafine powders at laboratory scale, such as flame (*2*), tube furnace (*3*), gas-condensation (*4*), thermal plasma (*5*), laser (*6*) , sputtering (*7*) and a variety of other aerosol processes named after the energy sources which are applied to provide the high temperatures during gas-to-particle conversion. However, until now, only flame processes have been scaled up to produce commercial quantities of ceramic particulates, such as silica, titania, etc., at low cost (about $ 1/lb).

Titania powders are extensively used in pigments, as catalyst supports, and more recently in synthesis of inorganic membranes and as photocatalyst in gas and water purification. Silica particles have found applications as optical fibers, fillers,

---

[1]Corresponding author

0097−6156/96/0622−0064$15.00/0

thickening and reinforcing agents, catalyst carriers, etc. (8). The annual production volume of these powders amounts to over 2 million tons worldwide (9). A large proportion of these powders are manufactured by flame aerosol processes such as the so-called "chloride process" where oxidation and hydrolysis of their respective chlorides (TiCl$_4$ and SiCl$_4$) take place in a flame reactor.

In flame synthesis processes, the control of particle size (both primary particle and aggregate size), morphology and phase composition is difficult and limited till now. The synthesis of silica in hydrocarbon flames was studied both theoretically and experimentally by Ulrich and co-workers (10-13). Silica particles were generated in flat flames with either propane/oxygen or methane/air as fuel/oxidant and SiCl$_4$ as precursor. They showed that the final silica aggregate and primary particle size were strongly affected by flame temperature and residence time while nucleation and surface reaction did not affect the final particle size. They also demonstrated that coalescence was the rate-controlling step in the growth of silica particles due to the high viscosity of the material. The effect of temperature was found to be important for the control of the particle growth: a decrease of only 2% (40 K) in flame temperature was sufficient to cause a surface area increase of up to 25% (13).

Silica particle formation was also studied in a counterflow diffusion flame reactors by Katz and coworkers (14, 15). In their work, silane was used as the source of silicon in H$_2$/O$_2$/Ar flames. The effects of silane loading, temperature on agglomerate size were investigated. Increasing flame temperatures tended to enhance homogeneous nucleation, leading to smaller agglomerates in higher numbers. Larger silane loading tended to accelerate particle nucleation, formed larger agglomerates and enhanced surface growth effects.

Formenti et al. (16) generated TiO$_2$ powders in H$_2$/O$_2$/N$_2$ co-flow diffusion flame using TiCl$_4$ as precursor. The influence of the precursor loading rate and the flame residence time on the primary particle size were investigated. They showed that the particle diameter increased with the concentration of TiCl$_4$ in the carrier gas and the residence time of particles in the flame. Almost pure anatase particles were synthesized under the employed flame conditions. However, the titania particles made by Hung et al (15) using the same reactants but a counter-flow diffusion flame were pure rutile, which was a totally different result, due to the different flame configuration they used.

George et al. (17) studied the formation of particulate TiO$_2$ by addition of small quantities of TiCl$_4$ vapor to a lean CO/O$_2$/N$_2$ flame with a maximum temperature of about 1400°C. According to their observation, chemical reaction was essentially complete 50 ms down stream of the CO flame front and that the particle growth continued for a further 200 ms by the coagulation mechanism. Good agreement with the theoretical prediction of Ulrich (10) was reached with respect to both the development of the mean particle size and the size distribution.

From the above studies, it is clear that flame temperature, residence time and precursor loading are key variables that determine final particle size and phase composition. However, our understanding for precise control of particle size, morphology and phase composition is still limited, and doping, electric charging and the control of reactant mixing are promising variables for tailoring the properties of flame-made particles.

Vemury and Pratsinis (18) investigated the effect of dopants on the characteristics of titania particles made by oxidation of TiCl$_4$ in diffusion flames. They found that introduction of Si^{4+} inhibited the anatase to rutile transformation and decreased the primary particle size of titania . However, the addition of Al^{3+} had the reverse results, presumably because Si^{4+} entered the lattice interstitially while Al^{3+} entered substitutionally into the titania lattice (3).

A gaseous electric discharge (corona) generated by two needle electrodes was applied in the synthesis of ultrafine titania particles recently (19). The presence of corona reduced the primary particle size and the level of crystallinity. For example, increasing the applied potential from 5 to 8 kV decreased the particle size from 50 to 25 nm and the rutile content from 20 to 8 wt%.

Pratsinis et al. (20) initially investigated the effect of reactant mixing in a diffusion flame reactor by altering the precursor $TiCl_4$, methane and air streams flowing through the reactor during synthesis of TiO_2 particles. They found that the better and the earlier the mixing between fuel and precursor vapor occurred, the coarser were the product particles and the higher was the rutile content. They were able to control the average primary particle size from 11 to 105 nm by merely altering the reactant mixing.

In the present study, flame synthesis of nanosize titania and silica is investigated. Since the sintering rate of silica is lower than that of titania, it is interesting to compare the effect of reactant mixing on the size evolution of these two materials. The effect of oxidant composition on product particle size and phase composition is also investigated. Using of pure oxygen as oxidant raises the flame temperature, and more importantly, changes the chemical environment in which particles are sintered. In the present study, we focus on investigating the specific surface area (primary particle size), morphology and phase composition of flame made titania and silica.

Background

Particle formation in flames follows the general gas-to-particle conversion mechanism. The flame temperature is usually between 1200 to 3000 K (16). This high temperature promotes rapid gas phase chemical reactions. The critical nucleus size of the product oxide species is usually smaller than a single molecule, thus, thermodynamically stable particles are formed directly by chemical reaction, and particles grow by Brownian coagulation and sintering (21). If the fusion of particles is faster than coagulation, spherical particles are obtained. However, as particles grow, the sintering rate decreases, particle coalescence cannot occur rapidly with respect to coagulation, so irregularly shaped aggregate particles are formed. In the diffusion flame, the maximum flame temperature usually occurs at the tip of the flame (22), where agglomerates fuse and leave the flame. Upon leaving the flame, the flame temperature drops quickly and particles continue to coagulate but no longer sinter, resulting in large agglomerates. Therefore, powders made in the flame reactors are agglomerates of primary particles. Agglomerates can be beneficial in synthesis of catalysts and catalytic substrates but they can be detrimental in fabrication of advanced ceramics.

Experimental

Apparatus. Titania and silica particle formation is studied in a diffusion flame reactor consisting of a series of concentric quartz tubes, similar to that described by Allendorf et al. (23). The central tube is 5 mm in diameter and the spacing between successive tubes is 1 mm. Only the three innermost rings of the reactor are used in the present study. The terminology of the diffusion flame reactor comes from the fact that the combustion processes are mainly determined by the rate of inter-diffusion of the oxidant and the fuel. The advantages of a diffusion flame burner are that it provides a stable flame over a wide range of operation conditions and it is extensively used in aerosol manufacture of particulate commodities (24). The experimental system is shown in Figure 1. Methane (Wright Brothers, 99%) is used as fuel while air or pure oxygen (Wright, 99.9%) is used as oxidant. Three flame configurations are investigated (Figure 2) in which Ar gas (Wright Brothers, 99.8%) initially saturated with the precursor vapor ($TiCl_4$ or $SiCl_4$) passes through the central tube of the burner (a) by passing argon through a bubbler (b) containing liquid $TiCl_4$ or $SiCl_4$ (Aldrich, 99.9%). The flowrates of CH_4 and oxidant (air or O_2) of each configuration are identical but the tube through which each reactant passes is varied. From flame A to C, the position of the oxidant stream is moved from the central tube to the 3rd tube, meanwhile methane moves from the 3rd tube to the 2nd tube (Figure 2). All reactant flow rates are metered by rotameters (c) while the temperatures are measured by thermocouples T2 and T3 or thermometer T1. The precursor loading is controlled by carefully controlling the bubbler temperature. The bubbler containing $TiCl_4$ is kept at 65 °C by an electrically heated mantel. The temperature of the burner and the manifold are kept 20°C above the bubbler temperature to prevent condensation of $TiCl_4$ in the lines. The liquid $SiCl_4$ is maintained at 0°C by keeping the bubbler in ice water.

Particles synthesized in the flame are collected on glass fiber filters with the pore size of 0.2 μm (Gelman Scientific; 143 mm) placed inside a stainless steel open-faced filter holder (e). The particle collection unit is placed 120 mm above the tip of the burner in all experiments. Since the luminous part of the flame is in the range of 20 ~ 90 mm, a distance of 120 mm between the collector and the flame suffices to quench particle sintering since the temperature drops very quickly downstream of the flame front. For example, the measured temperature at the filter is less than 300°C while at the flame front it can be as high as 1600°C. Sampling is facilitated by a vacuum pump (g) and corrosive byproducts such as Cl_2 and HCl are removed from the exhaust by passing them through a 1M NaOH aqueous solution (f) before being released to the fume hood.

Measurements. X-ray diffraction is used to determine the phase composition of the powders (D500, Siemens). Silica is predominately in amorphous phase while titania is in the crystal form of anatase and rutile. The weight fractions of anatase and rutile phases in the samples are calculated from the relative intensities of the primary peaks corresponding to anatase and rutile as described by Spurr and Myers (*25*). Transmission electron micrographs (TEM) are obtained on a Philips CM 20 microscope operating at 200 kV or an EM 400 operating at 100 kV. The powder specific surface area is measured by nitrogen adsorption (Gemini 2360, Micromeritics) at 77 K using the BET equation. Assuming spherical particles, the average grain size, d_p, can be calculated from the specific surface area, A, and particle density, ρ_p, by $d_p=6/\rho_p A$. The maximum temperature of each flame is measured with a 0.015" Pt-Rh R-type thermocouple (Omega Engineering) in the absence of precursor but with the carrier gas Ar flowing through the burner, otherwise the particles produced in the flame quickly coat the thermocouple causing measurement errors.

Results and Discussion

Table I lists the flame variables employed and the measured maximum flame temperatures without correcting for radiation losses. The latter depend on the actual flame temperatures, the reactant flowrates and the flame configurations. For the flame using air as oxidant (air flame), the radiation losses are 45 ~ 160°C under the present flame conditions. The radiation losses for the flame using oxygen as oxidant (oxygen flame) are greater due to the higher flame temperature.

The temperature of the oxygen flames are about 500°C higher than those for air flames (Table I). Combustion is diffusion controlled so that at the lower oxygen flowrate (1.6 l/min) incomplete combustion may take place resulting in lower flame temperatures compared to that at higher oxygen flowrates. As the CH_4 stream is moved closer to the oxygen stream (Flame B) or as it is enclosed by oxygen (Flame C) this effect is mitigated and flame temperatures increase. The temperatures of the oxygen flames increase with the increasing oxygen flowrate up to 3.8 l/min and then decrease because additional oxygen flow dissipates the generated heat very quickly. However, for air flames, the flame temperatures increase steadily with increasing air flowrate. Table I also shows that the maximum flame temperatures are quite constant with regard to flame configurations. However, it is worth mentioning that there are steep temperature gradients in oxygen flames A and B, that is, only a very thin high

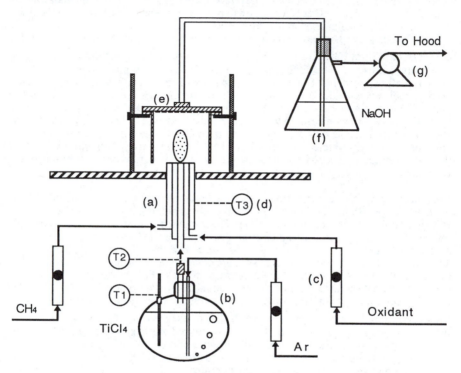

Figure 1. The schematic of the experimental set-up

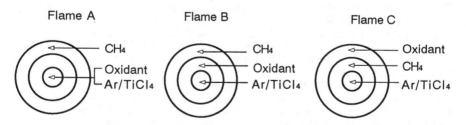

Figure 2. Reactant mixing configurations for Flames A, B and C

temperature layer exits at the fuel and oxygen front. The temperature gradient in flame C is relatively gentle because the fuel is totally enclosed by the supplied oxidant. Figure 3 shows the axial temperature profiles of air and oxygen flames along the center line in the flame configuration C with oxidant flowrate of 5.5 l/min. It can be seen that the luminous (high temperature) region of the oxygen flames is much closer to the tip of the burner. Furthermore this region is much shorter than that of the air flames because of the high combustion rates of the oxygen flames. Therefore, the residence time of particles in the oxygen flames at the highest temperature is shorter than that in corresponding air flames.

Table I. Experimental Conditions

CH$_4$: 0.4 l/min	Ar: 0.25 l/min	TiCl$_4$ or SiCl$_4$: 1.0 x 10^{-3} mol/min	
		Temperature (°C)[*]	
	Flame A	Flame B	Flame C
Air (l/min)			
2.5	971	1006	1020
3.8	1007	1017	1026
5.5	1023	1026	1061
O$_2$ (l/min)			
1.6	1356	1419	1423
2.5	1502	1507	1518
3.8	1553	1555	1568
5.5	1539	1543	1552

[*] The error bar for all the temperatures measured is within ±15°C.

The reaction yields of TiO$_2$ and SiO$_2$. The yield is defined as the ratio of the collected powder over the theoretically expected assuming that all precursor vapor is converted into powder. Figure 4 compares the yields of product oxides from precursor TiCl$_4$ and SiCl$_4$ using air or oxygen in flame configuration C, which provides the highest yields among all flames. To ensure that all particles generated are collected on the filter, a second filter is placed in series, however no particles were found on the second filter. Figure 4 shows that the reaction yields in the high temperature oxygen flame are higher than those in air flames for both precursors. However, most of the yields do not reach the complete conversion (100%) because of the short flame residence time (especially for oxygen flames) or the low flame temperature (especially for air flames). When the oxidant flowrate increases, the flame residence time decreases, as a result, the reaction conversion is further reduced.

In previous modeling work of flame synthesis of powders (*10, 26*), it was assumed that the rate of chemical reaction was rapid compared with particle formation processes (coagulation and coalescence) and no condensable material remained in the gas after the precursor gases pass through the flame. However, this is not always true since the chemical reaction and, subsequently, particle production depend on residence time and temperature. Thus, if the reactant gases are not exposed at high temperature complete conversion does not take place. This may be also one of the reasons that the theoretical prediction of the primary particle size was not in accordance with the experimental data of Wu et al. (*26*). TEM pictures of particles made in flame C clearly shows that tiny particles are attached to large particles (Figure 5e), indicating that some particles do not have enough time to grow in the flame or experience substantially different temperature histories.

The effect of flame configuration on particle size. Figure 6 shows that the

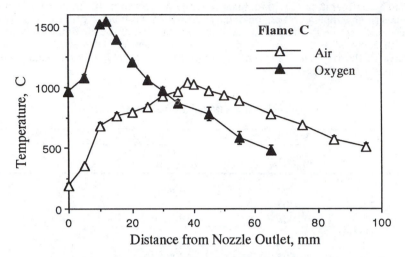

Figure 3. The axial temperature profiles at the center line of air and oxygen flames with flame configuration C

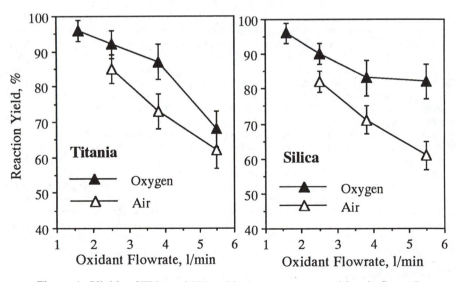

Figure 4. Yields of TiO$_2$ and SiO$_2$ with air or oxygen as oxidant in flame C

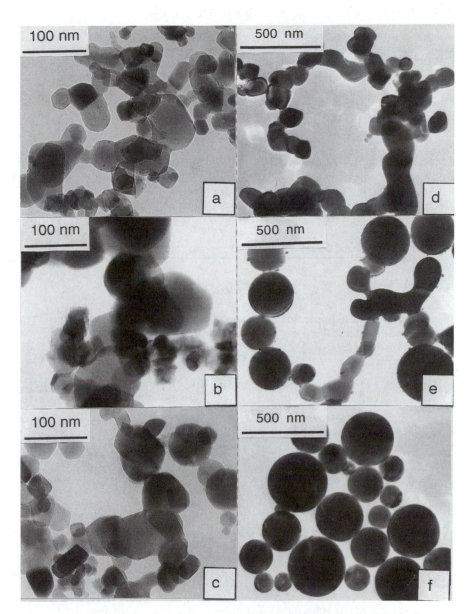

Figure 5. The morphology of titania particles synthesized with different O_2/N_2 molar ratio in the oxidant in (a) flame B, air; (b) flame B, 50% O_2; (c) flame B, pure O_2; (d) flame C, air; (e) flame C, 50% O_2; (f) flame C, pure O_2

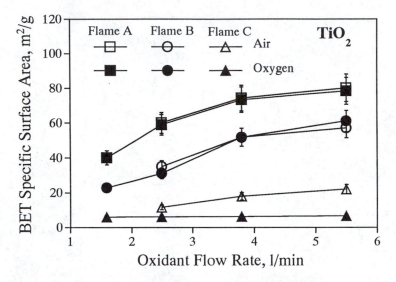

Figure 6. The BET specific surface area of titania powders made in 3 flame
configurations as a function of the oxidant flowrate

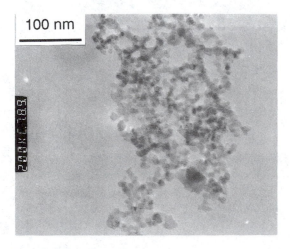

Figure 7. TEM picture of titania particles synthesized in flame A with air as
oxidant

titania specific surface area increases with increasing air and oxygen flowrate in flames A, B and C. Increasing the oxidant flowrate slightly increases the flame temperature (Table I), but reduces the flame height due to the more effective combustion (*20*). As a result, the residence time of $TiCl_4$ and particles at high temperature is reduced and, subsequently, less TiO_2 monomer is generated by $TiCl_4$ oxidation (Figure 4). This may indicate that there is not enough time for all $TiCl_4$ to be oxidized at the high temperature. Furthermore, the additional gas (oxidant) greatly dilutes the aerosol reducing its concentration and temperature. Therefore, the collision rate of newly formed titania as well as the effective sintering time decreases, which favor forming smaller particles and subsequently powders with high specific surface area as Figure 6 shows.

Whichever oxidant is used, flame A always produces the finest TiO_2 particles since the $TiCl_4$ stream is diluted with oxidant prior to its oxidation in the flame (6-22 times dilution for oxidant flowrate of 1.6 to 5.5 l/min, respectively). This decrease of precursor concentration in the flame reduces the initial particle number concentration, resulting in lower coagulation rates and small, loose agglomerates with less contact points between the primary particles. Since the sintering of the titania particles occurs on contact (*27*), the fusion rate among the primary particles decreases, resulting in small individual particles (Figure 7).

As to flame B, the dilution of the $TiCl_4$ stream with oxidant takes place further downstream from the burner mouth in the flame. Therefore, the dilution is limited to just prior to $TiCl_4$ oxidation resulting in larger particles than flame A. In addition, the dilution of the precursor is not as homogeneous as that in flame A, as a result, the particles synthesized in flame B have a wider primary particle size distribution as shown in Figure 5 a - c.

In flame C, the dilution of the $TiCl_4$-laden Ar stream with the CH_4 stream is not that significant since both streams have comparable flowrates 0.25 and 0.4 l/min, respectively. Furthermore, the mixing of precursor vapor and methane takes place much earlier than that in flame A and B, the newly formed titania particles stay longer in the flame and sinter, resulting in rather large primary particles (Figure 5 d - e).

As shown in Figure 6, titania particles synthesized in air and oxygen flames have comparable surface areas under same oxidant flowrates when flame A or B is applied, even though the maximum temperature of oxygen flames is much higher than that of air flames. As described above, oxygen flames A and B exhibit two diffusion flame fronts and have very steep temperature gradient. Aside from a very narrow high temperature region, the temperature of the rest of the flame is not so high, because the surrounding air may react also with the fuel and dissipates the heat; moreover, the height of oxygen flame is much lower (20 - 30 mm) than that of the air flame (55 - 85 mm). This means that the particle residence times in oxygen flames are much shorter than in the air flames. As a result, particles generated in both flames have similar sizes. In flame C, however, since the fuel is encircled completely by the supplied oxidant, there is only one diffusion flame front. Less energy is released to the surroundings, so the temperature gradient is relatively low; more importantly, even using oxygen as oxidant, the flame has a notable luminous height (40 - 55 mm) due to the less effective combustion than oxygen flames A and B, resulting in remarkable increase of sintering rates in oxygen flame, thus particles synthesized in oxygen flames have an average size of around 240 nm, 2 - 3 times larger than those in air flames.

The Effect of Material Property. Figure 8 shows the evolution of silica particle size with respect to oxidant flow rate in flames A and C. The flame conditions are listed in Table I and the precursor $SiCl_4$ loading rate is 1.0×10^{-3} mol/min identical to that of $TiCl_4$. Compared with Figure 6, it can be seen that though the silica particle size evolution has the same tendency as that of titania with respect to flame configuration. However, the effect of oxidant on particle size is more remarkable in silica than in titania. Even in flame A, silica particles synthesized in oxygen flames are about 1.3 ~

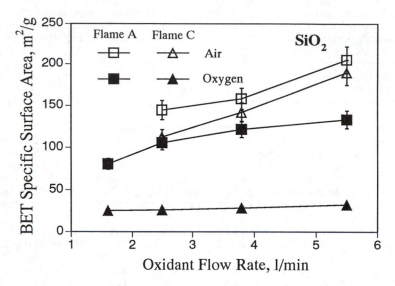

Figure 8. The BET specific surface area of silica powders made in flames A and C
as a function of the oxidant flowrate

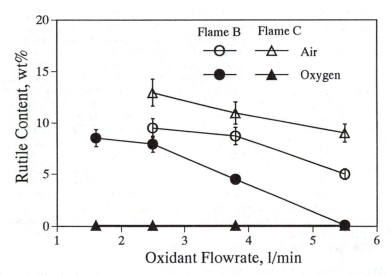

Figure 9. Rutile weight content of titania powders synthesized in flame B and C
with respect to oxidant (air or oxygen) flowrates

1.5 times larger than those in air flames. In flame C, the size difference increases to 3 ~ 5 times, again, larger than that of titania (1 ~ 2 times). Therefore, the sintering rate of silica is very sensitive to flame temperature, which is in agreement with Ulrich et al. (*13*).

Silica sinters by a viscous flow sintering mechanism while titania sinters by grain boundary diffusion. Kingery et al. (*28*) and Kobata et al. (*29*) have derived the characteristic sintering time, τ_f, for SiO_2 and TiO_2, respectively, which are related to sintering temperature, time and particle size. However, it is still difficult to calculate the τ_f here since the detailed temperature profiles of the current flames are not known. Nevertheless, since titania particle size is much larger than that of silica particles, it can be deduced that at the temperatures encountered in the present study, the sintering rate of titania is faster than that of silica due to the high viscosity of silica in agreement with the sintering expressions in the literature (*28, 29*).

The effect of oxidant and flame configuration on titania phase transformation and morphology. Figure 9 shows the rutile content of titania powders synthesized in flames B and C as a function of oxidant flowrate for both oxidants. The rutile weight fraction of particles generated in oxygen flames is lower than in the air flames at the same flowrate, especially in flame configuration C, in which pure anatase particles are synthesized in oxygen flames at all flowrates applied. The activation energy for transformation from anatase to rutile is of the order of 400 kJ/mol (*30*), so rutile is a more stable form than anatase. The rutile has a much closer atomic packing in its crystal pattern and the transformation of anatase to rutile involves volume contraction and a cooperative movement of titanium and oxygen ions, which increases with increasing temperature (*31*). Anatase to rutile transformation is usually accompanied by size enlargement from sintering. However, according to Figure 6, particles synthesized in flame C using oxygen as oxidant are the largest but their rutile content is the least. These two results contradict each other. Therefore, other mechanisms must be considered.

MacKenzie (*32*) investigated the effect of reaction atmosphere on the anatase-rutile transformation upon calcination of pure anatase powders at 1000°C in a furnace of periods 30 min each. He found that the greatest transformation rate and largest particle sizes were obtained in hydrogen atmosphere, while increasing oxygen partial pressure (from 2% O_2 mixed with Ar to 100% O_2) resulted in progressive retardation of the rutile formation from 14% to 7% rutile. He pointed out that creating defects such as oxygen vacancies during the firing was beneficial to the phase transformation of anatase to rutile. Iida and Ozaki (*33*) also found that the rate of anatase-rutile transformation decreased with an increase in the partial pressure of oxygen in the flowing gases when the samples were heated at 900°C for 3 hours. The rutile weight fraction of the sample sintered in flowing air was 70% and decreased to 52% in flowing O_2. Vemury and Pratsinis (*18*) studied the phase transformation of Al-doped titania powders in the flame reactor and they concluded that upon the introduction of Al^{3+} into the titania lattice, the rutile content of the powders increased attributing it to the creation of oxygen vacancies during the substitution of Al^{3+} for Ti^{4+}. In the present studies, the O_2 partial pressure in the oxygen flame is much higher than that in air flame, thus, the formation of oxygen vacancies may be inhibited resulting in the much lower rutile fraction in the powders made in O_2 than in air.

To prove the above statement, titania powder synthesis has also been conducted using oxidant stream with different oxygen molar ratio in flame configurations B and C. Table II lists the results. It can been seen from Table II that oxygen mole fraction in the oxidant has more marked effects on particle size and rutile content in flame C than in flame B, attributed to the total enclosure of fuel and precursor within the O_2 flow under configuration of C. With the increase of oxygen fraction in the oxidant, the rutile content of particles synthesized in flame C decreases until the pure anatase particles are formed in pure oxygen, which again proves that high partial pressure of oxygen retards the formation of oxygen vacancies, inhibiting thus, the formation of rutile.

Figure 10 shows the XRD pattern of titania particles synthesized in flame C with respect to oxygen ratio in the oxidant. It can be seen clearly that the (110) reflection of rutile ($2\theta = 27.5$) decreases as the O_2 ratio in the oxidant increasing. Form air to 100% O_2, the rutile content drops from 13.2 wt% to < 0.1%. Moreover, the peaks move about 0.3° toward the high 2θ angle when the oxygen ratio increases from 21% to 100%, indicating a reorganization of the titania lattice. The particle residence time in the flame is less than 100 ms. However, the effect of oxygen concentration on the phase transformation is still significant. This is the first time to point out that oxygen concentration affects the anatase-rutile transformation even in the flame synthesis processes.

Table II. The comparison of TiO_2 particle size and rutile content
synthesized under different oxygen concentration

CH4: 0.4 l/min	TiCl$_4$: 1.0 x 10^{-3} mol/min			Oxidant: 2.5 l/min		
	Flame B			Flame C		
O_2 ratio	Max.Temp. (°C)	Sur. Area (m^2/g)	Anatase wt%	Max.Temp. (°C)	Sur. Area (m^2/g)	Anatase wt%
21%	1006	34.7±2.1	90.7	1020	9.7±1.0	86.8
50%	1161	28.9±3.4	92.5	1196	6.2±0.6	94.3
75%	1343	31.9±1.8	91.8	1345	6.1±0.3	97.6
100%	1507	31.1±1.9	92.1	1552	5.9±0.3	99.9

The TEM pictures of particles made in flames B and C with respect to different oxygen ratio in the oxidant are shown in Figure 5 above. The morphology of particles generated in flame B (Figure 5 a, b and c) changes little with the oxygen ratio. The particles remained faceted whichever oxidant is used, indicating that sintering is not enhanced upon introducing more oxygen in flame B, in accordance with the results of Figure 6. When particles are made in flame C, things become more interesting. Figures 5 d - f nicely show the particle morphology evolution from faceted to spherical shape with increasing oxygen ratio in flame C. This means that particles are sintered more effectively at high oxygen ratios at this flame configuration.

Conclusions

Nanosized TiO_2 and SiO_2 are synthesized in a diffusion flame reactor. The effect of reactant mixing and oxygen concentration on product titania and silica particle size as well as on titania phase transformation were investigated. Titania sinters faster than silica, especially in low temperature flames, due to the different sintering mechanism of the two materials; the oxygen ratio in the oxidant has little effect on titania particle sizes when they are synthesized in flames A and B, however, it affects the size of silica powders sintered; the size of particles generated in flame C increased with oxygen ratio for both titania and silica. The oxygen concentration in the flame plays an important role for the generation of oxygen vacancies in the titania lattice and hence affects the anatase-rutile phase transformation; the higher is the concentration of oxygen in the flame, the less is the rutile content in the particles made; the short flame residence time (especially for O_2 flame) and low temperature (especially for air flame) limit the completion of the reaction of both $TiCl_4$ and $SiCl_4$.

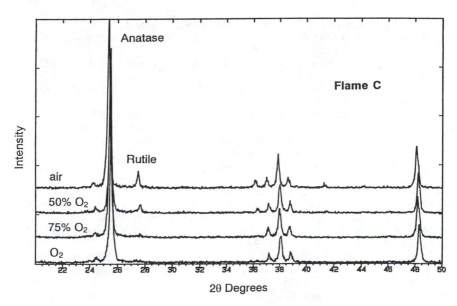

Figure 10. XRD pattern of titania particles synthesized in flame C as a function of oxygen ratio in the oxidant

Acknowledgments

This research was supported by the Advanced Fuel Research Co. and the National Science Foundation, CTS-8957042.

Literature Cited

1. S.E. Pratsinis and S.V.R. Mastrangelo, *Chem. Eng. Progress,* **1989**, *85*, p62.
2. M.R. Zachariah and S. Huzarewicz, *J. Mater. Res.,* **1991**, *6*, p264.
3. M.K. Akhtar, S.E. Pratsinis and S.V.R. Mastrangelo, *J. Am. Ceram. Soc.,* **1992**, *75*, p3408.
4. W-A. Saunders, P.C. Sercel, H.A. Atwater and R.C. Flagan, *Appl. Phys. Lett.,* **1992**, *60*, p950.
5. G.P. Vissokov, B.I. Stefanov, N.T. Gerasimov, D.H. Oliver, R.Z. Enikov, A.I. Vrantchev, E.G. Balabanova and P.S. Pirgov, *J. Mater. Sci.,* **1988**, *23*, p2415.
6. G.P. Johnston, R. Muenchausen, D.M. Smith, W. Fahrenholz and S. Foltyn, *J. Am. Ceram. Soc.,* **1992**, *75*, p3293.
7. H. Hahn and R.S. Averback, *J. Appl. Phys.,* **1990**, *67*, p1113.
8. G.D. Ulrich, *C & CE,* **1984**, *August* 6, p22.
9. P. Stamatakis, C.A. Natalie, B.R. Palmer, and W.A. Yuill, *Aerosol Sci.Technol.,* **1991**, *14*, p316.
10. G.D. Ulrich, *Comb. Sci. Tech.,* **1971**, *4*, p47.
11. G.D. Ulrich, B.A. Milnes and N.S. Subramanian, *Comb. Sci. Tech.,* **1976**, *14* p243.
12. G.D. Ulrich and N.S. Subramanian, *Comb. Sci. Tech.,* **1977**, *17*, p119.
13. G.D. Ulrich and J.W. Riehl, *J. Colloid Inter. Sci.,* **1982**, *87*, p257.
14. M.R. Zachariah, D. Chin, H.G. Semerjian and J.L. Katz, *Combustion and Flame,* **1989**, *78*, p287.

15. C.-H. Hung and J.L. Katz, *J. Mater. Res.*, **1992**, 7, p1861.
16. M. Formenti, F. Juillet, P. Meriaudeau, S.J. Teichner and P. Vergnon, *J. Colloid Inter. Sci.*, **1972**, *39*, p79.
17. A.P George, R.D. Murley and E.R. Place, *Farad. Symp. Chem. Soc.*, **1973**, 7 p63.
18. S. Vemury and S.E. Pratsinis, " Dopants in flame synthesis of titania", *J. Am. Ceram. Soc.*, **1995**, in press.
19. S. Vemury and S.E. Pratsinis, *Appl. Phys. Lett.*, **1995**, *66*, p3275.
20. S.E. Pratsinis, W. Zhu and S. Vemury, " The role of gas mixing in flame synthesis of titania powders", *Powder Technology*, **1996**, *86(1)*, in press.
21. K.A. Kusters and S.E. Pratsinis, *Powder Technology*, **1995**, *82*, p79.
22. A.G. Gaydon and H.G. Wolfhard, *Flames*, Chapman and Hall Ltd.: London, 1970.
23. M.D. Allendorf, J.R. Bautista and E. Potkay, *J. Appl. Phys.*, **1989**, *66*, p5046.
24. G.P. Fotou, S. Vemury and S.E. Pratsinis, *Chem. Eng. Sci.*, **1994**, *49*, p4939.
25. R.A. Spurr and H. Myers, *Analytical Chem.*, **1957**, *29*, p760.
26. M.K. Wu, R.S. Windeler, C.K.R. Steiner, T. Bors, and S.K. Friedlander, *Aerosol Sci. Tech.*, **1993**, *19*, p527.
27. W. Koch and S.K. Friedlander, *J. Colloid and Interface Sci.*, **1990**, *140*, p419.
28. W.D. Kingery, H.K. Bowen and D.R. Uhlmann, *Introduction to Ceramics*, Wiley: New York, NY, 1976.
29. A. Kobata, K. Kusakabe and S. Morooka, *AIChE J.*, **1991**, *37*, p347.
30. K.J.D. MacKenzie, *Trans. J. Brit. Ceram. Soc.*, **1975**, *74*, p77.
31. D.H. Solomon and D.G. Hawthorne, *Chemistry of Pigments and Fillers*, John Wiley & Sons: New York, NY, 1983; p51.
32. K.J.D. MacKenzie, *Trans. J. Brit. Ceram. Soc.*, **1976**, *75*, p121.
33. Y. Iida and S. Ozaki, *J. Am. Ceram. Soc.*, **1961**, *44*, p120.

RECEIVED December 1, 1995

Chapter 5

Synthesis of Nanostructured Materials Using a Laser Vaporization–Condensation Technique

M. Samy El-Shall[1], Shautian Li[1], Daniel Graiver[2], and Udo Pernisz[2]

[1]Department of Chemistry, Virginia Commonwealth University, 1001 West Main Street, Richmond, VA 23284–2006
[2]Dow Corning Corporation, 2200 West Salzburg Road, Auburn, MI 48611

A method which combines laser vaporization of metal targets with controlled condensation in a diffusion cloud chamber is used to synthesize nanoscale metal oxide and metal carbide particles (10 - 20 nm). The silica nanoparticles aggregate into a novel web-like microstructure. These aggregates are very porous and have a very large surface area (460 m^2/g). Bright blue photoluminescence from the nanoparticle silica has been observed upon irradiation with UV light. The photoluminescence is explained by the presence of intrinsic defects of the type $Si(II)^o$ in the amorphous silica. For the iron oxide nanoparticles, magnetic anisotropy constants were found to be one order of magnitude higher than the known bulk values. The synthesis of mixed nanoparticles of controlled composition has been demonstrated.

This chapter deals with the synthesis of nanoscale particles from supersaturated vapors generated by laser vaporization under controlled condensation conditions. The synthesis of nanoscale particles has received considerable attention in view of the potential for new materials with novel properties and the design of catalysts with specific dimensions and compositions (1-10). The reduced size and dimensionality of these atomically engineered materials are responsible for their unique electronic, magnetic, chemical and mechanical properties. The characterization of these properties can ultimately lead to identifying many potential uses, particularly in the field of catalysis. For example, particles of metal oxides and mixed oxides such as SiO_2/TiO_2 exhibit unusual acidic properties and can be used as acidic catalysts (11).

The chapter consists of three major sections. The first is a brief review of the processes of nucleation and growth in supersaturated vapors for the formation of clusters and nanoparticles. The second section deals with the application of laser vaporization for the synthesis of nanoparticles in a diffusion cloud chamber. In the third section, we present some examples of nanoparticles synthesized using this approach and discuss some selected properties.

Nucleation and Growth in Supersaturated Vapors

It is convenient to provide an abbreviated account of the theory of the nucleation of drops from the vapor. The rate of homogeneous nucleation in a one-compound system is usually expressed as J, the number of drops nucleated per cubic centimeter per second, and has the form (12,13).

0097–6156/96/0622–0079$15.25/0

$$J = K \exp(-W^*/kT) \tag{1}$$

where K is a slowly varying prefactor depending on temperature T, pressure P, density of the liquid, and several other parameters, while k is the Boltzmann constant. W^* is the free energy of formation of the critical nucleus, and can be shown to have the form:

$$W^* = 16 \pi v^2 \sigma^3 / 3(kT \ln S)^2 \tag{2}$$

where σ is the surface tension and v is the volume per molecule in the bulk liquid, while S, the supersaturation, is given by

$$S = P/P_e \tag{3}$$

where P is the actual pressure of the vapor and P_e is the equilibrium or "saturation" vapor pressure at the temperature of the vapor. For the one-component system, the free energy of formation, W, of a cluster of n molecules is assumed to be the sum of two terms; a positive contribution from the surface free energy and a negative contribution from the bulk free energy difference between the supersaturated vapor and the liquid. With smaller clusters the positive term dominates so that W increases with size. However, with increasing size the negative term overcomes this effect so that W passes through a maximum. The result is the free energy barrier illustrated qualitatively in Figure 1, where W is plotted versus n. The maximum corresponds to W^* where n is denoted as n^*. The cluster of size n^* is called a "condensation nucleus". Once formed it can grow spontaneously (with a decrease in free energy) into a macroscopic liquid drop. The value of n^* depends on S and is typically small (between 10 and 1000). The critical size (n^*) and radius (r^*) of the nucleus are given by Equations (4) and (5), respectively.

$$n^* = 32 \pi v^2 \sigma^3 / 3(kT \ln S)^3 \tag{4}$$

$$r^* = 2 v \sigma / kT \ln S \tag{5}$$

These equations reveal the very important rule that the larger the supersaturation ratio S, the smaller the critical size of the condensation nucleus (*13*).

The rate of nucleation J, is clearly very sensitive to the height of the free energy barrier W^*. Thus, if the vapor becomes sufficiently supersaturated, the barrier is reduced to the point where nucleation occurs at a high rate.

Preexisting surfaces (e.g. aerosol or dust particles), ions or large polymer molecules greatly accelerate the rate of nucleation by lowering W^*, defined by Eq. (2). Such surfaces accomplish this by reducing the amount of work required to provide the interface in nucleation (since a surface already exists) while ions accomplish this (especially with polar molecules) by dielectric polarization so that the barrier W^* can be lowered to the point where a single ion can induce the formation of a macroscopic liquid drop. This represents almost the ultimate in amplification and detection.

For a binary system it may be shown that the free energy hill in Figure 1 becomes a free energy surface (*14,15*) since now the free energy of formation of a cluster depends upon the number of molecules, n_1 and n_2, of both species in the drop. The clusters, on the way to becoming drops, flow over the surface and through a mountain pass. The saddle point marking the formation of this pass represents an energy barrier that embryos have to overcome in order to grow and become stable.

It should be noted that the application of the classical nucleation theory for the formation of nanoparticles from supersaturated vapors may not be appropriate at very high supersaturations (*16*). In these cases, the critical nuclei may be very small (less than ten atoms) and it is questionable to treat the nucleus as a macroscopic entity with macroscopic properties such as surface tension and density. Another problem in applying the nucleation theory arises when the particles are crystalline. Although, in principle, the capillarity approximation can be used when the clusters are crystalline, the bulk properties are usually unavailable at the low temperatures of nucleation (*16*). If the number density of the particles is high, they collide with one another and depending on the temperature of the vapor, they either coalesce to form larger particles or coagulate. However, it has been shown that coalescence is the dominant growth mechanism for nanoparticles (*17*). This process is usually described using a log normal distribution function (*18*).

Laser Vaporization of Metals

A wide range of scientifically interesting and technologically important nanoparticles has been produced by both chemical (*19,20*) and physical (*21-27*) methods. The most common physical methods involve gas condensation techniques where oven sources are usually used to produce metal vapors. In spite of the success of this method, there are, however, some problems and limitations, such as: possible reactions between metal vapors and oven materials, inhomogeneous heating which can limit the control of particle size and distribution, limited success with refractory metals due to low vapor pressures, and difficulties in controlling the composition of the mixed metal particles due to the difference in composition between the alloys and the mixed vapors. Laser vaporization techniques provide several advantages over other heating methods such as the production of high density vapor of any metal within an extremely short time (10^{-8}s), the generation of directional high-speed metal vapor from the solid target which can be useful for directional deposition of the particles, the control of the evaporation from specific spots on the target as well as the simultaneous or sequential evaporation of several different targets (*28*). Some of these advantages have been demonstrated in the synthesis of ultrafine metal particles and the laser surface treatments and film deposition (*29,30*).

In laser vaporization, a high energy pulsed laser with an intensity flux of about 10^{6}-10^{7} W/cm^2 is focused on a metal target of interest. The resulting plasma causes highly efficient vaporization and the temperature at the focusing spot can exceed 10,000 K. This high temperature can vaporize all known substances so quickly that the surrounding vapor stays at the ambient temperature. Typical yields are 10^{14} - 10^{15} atoms from a surface area of 0.01cm^2 in a 10^{-8} s pulse. The local atomic vapor density can exceed 10^{18} atoms/cm^3 (equivalent to 100 Torr pressure) in the microseconds following the laser pulse.

Experimental Technique

In the experiments, a modified upward thermal diffusion cloud chamber is used for the synthesis of the nanoscale particles (*10,31,32*). A sketch of the chamber with the relevant components necessary for the synthesis of nanoparticles is shown in Figure 2. This chamber has been commonly used for the production of steady state supersaturated vapors for the measurements of homogeneous and photo-induced nucleation rates of a variety of substances (*33*). Detailed description of the chamber and its major components can be found in several references (*33,34*). Here we only offer a very brief description of the modifications relevant to the synthesis of the

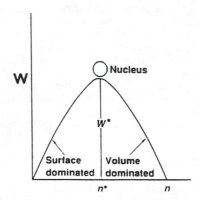

Figure 1. Free energy of cluster formation plotted vs. n, the number of molecules in the cluster.

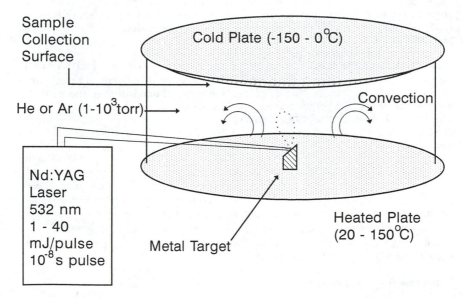

Figure 2. Experimental set-up for the synthesis of nanoparticles using laser vaporization in a convective atmosphere.

nanoparticles. The chamber consists of two, horizontal, circular stainless steel plates, separated by a circular glass ring. A metal target of interest sets on the lower plate (the top surface of the target is located about 0.5 reduced height of the chamber), and the chamber is filled with a carrier gas such as helium or Ar (99.99% pure) containing a known composition of the reactant gas (e.g. O_2 in case of oxides, N_2 or NH_3 for nitrides, CH_4 or C_2H_4 for carbides, etc.). The metal target and the lower plate are maintained at a temperature higher than that of the upper one (temperatures are controlled by circulating fluids). The top plate can be cooled to less than 120 K by circulating liquid nitrogen. The large temperature gradient between the bottom and top plates results in a steady convection current which can be enhanced by using a heavy carrier gas such as Ar under high pressure conditions (1-10^3 torr).

The metal vapor is generated by pulsed laser vaporization using the second harmonic (532 nm) of a Nd-YAG laser (1-40 mJ/pulse, 10^{-8} s pulse). The laser beam is focused on a small area on the surface of the metal target. The target surface is positioned to allow a 45° angle of incidence between the laser beam and the axis normal to the tilted target surface. Therefore, it is likely that the plasma plume above the target surface would expand toward the top plate of the chamber. The laser beam is moved on the metal surface in order to insure surface renewal and good reproducibilty of the amount of metal vapor produced. The hot metal atoms within the plasma plume react with O_2 (in the case of oxides syntheses) to form vapor phase metal oxide molecules which undergo several collisions with the carrier gas thus resulting in efficient cooling via collisional energy loss. Under the total pressure employed in the current experiments (800 torr), it is expected that the metal atoms and the oxide molecules approach the thermal energy of the ambient gas within several hundred microns from the vaporization target. If a small concentration of the reactant gas (O_2) is present, metal clusters will be formed first and then will react with O_2 to form surface oxidized clusters. The clusters are carried by convection to the top plate of the chamber where they are deposited.

The heat and mass flux equations in a convective transport system, as in the pulsed laser vaporization in a diffusion cloud chamber, are not simple. These equations have been solved for one dimensional plane parallel diffusion as in the conventional diffusion chamber (*34*). In this case, the solution of the boundary value problem yields the supersaturation and temperature profiles as a function of elevation within the chamber (*33*). In the present case, pulsed laser vaporization, reactive collisions, diffusion, convection and condensation processes must be considered in solving the flux conservation equations of mass, momentum and energy. The problem becomes much more complex than the one describing the conventional chamber within the one dimensional diffusion approximation. Some qualitative predictions could be made by considering the nature of the successive processes taking place in the chamber. Following the laser pulse, the ejection of the metal atoms and the eventual interaction with the ambient atmosphere will take place. Since the target surface where evaporation occurs is located near the middle of the chamber (about 0.5 reduced height) and because of the rapid decrease in the ambient temperature near the top plate, it is possible that a maximum supersaturation could develop within the upper half of the chamber above the surface target (perhaps closer to the target than to the top plate). This supersaturation can be increased by increasing the temperature gradient between the target and the top plate. As indicated by the Kelvin's equation (Equation 4 or 5), the higher the supersaturation, the smaller the size of the nucleus required for condensation (*13,16,33*). The role of convection in the present experiments is to remove the small particles away from the nucleation zone (once condensed out of the vapor phase) before they can grow into larger particles. The rate of convection will also increase by increasing the temperature gradient in the chamber. Therefore, by controlling the temperature gradient, the total pressure and the laser power (which determines the number density of the metal atoms released in the vapor phase), it is possible to control the size of the condensing particles. Nichrome heater wires are wrapped around the glass ring and provide sufficient heat to prevent condensation on the ring and to maintain a constant temperature gradient between the bottom and top plates. The particles formed in the nucleation zone are deposited on the top plate during the experiment (typically running at 10 Hz for about 1 - 2 hours) then the chamber is brought to room temperature and the particles are collected under atmospheric conditions. Glass slides or metal wafers can be attached to the top plate when it is desired to examine the morphology of the as-deposited particles. No particles are found on any other place in the chamber except on the top plate. This result supports the assumptions that nucleation takes place in the upper half of the chamber and that convection carries the particles to the top plate where deposition occurs.

Nanoparticles of Metal Oxides

Zinc Oxide. Scanning electron microscopy (SEM) of the ZnO particles synthesized using 20% O_2 in He at a total pressure of 800 torr and a top plate temperature of -100 $^{\circ}$C has indicated that the particle diameters are typically between 10 and 20 nm. The particle size is sensitive to the temperature gradient used in the experiment. For example, by setting the condensation plate (top plate) temperature at -40°C, the average particle size was found to be 50-60 nm.

Figure 3 (upper trace) displays the X-ray diffraction pattern of the ZnO nanoparticles synthesized by the laser vaporization method. For comparison we also include the pattern of the particles prepared in methanol (lower trace) according to the method of Henglein and coworkers (*35*). In both cases, the hexagonal Wurtzite structure is evident thus demonstrating that the bulk crystal structure is kept when the size of the semiconductor crystals falls into the nanometer scale. The results are consistent with the work of Spanhel and Anderson (*36*). The particle's size can be estimated from the X-ray diffraction data by using Scherrer's equation $D = K\lambda/(\beta - b)$ $\cos\theta$ where D is the particle's diameter, K is a constant taken as 0.89, λ is the X-ray wavelength, 1.54 $^{\circ}$A, β is the FWHM of the peak, θ is the angle of diffraction and b is a correction constant for the instrument (*37*). The results give D as 6.4 nm and 6.8 nm for the particles prepared by the laser and chemical methods, respectively.

The infrared spectrum of the bulk ZnO has two weak bands at 1350 cm^{-1} and 725 cm^{-1} and a very strong and broad band between 600 and 400 cm^{-1} (*38*). While the 725 cm^{-1} band is not observed in the nanoparticles, the intensity ratio $v_{1350}/v_{600-400}$ is larger as compared to the bulk (Figure 4-a). This suggests that the vibrational mode at 1350 cm^{-1} is enhanced in the nanoparticles due to specific surface processes. In fact, this intensity ratio does not change much from the bulk value in coated ZnO nanoparticles (Li, S.; El-Shall, M. S., unpublished data). For example, we prepared ZnO nanoparticles in a water / oil microemulsion (*39*) and modified the surface by coating it with a surfactant (undecylenic acid) and observed no significant change in the intensity ratio $v_{1350}/v_{600-400}$ from the bulk value (Li, S.; El-Shall, M. S., unpublished data). This indicates that the undecylenic acid layer covers the dangling ZnO bonds and thus reduces the specific surface processes. The ZnO nanoparticles also show small bands around 1570 cm^{-1} which are attributed to carboxylate species resulting from the surface reactions of ZnO with atmospheric CO_2 (*32*). In addition, because of the adsorption of water molecules on the large surface area of the particles, stretching vibrations of the surface OH groups are observed at 3400-3490 cm^{-1} (*32*). The Raman spectrum of the ZnO nanoparticles (Figure 4-b) shows two peaks at 438

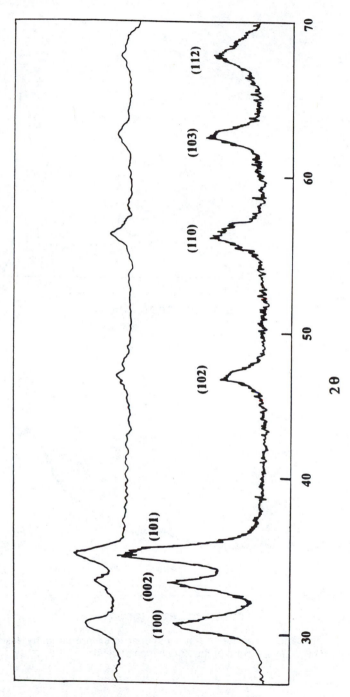

Figure 3. X-ray diffractions of ZnO nanoparticles prepared by laser vaporization (upper trace) and by a chemical method (*31*) (lower trace).

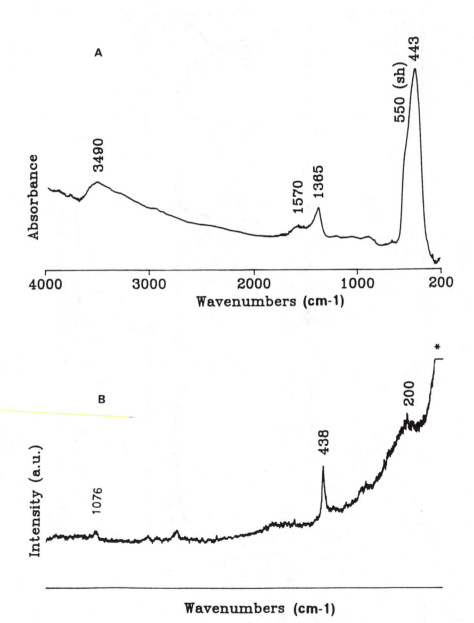

Figure 4. (A) Infrared spectrum of ZnO nanoparticles prepared by laser vaporization. (B) Raman spectrum of the ZnO nanoparticles.

cm^{-1} (E_2 nonpolarized) and 1076 cm^{-1} (due to multiphonon processes), consistent with the bulk spectrum (*40*).

From the UV absorption thresholds of the ZnO nanoparticles prepared by the chemical methods (*32*), the bandgap energies were found to be larger than that of bulk phase ZnO where the bandgap is 3.3 eV (*41*). This is consistent with the quantum size effect of the semiconductor nanoparticles (*4,42-44*).

Silica and Silicon. The SEM micrographs of the as - deposited silica particles on glass substrates reveal highly organized web-like structures characterized by micropores with pore diameters up to 1-2 µm and wall thicknesses of 10 - 20 nm (see Figure 5). The three dimensional arrays exhibit several remarkable features: (i) well-defined pore sizes and shapes (ii) fine control of particle size as indicated by the homogeneous particle diameter which is typically 10 - 20 nm as determined from the TEM analysis and (iii) a very high degree of pore ordering over micrometer length scales with regular interconnecting arrays (network). We found that these structures are more pronounced when the temperature of the condensation plate (upper plate) is kept lower than -40°C. At higher temperatures, a smoother, less porous deposit is obtained.

The observed web-like structure is different from the general structure of conventional silica prepared either by high temperature pyrolysis of a silicon precursor (silicone oxides or the more reactive chlorosilanes) or particle nucleation from supersaturated aqueous silicate solutions (*45*). In both these conventional preparation methods, nucleation yields primary, spherical particles which then aggregate into secondary structures consisting of a random three dimensional network. Variations in the process of both methods affect the nature of the aggregation and thus the structure of the network. Since many of the properties of silica are related to its aggregate structure, a great deal of effort has been made to understand and control the growth process of the primary particles, their linking to branch chains, and ultimately the formation of the aggregated network structure. The web-like aggregations with uniform large cavities, could provide special catalytic activities for the silica nanoparticles.

It is interesting to speculate on the possible growth mechanism of such self-arranged web-like structures. One possibility could arise from a secondary nucleation mechanism of the nanoparticles deposited from the vapor phase on the cold substrate surface. Structural defects at the interfaces of the nanoparticles could promote the nucleation of the observed agglomeration pattern. At lower temperatures, it is possible that the nanoparticles can identify these surface defects and thus self-arrange into the observed web-like morphology. At higher temperatures, the surface defects can be completely annealed and no such organized structures form. Another consideration relates to the size of the deposited particles. As the temperature of the condensation plate increases, the temperature gradient in the chamber decreases and the degree of supersaturation of SiO_2 (actual vapor pressure of SiO_2 at temperature T divided by the equilibrium vapor pressure of SiO_2 at the same T) decreases. Also, the convection current decreases and the residence time of the nucleating particles in the nucleation zone increases. This results in the formation of larger particles which may not be able to self-arrange to a web-like structure.

The X-ray diffraction analysis of silica nanoparticles prepared in 20% O_2 in He reveals a typical amorphous pattern very similar to that of the commercial fumed silica as shown in Figure 6.

The chemical composition of the silica nanoparticles was examined using FTIR. The spectrum shows strong absorption bands associated with the characteristic stretching, bending and wagging vibrations of the SiO_2 group. In addition, bands common to moisture-induced surface species on highly active silica (large surface area) were also observed. In particular, the band at 957 cm^{-1}, which can only be seen

X 9,800

Figure 5. SEM micrograph of web-like agglomeration of silica nanoparticles.

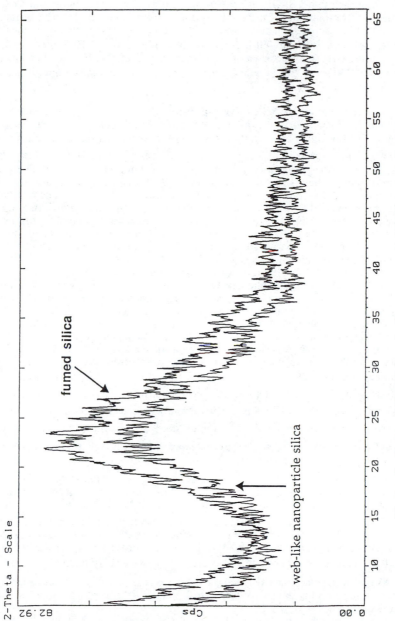

Figure 6. X-ray diffractions of silica nanoparticles prepared by laser vaporization and from fumed silica.

if the specific surface area of the silica is high enough, was observed in our sample. This band is directly related to the surface silanols and is assigned to the v (Si-O) stretching vibration in the Si-OH surface groups. Infrared surface analysis indicated that all the Si-OH groups are located on the surfaces of the nanoparticles. The surface area of the particles depends on the experimental parameters used during the synthesis and values between 380 and 460 m^2/g were determined by the BET method. In comparison, the largest surface area of a commercially available fumed silica has a surface area of 300 m^2/g ; the nominal particle size of this material is 7 nm (Cabosil EH-5, fumed silica).

Photoluminescence of Silica Nanoparticles. A striking feature of the SiO_2 nanoparticles is that they exhibit bright blue photoluminescence (PL) upon irradiation with UV light. The initial observation was made with a pulsed nitrogen laser ($\lambda = 337$ nm) with 0.12 mJ per pulse nominal energy at a pulse rate of 10 s^{-1}. Under these conditions the PL was clearly visible in standard room light. The PL spectrum of nanoparticle silica was measured at 77 K with a spectrofluorometer (SPEX Fluorolog-2) in emission and in excitation modes (see Figure 7) . The emitted photoluminescence (MPL) was excited with 300 nm light (spectral bandwidth 5 nm) and measured between 350 nm and 700 nm (with a spectral bandwidth of 1 nm). For the excitation spectrum (XPL), a wavelength of 476 nm (where one of the two emission maxima occurs) was selected for monitoring the intensity of the PL; the wavelength of the excitation light was varied from 200 nm to 425 nm.

The origin of this surprisingly strong blue photoluminescence is not yet completely clarified. A similar phenomenon was reported with oxygen-containing ultrafine particles of Si prepared by a gas evaporation technique and subsequently oxidized (46). The emission maxima were found at wavelengths at and below 470 nm depending on the oxygen treatment.

The MPL spectrum exhibits two broad peaks, one at 467 nm (2.64 eV) and the other one at 422 nm (2.94 eV). These peaks appear to be related also to defects in the SiO_2 structure. Several such defect models have been discussed in the literature (47,48). The emission at 2.65 eV has been assigned to a new intrinsic defect in amorphous SiO_2 for which a two-fold coordinated Si is proposed, i.e., a $Si(II)^0$ (neutral) center (48). Chemically, this is equivalent to the quasi-molecule $SiO_2{}^{2-}$ with a $^1A_1$ ground state, first excited $^1B_1$ singlet state, and a $^3B_1$ triplet state. The transition from the latter to the singlet ground state, $S_0 \leftarrow T_1$, consists of a radiative part at 2.65 eV in conjunction with a non-radiative transition to S_0 (48).

It is proposed that this $Si(II)^0$ center is present in significant amounts in oxygen-deficient silica. We have to assume, therefore, that the laser ablation of Si in oxygen as described here results in a silica nanostructure with a large concentration of this $Si(II)^0$ defect, especially at its internal surfaces, which in turn gives rise to the observed bright photoluminescence. Further reduction of oxygen due to incomplete oxidation could result in the fromation of Si-Si bonding across the oxygen vacancies. Indeed, the Raman spectrum of our SiO_2 particles shows a sharp band at 520 cm^{-1} (Figure 8) which indicates the existence of Si-Si bonds. The $Si(II)^0$ defect model is also compatible with the unusually large surface area determined for these silica nanoparticles.

The results provide strong evidence for the new form of silica which may become a promising material for future blue light luminescent devices.

Photoluminescence of Silicon Nanocrystals. Si nanocrystals have also been synthesized. In this case laser vaporization of a Si target takes place in a pure He

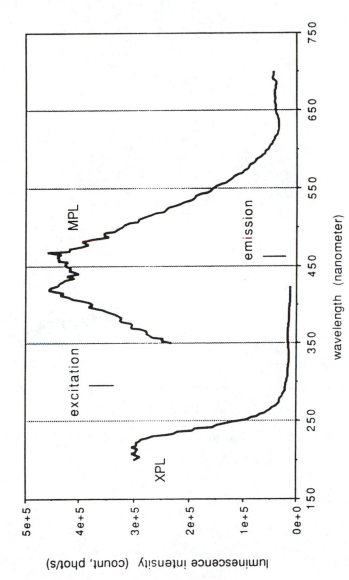

Figure 7. Emission and excitation photoluminescence spectra of the silica nanoparticles.

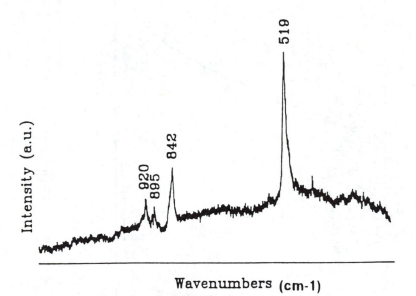

Figure 8. Raman Spectrum of silica nanoparticles.

or Ar atmosphere. The X-ray diffraction pattern of these particles is shown in Figure 9. The data conclusively shows the crystalline Si peaks at 47 and 56 degrees. These crystalline lines are not present in the amorphous silica as shown in Figure 6.

The Si nanocrystals exhibit photoluminescence upon irradiation with UV light at 230 nm. The MPL spectrum is shown in Figure 10. The spectrum is similar to that reported for 4 nm Si nanocrystals upon excitation with 350 nm at 20 K and also to that PL spectrum of Porous Silicon (*49*). In these systems the red luminescence is interpreted as a consequence of quantum crystallites which exhibit size-dependent, discrete excited electronic states due to a quantum effect (*6,50,51*). This quantum confinement shifts the luminescence to higher energy than the bulk crystalline Si (1.1 eV) band gap. This indirect gap transition is dipole forbidden in the infinite preferred crystal due to translational symmetry. By relaxing this symmetry in finite crystallite, the transition can become dipole allowed. As pointed out by Brus (*49*), the quantum size effect in Si nanocrystals is primarily kinetic mainly due to the isolation of electron-hole pairs from each other.

The similarity between the PL spectrum shown in Figure 10 and the spectrum obtained from 4 nm particles (*49*) may suggest that such small particles could be present in our sample.

Iron and Iron oxides. Nanoparticles of iron and iron oxides have been prepared by the laser vaporization of iron in a He atmosphere containing variable concentrations of O_2. The color of the samples goes from very dark brown, almost black (the color of FeO and Fe_3O_4) to reddish brown (the color of γ-Fe_2O_3) with increasing oxygen pressure. SEM/TEM micrographs were used to study the particle morphology. TEM bright field images showed the samples to consist of very small particles with a mean diameter of about 6 nm. Some bigger particles with size 50 - 500 nm were also present.

The diffuse electron diffraction patterns indicate the presence of FeO, Fe_3O_4 and γ-Fe_2O_3 phases. The diffraction patterns from different samples were much the same. In particular, no lines indicating α-Fe were observed in the Fe sample. This could arise from the Fe core being very small or having an amorphous structure and hence showing very broad lines.

The magnetic behavior of these particles was studied using DC- and AC-susceptibility measurements (*52*). All samples exhibited superparamagnetism (*53*) with blocking temperatures ranging from 50 K to above room temperature (*52*). Magnetic anisotropy constants were found to be one order of magnitude higher than the known bulk values. The mean particle size estimated from the magnetic data was found to be in perfect agreement with the TEM data which showed a mean particle diameter of about 6 nm.

Other Metal Oxides. Several other metal oxide nanoparticles have been synthesized using this technique. This includes Bi_2O_3, PdO, NiO, AgO, MgO, SnO, In_2O_3, CuO, Al_2O_3, ZrO_2, MoO_3, V_2O_5, TeO_2, Sb_2O_3, SnO_2 and Cr_2O_3. It is important to note that by controlling the experimental parameters such as the partial pressure of oxygen, the partial pressure of the metal (by varying the vaporization laser power) and total pressure of the carrier gas (He or Ar), it is possible to control the stoichiometry of the oxides formed. Several suboxides such as MoO_x (x < 3), V_2O_x (x < 5) and PdO_x (x < 1) were prepared and characterized using IR, Raman and XPS measurements.

Mixed Oxides. One of the significant advantages of the laser vaporization method is the ability to synthesize mixtures of nanoparticles of controlled composition or particles of mixed metals or metal oxides. This can be achieved by sequential or

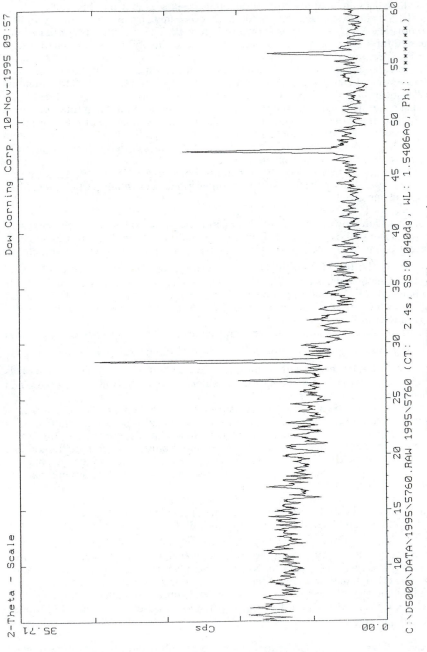

Figure 9. X-ray diffraction of Si nanocrystals.

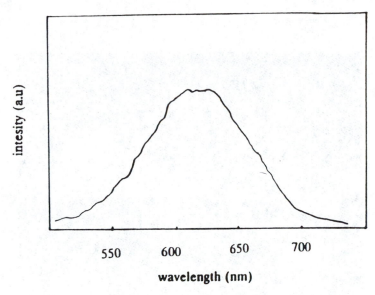

Figure 10. Emission photoluminescence spectrum of Si nanocrystals excited with $\lambda = 230$ nm.

simultaneous vaporization of different metal targets. Using sequential vaporization we prepared several mixtures of oxides such as CuO/ZnO, TiO_2/Al_2O_3 and $TiO_2/Al_2O_3/ZnO$. In these experiments, laser vaporization is alternated on the different metal targets of interest and by adjusting the number of laser shots on each target, we were able to control the composition of the oxides mixture. For example, in the case of CuO/ZnO mixtures, laser vaporization of Cu/Zn targets over successive duration of 1000 shots on each target, 1000 (Cu)/2000 (Zn), and 3000 (Cu)/1000 (Zn) resulted in mixtures containing CuO/ZnO with the ratios 20/80, 6/94, and 48/52, respectively as determined from the SEM analyses of the particles.

Mixed composite particles have been prepared by simultaneous vaporization of two or more metals. In this case beam splitters are used to focus the laser on the different targets. Figure 11 displays SEM micrographs obtained for the mixed Si/Al oxide. The particles appear to have a uniform size and shape. Energy dispersed analysis (shown in Figure 12) indicates that particles, within an area of 1 μm^2, contain both Si and Al with a composition that is identical throughout the sample. Other examples of composite particles prepared by this method include Si/B oxide, Mg/Fe oxide and Cd/Te oxide.

Metal Carbides and Other Carbonaceous Particles

The synthesis of metal carbide nanoparticles has been demonstrated by carrying out laser vaporization in the presence of a convenient source of carbon such as methane, ethylene or isobutene. Thus, silicon carbide nanoparticles were prepared by laser vaporization of Si in a He/isobutene mixture (*54*). Surface analysis using FT-IR has indicated that the nanoparticles consist of SiC cores covered with pure silica layers.

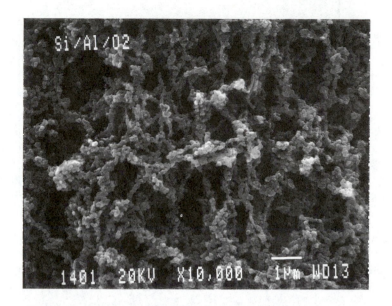

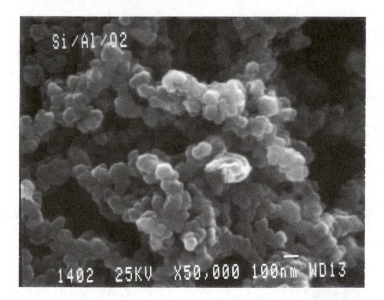

Figure 11. SEM micrographs of composite Al/Si oxide nanoparticles.

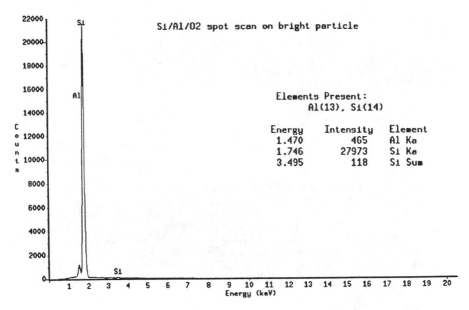

Figure 12. Energy dispersed analysis of composite Al/Si oxide nanoparticles.

The nanoparticles obtained by the laser vaporization of Ti in a He / isobutene mixture were found to contain Ti / C species such as $(TiC)_2$, $(Ti_2C)_2$ and $(TiC_2)_2$ as indicated from the mass spectrum obtained by laser desorption of the particles (*10*). When Mo was vaporized in the presence of a He / isobutene mixture, the mass spectra of the nanoparticles showed peaks corresponding to molybdenum carbide clusters of the stoichiometry $(MoC_4)_n$ with n = 1-4. These clusters have also been synthesized by laser irradiation of $Mo(CO)_6$ as recently reported by Jin et al (*55*). Interestingly, in our laser vaporization experiment, the mass spectra of the particles also revealed abundant ions corresponding to higher fullerenes (C_{80} - C_{150}) with small peaks corresponding to C_{60} and C_{70}. This suggests some catalytic effect for the Mo nanoparticles in the formation of higher fullerenes.

Acknowledgments

The authors gratefully acknowledge financial support from NSF Grant CHE 9311643. Acknowledgement is also made to the donors of the Petroleum Research Fund (2764-AC6), administered by the American Chemical Society, to the Dow Corning Corporation and to the Thomas F. and Kate Miller Jeffress Memorial Trust for the partial support of this research.

Literature Cited

1. See for example: "Frontiers in Materials Science", *Science* **1992**, 255, 1049.
2. Pool, R. *Science* **1990**, 248, 1186.
3. Siegel, R. W. *Annu. Rev. Mater. Sci.* **1991**, 21, 559, ; *Nanostructured Materials*, **1994**, 4, 121.
4. Henglein, A. *Chem. Rev.* **1989**, 89, 1861.
5. Steigerwald, M. L.; Brus, L. E. *Acc. Chem. Res.* **1990**, 23, 183.

6. Bawendi, M. G.; Steigerwald, M. L.; Brus, L. E. *Annu. Rev. Phys. Chem.* **1990**, 41, 477.
7. Wang, Y.; Herron, N. *J. Phys. Chem.* **1991**, 95, 525.
8. Weller, H. *Angew. Chem. Int. Ed. Engl.* **1993**, 105, 41.
9. Hanglein, A. *J. Phys. Chem.* **1993**, 97, 5457.
10. El-Shall, M. S.; Slack, W.; Vann, W.; Kane, D.; Hanley, D. *J. Phys. Chem,* **1994**, 98, 3067.
11. Stakheev, A. Yu.; Shapiro, E. S.; Apijok, J. *J. Phys. Chem.* **1993**, 97, 5668; Vishwanathan, B.; Tanka, B.; Toyoshima, L. *Langmuir* **1986**, 2, 113; Niwa, M.; Sago, M.; Ando, H.; Murakami, Y. *J. Catal.* **1981**,69, 69.
12. *Nucleation*; Zettlemayer, A. C., Ed.; Marcel Dekker: New York, 1969.
13. Abraham, F. F. *Homogeneous Nucleation Theory*; Academic Press: New York and London, 1974.
14. Reiss, H. *J. Chem. Phys.* **1950**, 18, 840.
15. Mirabel, P.; Reiss, H. *Langmuir* **1987**, 3, 228.
16. Reiss, H. In *Advances in Chemical Reaction Dynamics*; Rentzepisin, P. M.; Capellos, C., Eds; NATO ASI series 1985, vol. 184, pp. 115 - 133.
17. Granqvist, C. G.; Buhrman, R. A. *J. Appl. Phys.* **1976**,47, 2200.
18. Kurtz, S. K.; Carpy, F. M. A. *J. Appl. Phys.* **1980**, 51, 5725.
19. Shen, J.; Li, Z.; Yan, Q.; Chen, Y. *J. Phys. Chem.* **1993**, 97, 8504; Shen, J.; Li, Z.; Zhang, Q.; Chen, Y.; Bao, Q. Proceedings of 10th International Congress on Catalysis, Budapest, Hungary, 1992.
20. Wiley, J. B.; Kaner, R. B. *Science* **1992**, 255, 1093.
21. Siegel, R. W.; Hahn, H., In *Current Trends in the Physics of Materials*: Yussouff, M. Ed.; World Scientific: Singapore, 1987; p.403.
22. Granqvist, C. G.; Buhrman, R. A., *J. Appl. Phys.* **1976**, 47, 2200.
23. McHugh, K. M.; Sarkas, H. W.; Eaton, J. G.; Westgate, C. R.; Bowen, K. H., *Z. Phys. D.-Atoms, Molecules and Clusters,* **1989**, 12, 3.
24. Sarkas, H. W.; Kidder, L. H.; Eaton, J. G.; McHugh, K. M.; Bowen, K. H., *Mater. Res. Soc. Symp. Proc.* **1991**, 206, 277.
25. Iwama, S.; Hayakawa, K., Arizumi, T. J. *Cryst. Growth* **1982**, 56, 265.
26. Baba, K.; Shohata, N.; Yonezawa, M. *Appl. Phys. Lett.* **1989**, 54, 2309.
27. Hahn, H.; Averback, R. S. *J. Appl. Phys.* **1990**, 67, 1113.
28. Matsunawa, A.; Katayama, S. in *Laser Welding, Machining, and Materials Processing*; Abright, C., Ed.; Proc. ICALEO; IFS Publishing, 1985.
29. Matsunawa, A.; Katayama, S. *Trans. of JWRI* (Japan) **1985**, 14, 197.
30. Huibin, X.; Shusong. T.; Ngal, L. T. *Trans. of NF Soc.* **1992**, 2, 58.
31. El-Shall, M. S.; Slack, W.; Hanley, D.; Kane, D. in *Molecularly Designed Ultrafine/Nanostructured Materials*; Gonsalves, K. E.; Chow, G.; Xiao, T. D.; Cammarata, R. C., Eds.; Mat. Res. Soc. Symp. Proce.; 1994, 351; p. 369.
32. El-Shall, M. S.; Graiver, D.; Pernisz, U.; Baraton, M. I. *NanoStructured Materials* **1995**, 6, 297.
33. Wright, D.; Caldwell, R.; Moxely, C.; El-Shall, M. S. *J. Chem. Phys.* **1993**, 98, 3356.
34. Katz, J. L. *J. Chem. Phys.* **1970**, 52, 4733.
35. Koch, U.; Fojtik, A.; Weller, H.; Henglein, A. *Chem. Phys. Lett.* **1985**, 122, 507.
36. Spanhel, L.; Anderson, M. *J. Am .Chem. Soc.* **1991**, 113, 2826.
37. Kakudo, M.; Kasai, N. *X-Ray Diffraction by Polymers* ; Kodansha LTD and Elsevier Publishing Co.: 1972; P. 329.
38. Nyquist, R.; Kagel, R. *Infrared Spectra of Inorganic Compounds*; Academic Press: 1977; PP. 220-221.
39. Joselevich, E; Willner, I. *J. Phys. Chem.* **1994**, 98, 7628.
40. Damen, T. C.; Porto, S.; Tell, B. *Phys. Rev.* **1966**, 570.
41. Strehlow, W. H.; Cook, E. L. *J. Phys. Chem. Ref. Data* **1973**, 2, 163.

42. Brus, L. *J. Phys. Chem.* **1986**, 90, 2555.
43. Ekimov, A. I.; Efros, Al. L.; Onushchenko, A. A. *Solid State Communications* **1985**, 56, 921.
44. Wang, Y.; Herron, N. *Phys. Rev. B* **1990**, 42, 7253.
45. Iler, R. K. *The Chemistry of Silica. Solubility, Polymerization, Colloid and Surfaces Properties, and Biochemistry:*; Wiley-Interscience Pub.: New York, 1979; Chapter I.
46. Morisaki, H.; Hashimoto, H.; Ping, F. W.; Nozawa, H.; Ono, H. *J. Appl. Phys.* **1993**, 74, (4) 2977.
47. Griscom, D. L. *J. Non-Cryst. Sol.* **1985**, 73, 51.
48. Skuja, L. N.; Streletsky, A. N.; Pakovich, A. B. *Sol. State Comm.* **1984**, 50(12), 1069.
49. Brus, L. *J. Phys. Chem.* **1994**, 98, 3575.
50. Littau, K. A.; Szajowski, P. J.; Muller, A. J.; Kortan, A. R.; Brus, L. E. *J. Phys. Chem.* **1993**, 97, 1224.
51. Wilson, W. L.; Szajowski, P. F.; Brus, L. E. *Science* **1993**, 262, 1242.
52. Turkki, T.; Jonsson, J.; Strom, V.; Medelius, H.; El-Shall, M. S.; Rao, K. V. *J. Korean Magnetic Society,* Proc. 3rd Int'l Symp. on Physics of Magnetic Materials, 1995.
53. Bean, C. P.; Livingston, J. D. *J. Appl. Phys.* **1959**, 30, 120S.
54. Baraton, M. I. ; El-Shall, M. S. *NanoStructured Materials* **1995**, 6, 301.
55. Jin, C.; Haufler, R. E.; Hettich, R. L.; Barshick, C. M.; Compton, R. N.; Puretzky, A. A.; Dem'yanenko, A. V.; Tuinman, A. A. *Science* **1994**, 263, 68.

RECEIVED December 20, 1995

METAL COLLOIDS
IN POLYMERS AND MEMBRANES

Chapter 6

Nanodispersed Metal Particles in Polymeric Matrices

L. M. Bronstein[1], E. Sh. Mirzoeva[1], M. V. Seregina[1], P. M. Valetsky[1], S. P. Solodovnikov[1], and R. A. Register[2]

[1]Nesmeyanov Institute of Organoelement Compounds, Russian Academy of Sciences, 28 Vavilov Street, 117813, Moscow V–334, Russia
[2]Department of Chemical Engineering, Princeton University, Princeton, NJ 08544

Three routes to the preparation of nanodispersed metal or metal oxides particles in polymer matrices have been studied. Thermal treatment in air of W-, Mo- and Cr-carbonyl complexes with polyacrylonitrile (PAN) results in the formation of nanometer-size metal oxide particles in a polycyclic polymer. Co-containing polymers were prepared by mixing $Co_2(CO)_8$ with a PAN copolymer or an aromatic polyamide in dimethylformamide (DMF). $Co_2(CO)_8$ interacts with DMF giving the complex $[Co(DMF)_6]^{2+}[Co(CO)_4]^-_2$. Subsequent thermolysis converts this complex to nanodispersed Co particles. By ferromagnetic resonance and small-angle x-ray scattering, it was found that the average Co particle size depends on the type of polymeric matrix, the thermolysis conditions and the Co loading and varies from 1 to 10 nm.

Colloidal metal particles and metal clusters are intriguing because their behavior differs from both bulk metal and isolated metal atoms. The huge specific surface and confinement of charge carriers suggests potential applications as catalysts, as ferrofluids, and as materials for third-harmonic generation (*1,2*). The principal difficulty in preparing such metal colloids is developing a method to control the particle growth. This problem can be partly solved by carrying out the nucleation and growth process either in solid polymer matrices (*3*) or in cages (or pores, for example, zeolites) (*4*) or in organized media such as microemulsions, vesicles or micelles (*5*). Reactions in such microreactors work with the concept that growth processes are limited in nanostructured matrices by the size of the structures themselves. Here, we discuss another way to control the nucleation and growth process in polymeric matrices which might be especially appealing when the methods listed above cannot be applied, that is, when the polymeric matrix does not form vesicles or micelles. For this case, carrying out the metal colloid formation in solid polymeric matrices can provide size control by changing the type of polymeric matrix, complexation power of the matrix with the metal, conditions of colloid preparation, metal loading and so on. Two approaches to the formation of metal and metal oxide particles could be considered for such polymeric matrices as polyacrylonitrile (PAN). Since such polymers contain nitrile groups capable of complex formation, metal particles might form via complexation of organometallic compounds with PAN followed by decomposition of these complexes. Alternatively, organometallic compounds could be blended with polymers, although in this case miscibility between the two would be a paramount concern. Another important feature of PAN is that it produces polycyclic structures with conjugated double bonds under thermolysis at 200-300°C so materials derived from thermolyzed PAN may exhibit novel properties. For such polymeric matrices as aromatic polyamides (PA), which do not contain active groups, the preparation of blends with organometallic compounds is the most robust way for metal particle incorporation. Here, we present an overview of our recent work in the preparation of nanoscale metal and metal oxide particles embedded in polymeric matrices.

Experimental Section

PAN was synthesized by radical polymerization in DMF solution in the presence of boron alkyl catalysts. The resulting polymer was purified by reprecipitation from dimethylformamide (DMF) by methanol followed by extraction with ethanol for 2 hours in a Soxhlet extractor. An average molecular weight of $M_w=2.2\times10^5$ g mol^{-1} was determined using light scattering measurements of DMF solutions. The synthesis of tungsten, molybdenum and chromium carbonyl PAN with different metal contents was performed according to methods described elsewhere (6,7). Films were prepared by casting from DMF (PAN-W and PAN-Mo) and DMSO (PAN-Cr) solutions.

Thermolysis of Cr-, Mo-, W-containing, and initial PAN samples was carried out at 220°C in air. FTIR spectra were acquired with a Bruker IFR-25 spectrometer at 2 cm^{-1} resolution. Samples were prepared as films on KBr disks (polymer samples) or as Nujol mulls or solutions (model compounds).

A PAN copolymer containing itaconic acid (PAN-I) was synthesized by emulsion polymerization at the Research Institute of Synthetic Fibers (Tver, Russia) at feed ratio of 98.5/1.5 w/w acrylonitrile/itaconic acid, and has an intrinsic viscosity (dimethylsulfoxide, 25°C) of 2.06 dl/g. An aromatic polyamide (PA) derived from terephthalic and isophthalic acid dichlorides (tere/iso ratio of 85/15 w/w) and 9,9'-(4-aminophenyl) fluorene was synthesized at the Nesmeyanov Institute of Organoelement Compounds, and has [η]=0.98 dl/g (N-methylpyrrolidone, 25°C). $Co_2(CO)_8$ was recrystallized from pentane, while DMF (Aldrich) was used as received.

Thermolysis of the Co-containing polymer films was carried out in vacuum, at 220°C unless otherwise stated. Samples are identified by the Co content (in wt.%) in the precursor film before thermolysis, as determined by X-ray fluorescence (XRF) analysis; the suffixes "A" and "T" indicate samples as-cast from solution and after thermolysis, respectively.

Ferromagnetic resonance (FMR) and ESR spectra were registered on a E-12 Varian spectrometer employing 100 kHz field modulation and a 30.5 cm magnet. SAXS measurements were performed with a Philips XRG-3000 generator and a long-fine-focus Cu tube, an Anton-Paar compact Kratky camera, and a Braun one-dimensional position-sensitive detector (8). CuKα radiation was selected through a combination of nickel filtering (α/ß = 500) and pulse-height discrimination at the detector. Data were corrected for empty beam scattering, detector sensitivity and positional linearity, and system deadtime. Data for PA samples were normalized for sample thickness and transmittance. Data for PAN-I samples were normalized for sample transmittance, but the brittle nature of the thermolyzed films made accurate sample thickness measurements impossible. Data for the PAN-I samples are expressed on a consistent (assuming equal thicknesses) relative intensity scale, plotted against the scattering vector $q = (4\pi/\lambda)\sin\theta$, where θ is half the scattering angle and λ is the x-ray wavelength. The collimation system is such that the infinite-slit approximation is valid over the q range of principal interest ($q < 1.5$ nm^{-1}).

Preparation of Metal Particles via Complex Formation with VIB Group Metal Hexacarbonyls.

For preparation of polyacrylonitriles containing colloidal particles through complexation with organometallic compounds, we first studied the complex formation between PAN nitrile groups and VI B Group metal hexacarbonyls. From the literature on small molecule nitriles, it is known that reaction with VI B Group metal hexacarbonyls leads to the substitution of CO ligands by nitriles and usually a mixture of complexes (mononitrile pentacarbonyl complexes and dinitrile tetracarbonyl ones) is formed (9). Carrying out the reaction in CO, which is a reaction product, promotes the formation mainly of pentacarbonyl complexes.

The characterization of complexes immobilized on polymers was carried out by comparison of FTIR and NMR spectra of solid organometallic polymers with compounds prepared by complexing metal hexacarbonyls with CH$_3$CN (6,7). Relying on FTIR and NMR data complex formation of PAN nitrile groups with VI B Group metal hexacarbonyls can be envisioned as follows:

Where M = Cr, Mo, W; for M = Cr, m = 0.

Thermal treatment of PAN between 200 and 300°C is known (*10*) to result in reaction between the nitrile groups, leading to cyclization, with the formation of conjugated fragments. The main changes in the FTIR spectra of PAN films during thermolysis are the following: the decrease in the intensity of the $\nu(C\equiv N)$ band suggests the participation of nitrile groups in reactions involving the breakage of $C\equiv N$ triple bonds, and highly intense bands at 1580 and 1380 cm^{-1} are indicative of the formation of N-containing heterocyclic fragments. The decreasing intensity of the bands of CH_2 stretching vibrations and the appearance of a band at 807 cm^{-1} (attributed to nonplanar CH bending vibrations) suggests that aromatic fragments are also formed. The $\nu(C\equiv N)$ and $\nu(CH_2)$ bands do not vanish completely from the FTIR spectrum even upon prolonged thermolysis. This indicates that conjugated fragments of limited length and various structures are produced by thermolysis rather than a polymer having a regular structure. After 11 hours of thermolysis in air a new band at 2210 cm^{-1} appears in the FTIR spectrum. The 30 cm^{-1} decrease in the $\nu(C\equiv N)$ frequency is indicative of the formation of CN groups conjugated with olefin fragments. It is assumed that predominant interactions at 220°C are those between nitrile groups and the cyclization reaction. As the condensed cyclic fragments accumulate, the increased rigidity of the polymer chain poses some steric hindrance to further cyclization, and the initially secondary process of olefin structure formation becomes more probable.

In the case of metal-containing PAN the spectral picture is qualitatively the same (as compared with the initial PAN); however, the nature of the metal carbonyl introduced into the polymer somewhat influences the character of the transformations. By FTIR, ESCA and ESR, it was found that under thermal treatment in air at 220°C, first metal carbonyl complexes decompose (by FTIR spectroscopy tungsten and molybdenum carbonyl complexes lose carbon monoxide at 150-160°C, chromium carbonyl ones at 110°C) and chromium, molybdenum and tungsten oxide particles form.

Polymeric materials prepared by thermolysis of Cr carbonyl-containing PAN were investigated by ESR. The dispersity of Cr_2O_3 in the thermolyzed PAN sample manifests itself as a discrepancy between the temperature dependence of the ESR signals of Cr^{3+} ions in this sample and the same dependence in crystalline Cr_2O_3. The Curie point of the latter (the temperature of the transition from a paramagnetic to an antiferromagnetic state) is 40°C. In thermolyzed samples of Cr-containing PAN, decreasing the temperature to 77K does not lead to the disappearance of this signal.

The question arises as to the critical crystal size at which the destruction of the antiferromagnetic ordering of the spin system occurs. If the dimensional criterion is used, as was previously applied to ferromagnetic crystals by Vonsovsky (*11*), then a rough estimate gives a value about 3 nm. Thus, Cr-containing PAN after thermolysis in air contains Cr_2O_3 particles with a size less than 3 nm.

Complexation of the metal by the macromolecule produces very fine metal or metal oxide particles. However, the number of metal-containing polymers which can be prepared in this way is extremely limited.

Preparation of Metal Particles through Blending Organometallic Compounds with Polymers.

Another approach to producing polymeric materials with nanodispersed metal particles is blending organometallic compounds with polymers [12]. This approach is adaptable to any polymeric matrices since polymers need not contain active groups, although miscibility between components of the blend limits this method. As the polymeric matrix, we employed a copolymer of acrylonitrile and 1.5 wt.% itaconic acid (PAN-I), although all features observed with the copolymer are also found for PAN homopolymer. The presence of itaconic acid units in PAN-I accelerates thermolysis and permits successful thermolysis in vacuum for 5 hours whereas by FTIR thermolysis of PAN homopolymer lasted 46 hours.

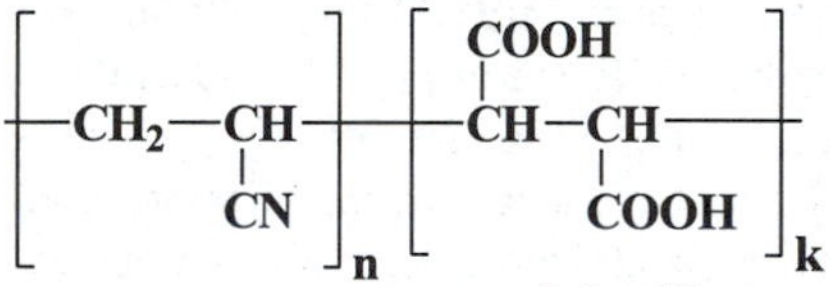

$$\left[-CH_2-\underset{\underset{CN}{|}}{CH}-\right]_n \left[\underset{\underset{COOH}{|}}{\overset{\overset{COOH}{|}}{CH}-CH}-\right]_k$$

1.5 wt%

Another polymeric matrix studied was the aromatic polyamide shown below:

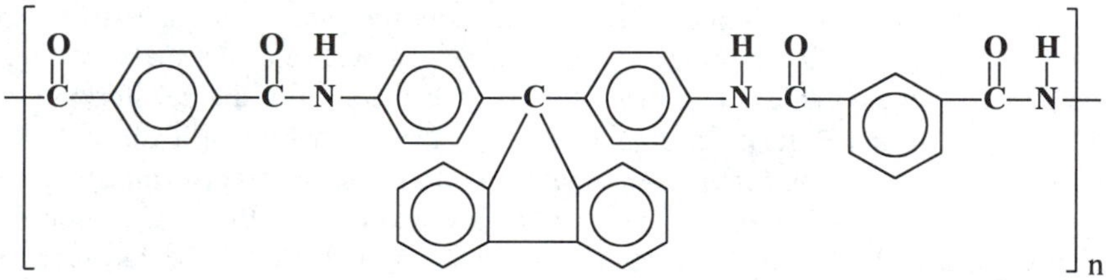

Unlike PAN, this polymer remains unchanged for thermal treatments below 420°C. The use of two types of polymeric matrices permits the study of the influence of the polymer nature on metal particle formation. Co octacarbonyl was chosen for incorporation into polymeric matrices because it can give ferromagnetic cobalt particles in mild thermal conditions; however, previously we have found that $Co_2(CO)_8$ is not compatible with polybutadiene, polystyrene and a poly(styrene-butadiene) block copolymer.

Nanodispersed Metal Particles in Thermolyzed PAN-I. To incorporate Co into PAN-I, the polymer and $Co_2(CO)_8$ were dissolved in DMF. Because DMF can react with $Co_2(CO)_8$, its interaction in the absence of polymer was studied first. The FTIR spectrum of the pink solution prepared by dissolution of orange $Co_2(CO)_8$ (4 wt.%) in DMF at room temperature for 20-30 minutes, exhibits a single band at 1890 cm^{-1}, which we assign to the CO stretch. The position of the band indicates that the species contains the anion $[Co(CO)_4]^-$, reported at 1886 cm^{-1} (*13*). When the DMF is evaporated in vacuum at 35-40°C, a solid pink product is obtained, whose FTIR spectrum exhibits two peaks at 1896 and 1880 cm^{-1}, which result from crystal splitting of the band observed in solution at 1890 cm^{-1}. Cyclic voltammetry of product formed in DMF showed a reduction potential Ep.c.=-1.50V and an oxidation potential Ep.a.=+0.24V, which are evidence for the formation of a cationic/anionic complex. Elemental analysis for this product yielded: Co - 20.85; C - 36.56; H - 5.26; N - 9.46 wt.%. For the structure $[Co(DMF)_6]^{2+}[Co(CO)_4]^{-}_2$, calculated values are: Co - 21.09; C - 37.19; H - 5.00; N - 10.01 wt.%. Therefore, we conclude that the interaction between $Co_2(CO)_8$ and DMF results in the formation of the cationic-anionic complex $[Co(DMF)_6]^{2+}[Co(CO)_4]^{-}_2$.

Co was incorporated into PAN-I by casting films of the blends of PAN-I and the Co complex from DMF solutions onto glass plates about 30 min after the addition of $Co_2(CO)_8$. PAN-I appears to be compatible with the cationic-anionic complex $[Co(DMF)_6]^{2+}[Co(CO)_4]^{-}_2$ until the Co content in the blends reaches 8 wt.%. At higher Co contents, this complex crystallizes on the surface of the films. FTIR spectra of these films also contain the two broad bands observed for the pure complex $[Co(DMF)_6]^{2+}[Co(CO)_4]^{-}_2$.

To convert the Co-containing PAN-I films to materials with nanodispersed Co metal particles, the films were thermolyzed under vacuum at 220°C. Progress of the thermolysis was monitored by FTIR. While heating in vacuum at 100°C for one hour is sufficient to remove the ligands from bulk $[Co(DMF)_6]^{2+}[Co(CO)_4]^{-}_2$, dispersing the complex in PAN-I evidently hinders this process; after 15 minutes of vacuum thermolysis at 220°C, sample 0.90T still showed substantial peaks at 1896 and 1880 cm^{-1}. Thermolyzing for 30 min causes these bands to vanish. During thermolysis, the FTIR bands arising from nitrile groups evolve similarly in pure PAN-I and in Co-containing PAN-I, but the presence of Co accelerates the thermolysis roughly tenfold. XRF analysis carried out by scanning through thermolyzed PAN-I films shows that the Co content is macroscopically uniform.

Because the size of the Co particles strongly influences their magnetic properties, ferromagnetic resonance (FMR) can be used as a convenient gauge of particle size. Bulk metal behavior appears at particle sizes > 10^3 nm (*11*). Monodomain particles form at sizes about 10^2 nm. Co particles about 10 nm in size exhibit superparamagnetic behavior; from the magnetization decay curve, obtained in zero magnetic field, a decay time constant related to particle size can be obtained (*14*). Ferromagnetic order of spins is broken at particle sizes less than 1 nm. Moreover, a correlation exists between the linewidth of the FMR signal and the particle size (*15,16*), although particle shape also influences the FMR spectra (*15*).

All FMR spectra of our samples show signals which are typical (*15*) for ß-Co, which has a face-centered-cubic structure (the position of signal does not depend on the examination temperature). For sample 0.90T, the width, shape, and position of the resonance line are very similar to those reported by Bean (*15*) for spherical Co particles with D = 5 nm; we thus infer a similar size for the Co particles in sample 0.90T (Table I). Unfortunately, the limited amount of data available for small spherical Co particles (*14,17*) limits the quantitative use of the ΔH values.

Table I. Characteristics of Thermolyzed PAN-I-Co Samples

Sample Identification	Weight Percent Co in Precursor, %	Linewidth ΔH (Oe), peak-to-peak	D (nm) from FMR	c (nm) from SAXS
0.40T	0.40	180/750	1-1.5[a]	2.95
0.90T	0.90	580	5[b]	3.41
1.85T	1.85	980	10[c]	3.67
5.92T	5.92	1250	-	5.75

[a]For narrow line; obtained from ΔH by comparison with data in (17).
[b]Obtained from ΔH by comparison with data in (15).
[c]Obtained from τ.

The FMR spectrum of the sample with 1.85 wt.% of Co after thermolysis at 220°C reveals superparamagnetic relaxation in the Co particles. From the magnetization half-time $\tau=2.3$ sec, D=10 nm is calculated (14). For the samples with less than 1.8 wt.% Co, the superparamagnetic fraction was too small to permit measurements of the relaxation half-time. The values of particle size for various samples identified by FMR data are shown in Table I. From these data, one can see that the Co particle size for these materials depends on Co loading: the particles become larger as the Co content increases.

It was found that the thermolysis temperature strongly influences the FMR signal. The increase of the thermolysis temperature from 220°C to 300°C results in the growth of signal intensity and linewidth [12]. The increase in signal intensity with increasing thermolysis temperature suggests that very small particles (<1 nm), which do not contribute to the FMR signal, are formed at low thermolysis temperatures. More intensive thermal treatment increases the linewidth, and by inference the particle size. The weak narrow signal results from free radicals in the polymer (18), so this signal should have $g = 2.003$. Comparison of the positions of the Co and free-radical signals shows that the value of g_{eff} for the Co particles is constant for the different thermal treatments, implying that no substantial change in particle shape occurs; moreover, the symmetric shape of the signal implies that the particles are spherical (15).

Figure 1 illustrates how the form of FMR spectra changes when the thermolysis temperature is greatly increased. Figure 1(a) is characteristic for samples thermolyzed at 220°C in vacuum for 5 hours. The spectra in Figures 1b, c, d were recorded after carbonizing treatment at 980°C in Ar for 30 min. Distinct fluctuations of magnetization are seen at low magnetic field (the Barkhausen effect) [11] and these fluctuations disappear in high fields. The different registration time constants show a distribution of domain wall motion velocities. The presence of the Barkhausen effect is also evidence of the formation of large cobalt particles containing many magnetic domains.

X-ray diffraction showed no clear peaks in the range expected for fcc Co. This is not surprising, given the low Co loadings and the thinness of the films used (transmission for CuKα > 85%). Thus, small-angle x-ray scattering (SAXS) was employed to assess the size of the particles formed. Slit-smeared SAXS patterns for samples 0.40A and 0.40T are shown in Figure 2. In the as-cast film, very little scattering is observed, while the scattered intensity in the same region is quite substantial in the thermolyzed sample. The change in the SAXS pattern is a direct consequence of the formation of Co particles during the thermal treatment. Similar results were observed for all four PAN-I materials examined, before and after thermolysis. Note that all the SAXS patterns exhibit a strong background due to Co fluorescence (K edge = 7710 eV) in the CuKα x-ray (8042 eV) beam. Fluorescence should appear as a constant intensity contribution, independent of q over the small angular range probed by SAXS. This contribution was estimated from the intensity at high q and subtracted from the data prior to the analyses described below.

A Guinier analysis of the SAXS data was unsuccessful in estimating the particle size in the thermolyzed PAN-I samples, as no substantial linear region could be found in the Guinier plots. This observation, coupled with the continuous nature of the scattering in Figure 2, suggests that the particle size distribution is broad, with a substantial fraction of the particles having $qR_g > 1$. Thus, a model appropriate to systems with a more random distribution of size scales was selected: the Debye-Bueche random-two-phase model (*19*). The parameter extracted from the analysis is a correlation length c, which reflects an average size scale in the material. The equation describing the model, for infinite-slit collimation, is (*20*):

$$I_c = K(1 + c^2q^2)^{-3/2} \tag{1}$$

where the constant K reflects both instrumental parameters and the mean electron-density fluctuation. I_c is the net intensity above background. A plot of slit-smeared data as $(I_c)^{-2/3}$ vs. q^2 should produce a straight line, with c^2 given by the ratio of slope to intercept. A representative Debye plot is shown in Figure 3. With all the thermolyzed PAN-I samples, the Debye-Bueche model gave a good description over most of the q range; however, at very low q, the data exhibit scattered intensity in excess of that predicted by the model. This discrepancy between model and data at low q, which was observed for all four thermolyzed samples, has two possible origins: a small population of distinctly larger particles (recall that the FMR spectrum for sample 0.40T showed both narrow and broad signals), or voids formed during thermolysis. Correlation lengths obtained from the best fits to the data are given in Table I. Clearly, the average particle size increases with increasing Co content in the precursor blends.

By comparing the $q \to 0$ limits of Guinier's Law and the Debye-Bueche equation, we find $R_g = (9/2)^{1/2}c$. This relation then allows comparison of the sizes obtained by SAXS and FMR; assuming the particles are roughly spherical, then $D = (20/3)^{1/2}R_g = (30)^{1/2}c$. Applying this approximate factor to the SAXS-determined c values in Table I, we see that the sizes determined by FMR and by SAXS are of the same order of magnitude, although the SAXS values are larger; this

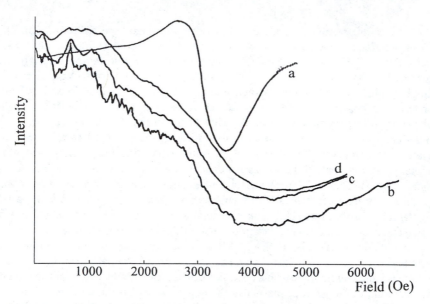

Figure 1. FMR spectra of sample 1.85T (containing 1.85 wt.% Co in precursor) after thermolysis (a) and carbonization (b, c, d); registration time constant $\tau=0.3$ s (b), $\tau=1$ s (c), $\tau=3$ s (d).

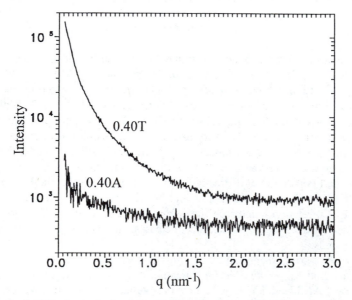

Figure 2. SAXS data obtained for PAN-I-Co samples with 0.4 wt.% Co (in precursor) before (0.40A, bottom) and after (0.40T, top) thermolysis at 220°C for 5 hours.

discrepancy could well result from how each technique perceives the "average" size in a broad distribution, as well as from the assumptions made in deriving sizes from the data.

Nanodispersed Metal Particles in Polyamides. As was mentioned above, in principle any polymeric matrix can be employed for preparing nanoscale particles through blending with organometallic compounds. To explore the influence of the polymeric matrix on metal particles formation we have studied an aromatic polyamide (PA) as the matrix for preparation of Co colloids. Co-containing PA films were prepared in the same way as Co-containing PAN-I, that is, through dissolution of $Co_2(CO)_8$ in DMF solutions of PA. FMR data for the samples with 1.00 and 1.50 wt.% Co in the precursor are given in Table II.

Table II. FMR and SAXS Data for Thermolyzed PA-Co Samples

Sample Identification	Weight Percent Co in precursor	Linewidth ΔH (Oe), peak-to-peak	A (relative)	R_g (nm)
1.50T	1.50	500	9.32	0.98
1.00T	1.00	500	8.98	1.00

It can be seen that the FMR linewidth is the same for both Co loadings. According to (*15*), such a linewidth corresponds to particles about 5 nm in size. It was also found that thermolysis time (from 15 minutes to 5 hours) does not influence the FMR linewidths; however, increasing the thermolysis temperature from 220°C to 270°C results in an increase of the FMR linewidth (from 500 to 1350 Oe) and, therefore, the particle size.

Slit-smeared SAXS patterns for sample 1.00A (as-cast) and 1.00T (thermolyzed) are shown in Figure 4. As in the PAN-I-Co case, the change in the SAXS pattern after thermolysis reflects Co particle formation. The net scattering (above the background) for samples 1.00T and 1.50T was well-described by Guinier's Law (*21*), as shown in Figure 5:

$$\ln(I_{net}) = A - (R_g^2/3)q^2 \qquad (2)$$

where A is a constant describing the overall intensity and R_g is the particle radius of gyration. Note that the form of Guinier's Law is identical (*19*) for both point and infinite-slit collimation; only the value of the constant A differs. Results of the Guinier analysis are given in Table I. Both PA-Co specimens exhibit $R_g \approx 1.0$ nm, with the main difference between the two specimens being the overall intensity (number of particles). The value of exp(A) should be proportional to the average particle volume. The ratio of $exp(A)/R_g^3$ values between samples 1.00T and 1.50T (1.5) as well as the ratio of background intensities (1.5) is in good agreement with the

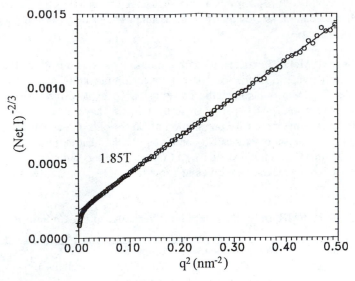

Figure 3. Debye plot of SAXS intensity (after fluorescent background subtraction) for PAN-I-Co sample 1.85T, showing data (O) and best-fit line.

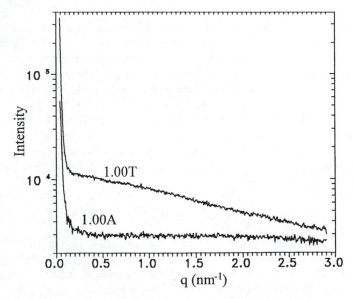

Figure 4. SAXS data obtained for PA-Co samples with 1.00 wt.% Co (in precursor) before (1.00A, bottom) and after (1.00T, top) thermolysis at $220°C$ for 5 hours.

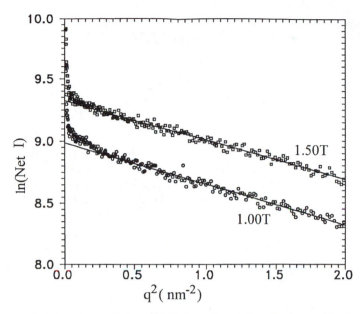

Figure 5. Guinier plots of the SAXS intensity (after fluorescent background subtraction) for PA-Co samples 1.50T (□, top) and 1.00T (O, bottom). Best-fit lines are shown; parameters are given in Table 2.

relative Co loadings of the two films (1.5). The range of q^2 used in the fits to equation (2) corresponds to $0.44 < qR_g < 1.41$; Guinier's Law should hold well for $qR_g < 1$, but is often observed to hold to somewhat larger q values, as is the case here.

So, the particle size in thermolyzed PA was found by FMR and SAXS to be independent of Co loading, at least at the two levels investigated; an increase of the Co content in the precursor leads only to an increase in the number of particles. Previously we have found that iron nanoparticles formed via the organometallic polymer route are quite different for solid polymeric matrices than when the matrix is able to melt (22). Comparing PAN and PA, polyacrylonitrile takes part in chemical reactions during thermolysis at 220°C and though PAN has no real Tg, constant changes take place when Co particles form that can facilitate diffusion not only of atoms, but also of bigger particles. PA, by contrast, does not change its properties during thermolysis at 220°C or 270°C, and diffusion of large metal clusters can be difficult.

Two processes of crystallite growth can be surmised [reactions (3) and (4)] and we presume that both occur when Co nanostructured particles form in PAN-I, leading to the formation of very irregular and highly polydisperse particles:

$$M_n + M \longrightarrow M_{n+1} \tag{3}$$

$$M_n + M_m \longrightarrow M_{n+m} \tag{4}$$

In parallel with crystallite growth there can also be interactions of the metal particles with polymeric molecules S, terminating particle growth:

$$M_n + kS \longrightarrow \begin{array}{ccc} & S & \\ S & & S \\ & M_n & \\ S & & S \\ & S & \end{array}$$

Because the nitrile groups in PAN are reacting to form polycyclic structures, they may be less likely to complex with surfaces of the growing Co particles. By contrast, the PA is thermally stable under the thermolysis conditions used; we speculate that the amide groups in PA act to stabilize the small colloidal particles.

Conclusions

We have demonstrated that thermolyzed PAN consisting of conjugated heterocycles and including metal or metal oxide particles, can be prepared by two basic ways. The first involves complex formation of PAN nitrile groups with corresponding metal carbonyls, followed by thermolysis of the organometallic polymers. This route is acceptable for preparation of thermolyzed PAN with Cr, Mo, and W oxides (or metal particles) distributed throughout the polymer. Complexation by the macromolecule yields nanometer-size metal or metal oxide particles. The second route employs blends of polymers with organometallic compounds; the compatibility between the two can be improved by the use of a donor solvent. DMF reacts with $Co_2(CO)_8$ to form $[Co(DMF)_6]^{2+}[Co(CO)_4]^-_2$, increasing miscibility with the polymer. Nanodispersed ß-Co particles form in a polymeric matrix, by thermolyzing films cast from DMF solutions of PAN-I and Co complex. A portion of the Co particles formed after thermolysis exhibit superparamagnetic behavior. It was also found that for this polymeric matrix the size of the Co particles could be controlled either by varying the thermolysis conditions or the loading of Co in the PAN films before thermolysis. In the aromatic polyamide (PA), the Co particle size seemed to be independent of Co loading; an increase in the Co content in the precursor leads only to an increase in the number of particles. The differences between the two cases may be due to interactions between the colloidal Co particles and functional groups in the polymer matrix.

Acknowledgments

We thank Prof. Ya.S. Vygodsky for samples of the aromatic polyamide. The research carried out at Nesmeyanov Institute was made possible in part by the Grant No. 93-03-4317 from the Russian Science Foundation, by Grant No. M3U000 from the International Science Foundation and by Grant No. M3U300 from the International Science Foundation and Russian Government. Research at Princeton was partially supported by the National Science Foundation, Polymers Program [DMR-9257565].

Literature Cited

1.   Henglein, A. *Chem. Rev.* **1989**, *89*, 1861.
2.   Ogawa, S; Hayashi. Y.; Kobayashi, N.; Tokizaki, T.; Nakamura, A. *Jpn. J. Appl. Phys.* **1994**, *33*, L331.
3.   Sur, G.S; Mark, J.E. *Polymer Bulletin.* **1987**, *18*, 369.
4.   Kawi, S; Gates, B.C. In *Clusters and Colloids*; Schmid, G., Ed.; VCH: Weinheim, 1994, Chapter 4, pp. 299-316.
5.   Bradley, J.S. In *Clusters and Colloids*; Schmid, G., Ed.; VCH: Weinheim, 1994, Chapter 6, pp. 459-537.
6.   Loginova, T.P.; Bronstein, L.M.; Valetsky, P.M.; Ezernitskaya, M.G.; Lokshin, B.V.; Lependina, O.L.; Bakhmutov, V.I.; Vinogradova, S.V. *Organometallic Chemistry in the USSR.* **1990**, *3*, 87.
7.   Loginova, T.P.; Bronstein, L.M.; Valetsky, P.M.; Ezernitskaya, M.G.; Dyumaeva, I.V.; Lokshin, B.V.; Lependina, O.L.; Vinogradova, S.V. *Organometallic Chemistry in the USSR.***1990**, *3*, 222.
8.   Register, R.A.; Bell, T.R. *J. Polym. Sci. B: Polym. Phys.* **1992**, *30*, 569.
9.   Dobson, G.R.; Amr, M.F.; Sayed, E.L.; Stolz, I.W.; Sheline, K.K. *Inorg. Chem.* **1962**, *1*, 526.
10.   Henrici-Olivé, G.; Olivé, S. *Adv. Polym. Sci.* **1983**, *51*, 1.
11.   Vonsovskii, S.V. *Magnetism*; Nauka: Moscow, 1971; pp. 800-812.
12.   Bronstein, L.M.; Mirzoeva, E. Sh; Valetsky, P.M.; Solodovnikov, S.P.; Register, R.A. *J. Mater. Chem.*, **1995**, *8*, 1197.
13.   Edgell, F.; Huff, J.; Thomas, J.; Lehman, H.; Angell, C.; Asato, G. *J.Amer.Chem. Soc.* **1960**, *82*, 1254.
14.   S.P. Solodovnikov, S.P.; Vasil'kov, A.Yu.; Titova, E.F.; Sergeev, V.A. *Dokl. Akad. Nauk SSSR.* **1989**, *310*, 911.
15.   Bean, C.P.; Livingston, J.D.; Rodbell, D.S. *Acta Metall.*, **1957**, *5*, 682.
16.   Sharma, V.K.; Baiker, A. *J. Chem. Phys.,* **1981**, *75*, 5596.
17.   Kuznetsov, V.L.; Lisizin, A.C.; Golovin, A.V.; Aleksandrov, M.N.; Ermakov, Yu.U. *5th Intern. Symp. on Connection Between Homogeneous and Heterogeneous Catalysis, Novosibirsk.*1986, III, part II, 83.
18.   Solodovnikov, S.P.; Bronstein, L.M.; Loginova, T.P.; Mirzoeva, E.Sh.; Lependina, O.L.; Valetsky, P.M. *Polym. Sci.* **1993**, *35*, 128.
19.   Debye, P.; Bueche, A.M. *J. Appl. Phys.,* **1949**, *20*, 518.
20.   Chu, B.; Wang, J.; Li, Y.; Peiffer, D.G. *Macromolecules.* **1992**, *25*, 4229.
21.   Guinier A.; Fournet, G. *Small-Angle Scattering of X-Rays.* Wiley: New York. 1955, p.25.
22.   Bronstein, L.M.; Solodovnikov, S.P.; Mirzoeva, E.Sh.; Baukova, E.Yu.; Valetsky P.M. In Preprints of the ACS; PMSE Division; Fall Meeting, Washington, 1994, Vol.71, pp.397-398.

RECEIVED November 9, 1995

Chapter 7

Stabilization of Gold Colloids in Toluene by Block Copolymers of Polystyrene and Poly(ethylene oxide)

Arno Roescher[1] and Martin Möller[2]

[1]Department of Chemical Technology, University of Twente,
P.O. Box 217, NL–7500 AE Enschede, Netherlands
[2]Department of Organic Chemistry III, Macromolecular Chemistry,
University of Ulm, D–89077 Ulm, Germany

Gold nanoparticles were prepared in the core of micelles of polystyrene-block-polyethyleneoxide. For this purpose the block copolymer was dissolved in toluene and the glycolether units were complexed by $LiAuCl_4$. Light scattering and size exclusion chromatography demonstrated a strong dependence of the micelle formation on the length of the polyethyleneoxide block and on the complexation of lithium cations. Hydrazine hydrate was shown to be a suitable reducing agent to convert the micelle bound $AuCl_4^-$ into small gold particles. Thin films of the micellar solutions were cast and studied by transmission electron microscopy, which permitted determination of the size of the gold particles. Long equilibration times of the block copolymer/toluene/$LiAuCl_4$ turned out to be of great importance in order to achieve narrow size distributions of the final gold particles. However, in most cases particles were formed which are too large to be generated only by the $AuCl_4^-$ ions of one initial micelle. This is explained by reorganization of the micelles during the reduction reaction.

Reduction of a noble metal salt in polymer solution is a widely used method to prepare noble metal colloids (*1*). Stabilization is based on the fact that the polymer adsorbs at the surface of the metal particles and thus prevents aggregation by forming a protective layer. It is therefore commonly assumed that strong adsorption of the polymer at the metal particle surfaces is desirable (*2*). While strong adsorption requires high affinity to the metal surface and lower affinity to the medium (solvent), the protective polymer layer has also to be extended enough in order to keep the particles separated (*3*). Thick layers are, however, more likely when the polymer is well soluble. The two opposing requirements can be met either by proper adjustment of the polymer solubility and molecular weight or by use of block copolymers, where one block has a high affinity to the solvent, and the second block has a high affinity to the particle surface (*2*). For example, block copolymers of polystyrene and poly(2-vinylpyridine) have been shown to be effective stabilizers for gold and palladium nanoparticles in toluene (*4,5*).

In addition to the topological separation of the anchoring site and the solubilizing tails, block copolymers have also the potential to achieve a purely steric stabilization of the colloidal particles because of their ability to form micelles (*6, 7*). Within the core of such a micelle, a metal salt can be bound or preferentially dissolved. Formation of colloidal metal particles by reduction will consequently occur within these cores. Provided sufficient stability of the micelles, the particle might get entrapped within the core independently of its actual interaction with segmental units of the macromolecules. Thus, the structure of the block copolymer, i.e., block length and degree of association, and the amount of metal salt by which the micelle is loaded might control the size of the colloidal particle formed within such a micelle.

In the case of steric stabilization without chemisorption of the polymer being necessary, electronic properties can be tuned by adding a suitable adsorbent in deliberate amounts. Such a concept could also be very helpful for the tailoring of catalytic activities in an organic environment.

In order to study the stability of gold colloids in solutions of poorly interacting polymers, we chose block copolymers of styrene and ethyleneoxide (PS-b-PEO). Polyethyleneoxide does not interact strongly with transition metal surfaces and is thus a rather poor stabilizer for colloidal gold (*8*). It can, however, complex lithium cations and can also be protonated (*9*). In this way, PEO can bind compounds like $LiAuCl_4$ and $HAuCl_4$, and we attempted to load inverse PS-b-PEO micelles by these gold salts. Subsequent reduction was expected to yield small gold crystallites entrapped in the micellar cores.

Experimental

Materials and solvents: Toluene (Merck, p.A.), chloroform (Merck, p.A.), LiCl (Fluka, p.A.), $HAuCl_4$ (Fluka), sodium borohydride (Merck) and hydrazine hydrate (Fluka, 24% in water) were used as received.

Polymers: Polystyrene-block-polyethyleneoxide samples of different compositions were prepared by anionic polymerization in THF in inert atmosphere, using high vacuum techniques (*10*).

Size exclusion chromatography (SEC): Measurements in toluene or chloroform were performed on a system equipped with a Waters 590 programmable HPLC pump coupled with μ-Styragel columns (10^5, 10^4 and 10^3Å; in the case of chloroform an extra 10^6Å column was added). 60 μL polymer solution (4 mg/mL) was injected, and elution was monitored with a Waters 410 differential refractometer and a Viscotek H502B viscometer. When chloroform was used as solvent, a Waters 486 tunable absorbance detector (254 nm) was employed additionally. Elution volumes were converted to molecular weights by means of calibration curves obtained with narrow molecular weight polystyrene standard samples (PSS, Mainz). For measurements in the presence of LiCl, the polymer solution in toluene was stirred over LiCl (1 mole LiCl per mole EO-polymer-groups) for 2 days. Subsequently, the solution was centrifuged at 5000 rpm for 15 minutes. The clear solution was separated from the precipitate (LiCl).

Static light scattering: Experiments were performed in glass cuvettes (Ø=21 mm) at 25 C on a PL-LSP light scattering apparatus equipped with a 10 mW laser (633 nm), and LSP Version 4.0 software (Polymer Laboratories). Measurements were performed in a matching bath of di-n-butyl phthalate. Scattering between angles of 20 and 150° was measured and related to scattering of a toluene standard at 90° ($R_{90}=14.09*10^{-6}$ cm^{-1}). Refractive index increments (dn/dc) were measured with an Photal RM-102 Differential Refractometer (Otsuka Electronics) at $\lambda_o=633$ nm at 25°C. The solvent was filtered 3 times using Whatman Anodisc 47 filters (poresize 0.02 μm), and polymer solutions were filtered 3 times using Spartan 13 (poresize 0.5 μm). LiCl-saturated polymer solutions were centrifuged.

LiAuCl$_4$ was synthesized by mixing equimolar amounts of aqueous LiCl and HAuCl$_4$ solutions after which the water was removed by evaporation (*11*). The yellow product was dried in a desiccator above potassium hydroxide in vacuum.

Preparation of Au-colloids: An 1 wt% solution of PS-b-PEO block copolymer was prepared in toluene and stirred for two days. Eventually the solution was saturated with LiCl by stirring overnight with excess LiCl, centrifuged (5000 rpm) and filtered. A defined amount of LiAuCl$_4$ was added. The solution was stirred in the dark to avoid reduction of the gold salt. A clear yellow solution remained after filtration (Spartan 13, 0.5 μm). Reduction was carried out either by exposure to UV-light (366 nm, 6W) at room temperature, or by addition of a slight excess of reduction agents such as hydrazine hydrate or BH$_3$/methanol at -10°C (typical 0.1 mL reduction solution in 10 mL polymer/salt solution). The BH$_3$/methanol reduction solution was freshly prepared by saturating methanol with excess NaBH$_4$.

UV-Vis Spectroscopy: Spectra were recorded on a Perkin Elmer Lambda 16 UV/VIS spectrometer. Samples of gold colloids were taken and diluted to suitable concentrations, where the absorbance of the plasmon peak amounted to approximately 2. The solutions were measured in 10 mm quartz glass cuvettes against a cuvette filled with toluene as the reference, at a rate of 480 nm/min.

Transmission electron micrographs: bright field micrographs were obtained using a Philips EM-401 or EM-301 electron microscope, both operating at 80 keV. Calibration was done with catalase crystals. 30 minutes after colloid preparation, samples were prepared by putting a drop of polymer protected colloid solution on a carbon-coated grid and immediately soaking away the fluid in an underlying paper.

Results

Polymerization: Polystyrene-block-polyethyleneoxide samples were prepared by sequential anionic polymerization in THF with cumyl potassium as the initiator. The PEO block lengths were varied from 1 to 25 mole%. Samples were characterized by SEC and ^{1}H-NMR. Results are summarized in Table I. The monomer to initiator ratio was chosen so that molecular weights of about 50.000 g/mole could be expected for all polymers. The actual molecular weights which were found by SEC were higher. This can be explained by partial precipitation of the initiator in the stock solution. It must be noted also, that SEC elution volumes had been transformed to molecular weights by means of a polystyrene calibration curve. For this reason, some deviation of the actual molecular weight of the styrene/ethyleneoxide block copolymers is expected.

Table I: Molecular weights and composition of the PS-b-PEO block copolymers

Polymer[1]	mole% EO[2]	M_n (g/mole)[3]	M_w (g/mole)[3]	M_w/M_n
PS(540)-b-PEO(220)	29	45000	66000	1.46
PS(610)-b-PEO(80)	12	58000	67000	1.15
PS(580)-b-PEO(45)	7	51000	62000	1.22
PS(480)-b-PEO(5)	1	39000	50000	1.28

[1]The numbers in brackets refer to the numbers of monomer units in the constituent blocks, based on proton-NMR and SEC [2]From proton-NMR, [3]from SEC relative to polystyrene in chloroform

All polymers had similar molecular weights and narrow enough molecular weight distributions to allow studying the influence of the ethylene oxide block length on solution properties of the block copolymers systematically. Only in the case of PS(540)-b-PEO(220), the sample contained some PEO homopolymer. The block copolymer did not precipitate in methanol. Therefore, the polymer was difficult to isolate free of the small amount of polyethyleneoxide homopolymer which also was formed during the polymerization reaction.

Micelle Formation: Micelle formation occurred when the block copolymers were dissolved in toluene which is selective for PS. Formation of micelles can be enhanced further by addition of a lithium salt. Lithium cations are complexed by PEO segments, and the solubility of the resulting polyionic block in toluene is even less than that of the PEO.

A first indication of the micelle formation of the block copolymers in toluene could be obtained from SEC experiments. Figure 1 depicts the SEC elution diagrams for PS(540)-b-PEO(220) in toluene. The SEC curve of an one day old solution of the block copolymer showed a peak for the free polymer chains and a shoulder for aggregated polymer chains. When the same polymer solution was allowed to equilibrate for 10 days, the micelle peak was more pronounced while the initial free polymer peak had reduced to a shoulder. Clearly, the association equilibrium is reached only after longer times, i.e., several days (see below), indicating a slow exchange between the associates and the free state of the block copolymer. This allowed separation of the micelles and unimeric macromolecules by chromatography. The same trend was observed for LiCl-containing PS-b-PEO solutions. However, in this case the micelles dominated already one day after preparation of the solution, indicating enhanced micelle formation.

Further insight into the micelle formation can be obtained by light scattering experiments. For the evaluation of light scattering data, we assumed that all polymer molecules in the solution possessed the same refractive index increment, i.e., an average refractive index increment dn/dc including a contribution of the polystyrene block and the polyethyleneoxide block. Because the refractive index increment is not equal for polystyrene and polyethyleneoxide (*12*), the obtained molecular weights will deviate from the real molecular weights and must be regarded as apparent values which only allow relative comparison (*13*).

Table II: Apparent weight-average molecular weight (g/mole) of PS-b-PEO block copolymers in chloroform, toluene and after complexation of LiCl to the PEO block in toluene

Polymer	$M_{w,app}$(chloroform)	$M_{w,app}$(toluene)	$M_{w,app}$(toluene/LiCl)
PS(480)-b-PEO(5)	67000	79000	74000
PS(580)-b-PEO(45)	71000	97000	117000
PS(610)-b-PEO(80)	79000	140000	241000
PS(540)-b-PEO(220)	82000	289000	5200000

Table II summarizes the light scattering results for the different block copolymers. In chloroform, which is a good solvent for both blocks, molecular weights were around 75000 for all polymers. These values are only little higher than the molecular weights found by SEC in chloroform according to PS calibration (Table I). In toluene, the samples with a longer PEO block yielded higher molecular weights indicating association. Obviously the PEO block of PS(480)-b-PEO(5) is too short to cause association in toluene. Even when LiCl was added to this polymer solution, no significant change of the molecular weights were found. In the case of longer LiCl-complexed PEO blocks, the increased apparent molecular weights indicated strong association.

Loading of the Micelles by AuCl₄⁻: Tetrachloroauric acid (HAuCl₄) is an often used precursor for the preparation of polymer-protected gold colloids (*14*). Although it is insoluble in toluene, the acid was slowly taken up when it was added to a firmly stirred solution of PS-b-PEO in toluene, provided the toluene and the acid were dried carefully. Fast uptake was achieved when the gold acid was added as a solution in diethylether. After addition of the acid/ether solution, the ether could easily be removed by evaporation (*6*). However, the originally yellow solutions of the obtained polymer-acid complexes turned brown within a few hours, indicating reduction of the gold salt even when light was excluded. Apparently, the polyethyleneoxide block was cleaved and oxidized in toluene (*15*).

Better results were obtained with lithium tetrachloroaurate. LiAuCl₄ did not attack ether bonds and is also not as hygroscopic as HAuCl₄. High amounts of LiAuCl₄ could be brought into the polymer solution just by adding the salt while the solution was vigorously stirred. The tetrachloroaurate anions get bound as counterions as the lithium cations are complexed to the polyethyleneoxide chain (*10*):

The uptake of LiAuCl$_4$ was quantitative in the range of 0.01-0.30 mole LiAuCl$_4$/mole EO.

Reduction of the Polymer-Salt Complexes in Solution: Reduction was carried out either by exposure to UV-light, or by addition of a slight excess of hydrazine hydrate or BH$_3$/methanol. Solutions which were reduced by UV light led to unstable colloids. A black precipitate could be observed within a few hours. In 2 days, all the gold had precipitated as indicated by a colorless solution. Better results could be obtained when the reduction was performed with BH$_3$/methanol or hydrazine hydrate. Solutions obtained by reduction with BH$_3$/methanol were clear and red/violet. Only after a few days, the colloidal solution became turbid brown, and a dark precipitate could be observed. Reduction with hydrazine hydrate yielded the most stable colloids. Precipitation of gold was only observed at high salt loadings >0.2 mole LiAuCl$_4$/mole EO units. This is likely due to uncontrolled reduction of uncomplexed salt.

Only PS(480)-b-PEO(5) appeared to be unable to stabilize gold colloids, independent of the applied reduction method. Within a few hours, the first violet/blue precipitate was seen. This is consistent with the light scattering experiments, which did not show micelle formation.

Figure 2 depicts the plasmon absorbance of gold colloids obtained by reduction of LiAuCl$_4$ by hydrazine hydrate in toluene, in the presence of different block copolymers. Spectra were recorded directly after reduction and after 700 hours (29 days). It is well established that the plasmon absorbance peak is sensitive to the size of the Au-particles (*16*). Right after reduction, λ_{max} was found to increase with increasing PEO block length of the block copolymer; for PS(580)-b-PEO(45), PS(610)-b-PEO(80), and PS(540)-b-PEO(220) respectively 518, 533 and 542 nm. In the case of PS(580)-b-PEO(45), a new shoulder around 620 nm had appeared after 700 hours beside the plasmon absorbance peak which had shifted to 535 nm. For this sample, a gradual blue coloration and a dark precipitate was observed after several days. In the case of the other block copolymer/gold solutions, the shift of the plasmon absorbance was less pronounced. The absorbance band remained rather narrow and λ_{max} did not exceed 545 nm.

TEM Characterization of the Particle Size in Thin Films cast from the Colloidal Solutions: Because the UV/Vis spectra did not give direct evidence on the particle size, transmission electron microscopy was done on thin films which had been casted from the micellar solution. It has been demonstrated recently that the micellar structures can be well preserved when a block ionomer solution is evaporated to yield a solid film (*17,18*).

Figure 3 shows typical TEM pictures of such gold colloid containing block copolymer films which were obtained after reduction of PS-b-PEO/LiAuCl$_4$ complexes in toluene either by BH$_3$/methanol or by N$_2$H$_4$(aq). It was generally observed that reduction with BH$_3$/methanol led to bimodal products containing very small (1.5 nm) and larger (ca. 15 nm) particles in each micelle core (Figure 3A). In contrast, reduction with hydrazine hydrate led to particles with sizes mostly between 6 and 15 nm (Figure 3B). Size and location of the particles was, however, not directly related to a micellar film structure.

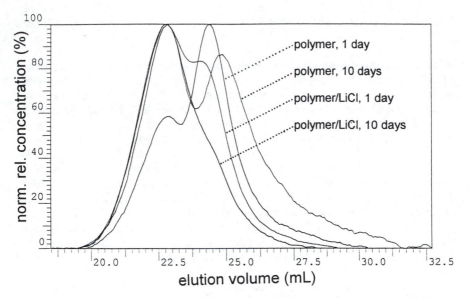

Figure 1: SEC curves of PS(540)-b-PEO(220) in toluene and after saturation with LiCl, measured 1 and 10 days after solution preparation

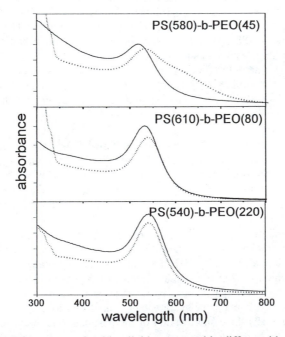

Figure 2: UV/Vis spectra of gold colloids protected in different block copolymer solutions (1 wt%) in toluene (0.2 mole Au/mole EO units). —— = just after reduction, ---- = after 700 hours.

When reduction with hydrazine hydrate was carried out within a short time (48 hrs) after the lithium gold salt had been added to the polymer solution, colloids with broad particle size distributions were found (Figure 3B). To investigate the time dependence more quantitatively, a 0.4 wt% solution of PS(610)-b-PEO(80) in toluene was saturated with LiCl, loaded with 0.1 mole LiAuCl$_4$ per mole EO units and stirred in the dark at room temperature. After different times of stirring, a sample was taken from the solution and reduced with N$_2$H$_4$(H$_2$O). The average particle sizes of the products were calculated from TEM pictures. The standard deviation of the particle size was taken as a measure for the width of the particle size distribution. In Figure 4A, the average particle size and the standard deviation are plotted as a function of stirring time. Figure 4B depicts the particle size distributions of colloids obtained after different stirring times. The average particle size stayed rather constant around 10-11 nm. The standard deviation dropped, however, drastically in the first 100 hours. From the particle size distributions shown in Figure 4B, it is evident that narrowing of the particle size distribution with longer stirring time is mainly due to decreased formation of larger particles (>15 nm). Already after 1 hour of stirring the polymer/salt solution, reduction led to a particle size distribution with a peak for particles of 7 nm. The position of this peak did not change when the solution was stirred longer before reduction was carried out.

Figure 5 shows the influence of the PEO block length and of the LiAuCl$_4$ loading on particle size (distribution). The solutions were stirred for about 300 hours before reduction with hydrazine hydrate was carried out in order to allow the gold salt to be properly dissolved into the micelle cores. Increased particle sizes were observed for all block copolymers when low salt loadings (≤6%) were applied (Figure 5A). Also the size distribution became broader at lower salt loadings, as it can be seen in figure 5B for colloids stabilized by PS(610)-b-PEO(80). The most well-defined colloids were found in the case of 6 and 10% loading, where the spreading in the particle size was minimal. A minimum particle size could be observed at a salt loading of about 0.1 mole LiAuCl$_4$ per mole EO units. At a loading of 30%, the particle size distribution was very broad, mainly due to a shoulder for larger particles as shown for PS(610)-b-PEO(80) in figure 5B. This broad shoulder can be ascribed to uncontrolled reduction of larger gold salt crystallites, which might be due to overloading of the block copolymer solution.

Figure 6 shows TEM micrographs of a film from PS(610)-b-PEO(80) with 0.2 equivalent LiAuCl$_4$. The micrographs were recorded directly after colloid preparation and after 2900 hours (120 days). Average particle sizes which were calculated showed only a small increase from 13 nm right after reduction to 15 nm after 2900 hrs. As discussed before (Figure 4), the particle size did not increase significantly with time. This was not the case when the stabilizing polymer was PS(580)-b-PEO(45), where larger aggregates were observed after extended time.

Discussion

In general, the sizes of the gold particles were not consistent with the size of the micelles, as determined by light scattering or electron microscopy (TEM). In the case of PS(540)-b-PEO(220) for example, the number of associated polymer chains in a micelle loaded with a lithium salt can be roughly estimated to be around 10^2 according

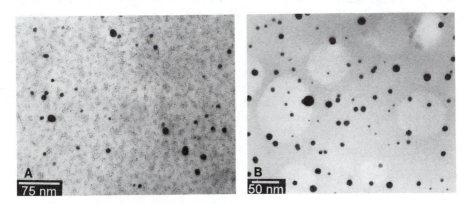

Figure 3: TEM micrographs of PS(540)-b-PEO(220) with 0.10 mole LiAuCl₄, after reduction with BH₃/methanol (A), and N₂H₄(aq) (B)

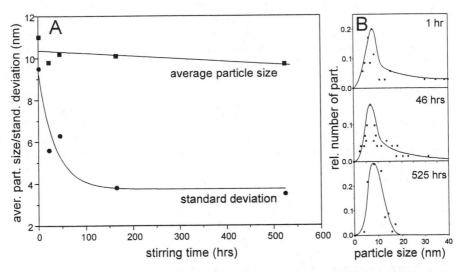

Figure 4: average particle size and standard deviation obtained by reduction of PS(610)-b-PEO(80) loaded with 0.1 mole LiAuCl₄ per mole EO units at different times after solution preparation (A), and particle size distributions (B)

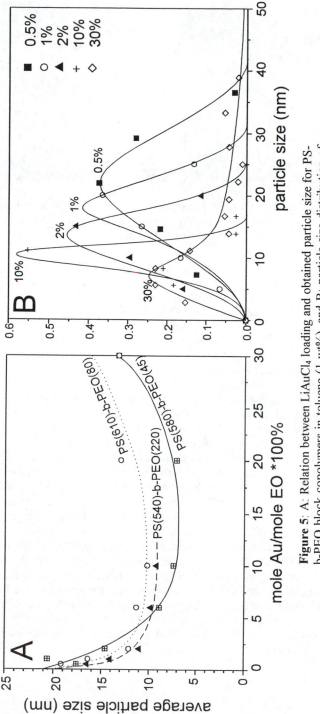

Figure 5: A: Relation between LiAuCl$_4$ loading and obtained particle size for PS-b-PEO block copolymers in toluene (1 wt%), and B: particle size distribution of colloids protected by PS(610)-b-PEO(80) for different salt loadings.

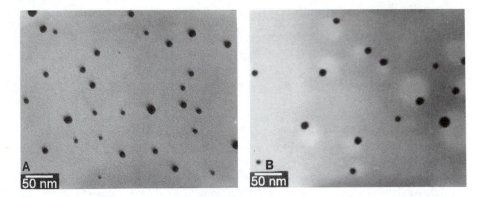

Figure 6: TEM pictures of films casted from gold colloids in solutions of PS(610)-b-PEO(80) in toluene (1 wt% polymer, 0.2 mole Au/mole EO units). A: immediately after reduction, B: idem after 2900 hours

to light scattering experiments. At a loading of 0.1 mole LiAuCl$_4$/mole EO, this involves approximately 2000 gold atoms present in a micelle. Assuming a fcc packing of the gold atoms in spherical particles (*19*), this implies particles with a diameter of about 4 nm. In the case of PS(540)-b-PEO(220), at a loading of 0.1 mole LiAuCl$_4$/mole EO units, no particles larger than 4 nm should be expected. The block copolymers with the shorter PEO blocks should even yield smaller gold particles because of the smaller micellar associates which they form.

The much larger particles which result in the case of reduction of PS-b-PEO/LiAuCl$_4$ complexes in toluene (Figure 5A) might be explained by reorganization of micelles during reduction and/or film formation. The driving force for this process is the formation of energetically favored larger gold crystallites and eventual reorganization of the micellar structure upon solvent evaporation.

The increasing particle size with decreasing salt loading is in agreement with reorganization of the micelles. At lower salt loadings, the interaction between the different core blocks will be weaker. Furthermore, the micelle stability can expected to be larger in the case of longer PEO blocks. This means that at a particular low loading smaller particles should be formed when the PEO block is longer, because reorganization is more difficult. In case of 0.01 and 0.02 mole LiAuCl$_4$/mole EO, increasing particle size was indeed observed for decreasing PEO block length. At higher loadings (>0.10 LiAuCl$_4$/mole EO), the stability of micelles with shorter blocks as well will be higher and the influence of the PEO block length may vanish. In this case, the particle size can be expected to be more strongly related to the loading of the micelle cores. In fact, at 0.2 LiAuCl$_4$/mole EO smaller particles were obtained in PS(580)-b-PEO(45) solution compared to the PS(610)-b-PEO(80) solution. Consistently, the particle size distribution became broader with decreasing salt loading for the same block copolymers.

Conclusions

Although polyethyleneoxide is a bad stabilizer for colloidal noble metals, it appeared possible to prepare rather stable colloidal gold solutions in toluene with polystyrene/polyethyleneoxide block copolymers by reduction of $LiAuCl_4$. When polymer/salt complexes were allowed to approach their equilibrium during longer times, a more homogeneous distribution of gold salt in the micelles resulted in more narrow particle size distributions.

The size of the formed gold particles was too large, to be related to reduction of defined amounts of salt in polymer micelles. Apparently, reorganization processes took place. Although the micelles did not directly control the size of the formed gold particles, the particle sizes were related to the stability of the micelles.

References

(1)   Schmid, G. *Chem. Rev.* **1992,** *92,* 1709
(2)   Tadros, Th. F. *Polymer J.* **1991,** *23*, 683
(3)   Napper, D.H. *Polymeric Stabilization of Colloidal Dispersions,* Academic Press, London, 1983
(4)   Roescher, A.; Möller, M. *Adv. Mat.* **1995,** *7*, 151
(5)   Roescher, A.; Möller, M. in preparation
(6)   Tuzar, Z.; Kratochvíl, P. in *Micelles of Block and Graft Copolymers in Solutions;* Matijevic, E., Surface and Colloid Science, Plenum Press, New York, 1993, vol. 15, 1-83
(7)   Chu, B. *Langmuir* **1995,** *11*, 414
(8)   Hirai, H.; Nakao, Y.; Toshima, N. *J. Macromol. Sci.-Chem.* **1979,** *A13(6),* 727
(9)   Searles Jr., S.; Tamres, M. in *Basicity and complexing ability of ethers;* Patai, S., The chemistry of the ether linkage, Interscience Publishers, London, 1967, 243-308
(10)   Roescher, A. *Stabilization of colloidal noble metals by block copolymers,* thesis, University of Twente, Enschede, 1995
(11)   *Gmelins Handbuch der Anorganischen Chemie, "Gold",* Achte Auflage, Verlag Chemie GmbH, Weinheim, 1954, 731
(12)   J. Brandrup and E.H. Immergut, *Polymer Handbook,* 3rd edition, John Wiley&Sons, New York, 1989
(13)   Tuzar, Z.; Kratochvíl, P.; Straková, D. *Eur. Polym. J.* **1970,** *6*, 1113
(14)   Harada, M.; Asakura, K.; Toshima, N. *J. Phys. Chem.* **1993,** *97*, 5103
(15)   T.W.G. Solomons, *Organic Chemistry,* 3rd edition, John Wiley&Sons, New York, 1984, 674.
(16)   Fragstein, C.v.; Römer, H. Z. *Physik* **1958,** *151,* 54
(17)   Zhang, L.; Eisenberg, A. *Science* **1995,** *268,* 1728
(18)   Spatz, J.P.; Roescher, A.; Sheiko, S.; Möller, M. *Adv. Mater.* **1995,** *7*, 731
(19)   Toshima, N. *J. Macromol. Sci.-Chem.* **1990,** *A27(9-11),* 1225

RECEIVED January 4, 1996

Chapter 8

Spontaneous Formation of Gold Particles in Aqueous Polymeric Solutions

L. Longenberger and G. Mills[1]

Department of Chemistry, 179 Chemistry Building, Auburn University, Auburn, AL 36849

The spontaneous formation of Au particles at room temperature in air-saturated aqueous solutions of poly(ethylene glycol)s was investigated using optical, potentiometric and conductivity techniques. The kinetic information is consistent with a mechanism in which Au(III) complexes bind through ion-pairs to pseudocrown ether structures of the polymers. Reduction of the metal centers follows through their reactions with the oxyethylene groups that form these cavities. The particle size of the metal crystallites is controlled by the molar mass of the polymers. Agglomerates of small Au particles are formed as final products when polymers of low molar mass are used in the synthesis.

Interest on small metal particles has increased lately because of their unusual physical and chemical properties (*1-3*). Variations in the optical properties and chemical reactivity of the crystallites have been correlated to size and surface effects. In the case of Au particles changes in the surface plasmon resonance of the crystallites have been used extensively to study these effects (4-13). The development of new procedures for the synthesis of small Au particles in a variety of matrices has facilitated determinations of size and surface effects (*9-13*).

Recently, we have reported a simple method for the formation of small colloidal Au, Ag and Pd particles (*14*). The crystallites are formed via the spontaneous reduction of Au(III), Ag(I) and Pd(II) complexes by reaction with poly(ethylene glycol)s and poly(vinyl alcohol)s in air-saturated aqueous solutions. The particle formation processes were faster with former polymers as compared with the formation processes initiated by poly(vinyl alcohol)s (*15*); formation of Au particles occurred with the highest reaction rate. These results suggested that Au complexes, which are bound to pseudocrown ether cavities formed by the interaction of poly(ethylene glycol)s with cations (*16,17*), were reduced by the oxyethylene groups of the macromolecules (*14*). Further investigations on the formation of Au particles have been conducted and the results are presented here. The kinetic data gathered in this study supports the previously formulated mechanism. In addition, optical results suggest that the size of the Au particles is controlled by the molar mass of the polymers.

[1]Corresponding author

0097–6156/96/0622–0128$15.00/0

Experimental Section

NaAuCl$_4$ · 2H$_2$O (Aldrich) and poly(vinyl alcohol) with average molar mass of 2.5 x 10^4 g mol^{-1} (Polysciences) were used as received. The poly(ethylene glycol)s used were Carbowax polymers from Union Carbide, with average molar masses (in g mol^{-1}) of: 1.45 x 10^3 (PEG 1450), 3.35 x 10^3 (PEG 3350), 8 x 10^3 (PEG 8000) and 1.75 x 10^4 (PEG 20M). They were recrystallized 3 times from hot 2-propanol prior to use, and the resulting samples were dried and stored under vacuum. Peroxides or aldehydes were not found in these polymeric samples (*14*). However, the Carbowax materials contained base since the following pH values and concentrations of basic groups were determined by titration (under Ar) of solutions with 3 x 10^{-3} M of the purified polymers: 6, 6.7 x 10^{-6} M base (PEG 1450); 6.9, 2.4 x 10^{-5} M base (PEG 3350); 7.7, 9.6 x 10^{-5} M base (PEG 8000); and 10, 9.7 x 10^{-4} M base (PEG 20M). A pH of about 5.5 was typical of PEG-free solutions. The nature of the base present in the PEG samples is not known; we speculate that the base is either NaOH or KOH since these compounds are used in the preparation of the polymers (*17*). It should be noted that the rates of reaction remained unchanged when unpurified PEG materials were used. Also, similar results were obtained by using PEG polymers from Fluka or Polysciences, which do not contain base.

Colloids were prepared by adding 2.5 mL of a fresh 2 x 10^{-3} M NaAuCl$_4$ solution to approximately 40 mL of a polymer solution, followed by adjusting the volume to 50 mL with water while stirring. The final concentrations were 1 x 10^{-4} M NaAuCl$_4$ and 3 x 10^{-3} M PEG. To avoid excessive ligand substitution reactions of the AuCl$_4^-$ ions with the solvent a different procedure was adopted for determinations of conductivity or Cl$^-$ ion concentrations. In these experiments the required amount of solid NaAuCl$_4$ was dissolved directly, under vigorous stirring, into the solutions containing polymers. Dissolution of the metal salt was very fast (less than 20 sec); optical determinations showed that both methods yielded similar results. Experiments were conducted at room temperature and without ambient light to prevent photoreactions of the gold complex (*10*). Changes in the chloride ion concentrations were followed using potentiometric methods by means of a Radiometer K601 mercurous sulfate reference electrode, a Cl$^-$ ion selective electrode (Orion), and a Radiometer PHM 95 pH meter. Conductivity measurements were carried out with a Radiometer CDM 83 instrument. UV-Vis spectra were collected using a Hitachi U-2000 spectrophotometer.

Results

Figure 1 shows absorption spectra collected at different times after mixing AuCl$_4^-$ ions with neutralized PEG 20M (pH = 5.5). Similar results were obtained using HNO$_3$ or HCl to neutralize the polymer samples. For times shorter than 15 min, the plasmon band of Au particles located at $\lambda \geq 500$ nm increased rapidly. During this time, a broad absorption band centered at 534 nm was detected at 3 min, but the maximum shifted to 541 nm. A second and slower step that lasted for about 24 h resulted in an increase in the intensity of this band without a change in λ_{max}. These results differ considerably from the observations made when PEG 20M was not neutralized (*14*). In this case, the two reaction steps were faster and the formation process was completed in about 8 h. Also, only a small blue shift of λ_{max} (from 525 to 518 nm) was noticed during the first step. Furthermore, the colloids exhibited weaker plasmon bands than those of Au particles formed with neutralized PEG 20M.

An important difference between the spectra of solutions containing neutralized and non-neutralized PEG 20M is that in the former case the spectra measured within 3 min of reaction contained an absorption band centered at 383 nm. This absorption decayed in a synchronous fashion with the formation of the particle surface plasmon at $\lambda \geq 500$ nm, meaning that the initial species absorbing at 383 nm are precursors of the metal particles.   In view of the similar optical properties of these species and of transient Au clusters (*6a*), it seems reasonable to assume that the initial species consist of Au particles containing only a few metal atoms. If this assumption is correct, it implies that these clusters are relatively stable toward oxidation by O_2, which may allow for their isolation and characterization.

Presented in Figure 2 is a comparison of the changes in [Cl$^-$] and in optical density during the reactions of Au(III) complexes in solutions containing 1×10^{-4} M NaAuCl$_4$ and 3×10^{-3} M non-neutralized or neutralized PEG 20M. Included in Figure 2a are data of the reaction of AuCl$_4^-$ ions in pure water; the slow liberation of coordinated chloride ions resulted from the ligand substitution reactions of the metal complexes with water (*18*). Since attempts to fit the kinetic data with simple kinetic models were unsuccessful, initial rates of reactions were used. The initial rates ($R_0(Cl)$) are expressed as the change in the concentration of chloride ion per min (Δ[Cl$^-$] min^{-1}). Chloride ions were released during the ligand exchange of AuCl$_4^-$ ions in pure water ($R_0(Cl) = 5.6 \times 10^{-5}$ Δ[Cl$^-$] min^{-1}). In contrast, the first step in the reaction of the system containing Au(III) complexes and non-neutralized PEG 20M was so fast that it was not possible to determine a rate of reaction. The fast release of Cl$^-$ ions is related, in part, to the efficient ligand substitution of chloride ions induced by OH$^-$ ions in basic solution (*18*),

$$AuCl_4^- + xOH^- \longrightarrow [AuCl_{4-x}(OH)_x]^- + xCl^- \qquad (1)$$

As shown in Figure 2a the second step was finished in about 12 min, at which time all Cl$^-$ ions initially coordinated to the metal centers were found in solution. However, the release of coordinated chloride ions during this step was not due to ligand exchange processes alone. The results of Figure 2b show that the absorbance of the particles during this period of time was about 70% of the optical signal obtained after formation of the Au crystallites was completed. Obviously, a significant number of metal particles were generated during the fast release of Cl$^-$ ions. Thus, the increases in chloride ions concentration are related to two simultaneous processes: reaction 1 in competition with the reduction process of Au(III) complexes, which is represented by the overall reaction

$$AuCl_4^- + 3RH \longrightarrow Au + 4Cl^- + 3H^+ + 3R \qquad (2)$$

where RH represent functional groups of PEG that undergo oxidation by reaction with the metal ions, and R represent oxidation products. While reaction 2 predominated at short times, a slower formation of Au particles was noticed at longer times (Figure 2b). This is probably due to the fact that aquo-complexes of Au(III) are harder to reduce than the chloro-complexes (*19*), and because of the inhibiting effect of Cl$^-$ ions in the reduction process (see bellow).

A slower release of Cl$^-$ ions was determined in solutions with neutralized PEG 20M, but the initial rate ($R_0(Cl) = 1.2 \times 10^{-4}$ Δ[Cl$^-$] min^{-1}) was higher by a factor of 2 than the rate measured in the absence of the polymer (Figure 2a). These results, in conjunction with the optical data of Figure 2b indicate that reduction of metal ions predominates in the presence of neutralized PEG 20M. Further evidence for reaction 2 was obtained from conductivity experiments, which, in agreement with the results of Figure 1, showed that the reduction process is over after 24 h. Additional optical and

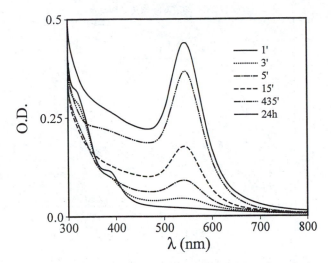

Figure 1. Evolution of the absorption spectra with time of a solution with 1 x 10^{-4} M NaAuCl$_4$ and 3 x 10^{-3} M neutralized PEG 20M.

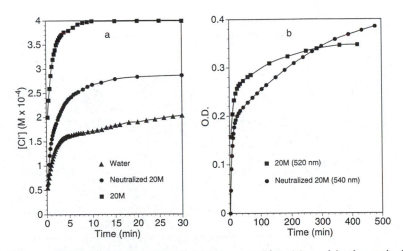

Figure 2. Changes in the chloride ion concentration (a), and in the optical density (b) during the reaction of 1 x 10^{-4} M AuCl$_4^-$ ions in: (▲) pure water, (●) a solution with 3 x 10^{-3} M neutralized PEG 20M, (■) a solution with 3 x 10^{-3} M non-neutralized PEG 20M.

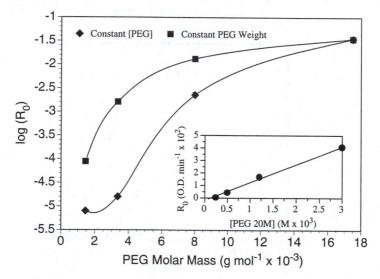

Figure 3. Effect of the polymer molar mass on the logarithm of the initial rate of reaction for non-neutralized polymers. The solutions contained 1×10^{-4} M $NaAuCl_4$ and: ($\blacklozenge$) 3×10^{-3} M PEG, ($\blacksquare$) 60 g L^{-1} PEG. The insert is a plot of the initial reaction rate as function of the concentration of PEG 20M.

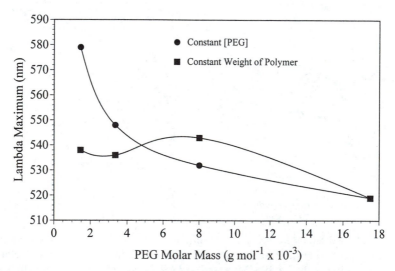

Figure 4. Dependency of the wavelength of the absorption maximum on the molar mass of non-neutralized polymers in solutions with 1×10^{-4} M $NaAuCl_4$ and: ($\bullet$) 3×10^{-3} M PEG, ($\blacksquare$) 60 g L^{-1} PEG.

potentiometric data were obtained in solutions containing poly(vinyl alcohol). Colloid formation was very slow, lasting for about 6 days. During this process, a shift of λ_{max} from 584 to 520 nm took place, and chloride ions were released with an initial rate $(R_0(Cl) = 1.4 \times 10^{-5} \Delta[Cl^-] min^{-1})$ that was four times smaller than that of solutions without polymer.

Variations in optical density at a wavelength close to the final λ_{max} of the colloids were utilized to follow the particle formation process in systems containing non-neutralized polymers. In this case initial rates of reaction (R_0) are expressed as the change in optical density per min $(\Delta O.D. min^{-1})$. Depicted in Figure 3 is the dependence of log (R_0) on the molar mass of the PEG polymers. Two sets of experiments are presented in this Figure. In the first series the polymer weight per volume was held constant at 60 g L^{-1}; the second set of initial rates was obtained by using a constant molar concentration of polymer, $[PEG] = 3 \times 10^{-3}$ M. Large increases in R_0 occurred with increasing PEG molar mass in both cases. In the first series R_0 varied from 8 x 10^{-5} $\Delta O.D. min^{-1}$ to 4 x 10^{-2} $\Delta O.D. min^{-1}$ when the molar mass changed from 1.45 x 10^3 to 1.75 x 10^4 g mol^{-1}. At constant molar concentration of polymer it varied from 8 x 10^{-6} $\Delta O.D. min^{-1}$ to 3.5 x 10^{-2} $\Delta O.D. min^{-1}$. The insert in Figure 3 shows that the initial rate increased linearly with increasing molar concentration of PEG 20M. This linear relationship obeyed the empirical equation $R_0 = -2.4 \times 10^{-3} + 14.5[PEG \ 20M]$. A few experiments were carried out with solutions of PEG 20M containing excess Cl^- ions. The initial rate decreased by a factor of 2 in the presence of 0.1M NaCl $(R_0 = 1.8 \times 10^{-2} \Delta O.D. min^{-1})$ as compared with the initial rate determined in the absence of the salt. Also, an induction period of 7 min was noticed in the former system, and the plasmon band of the resulting colloid was stronger and broader with $\lambda_{max} = 525$ nm. Reduction of the Au complexes was much slower in solutions with 0.1 M HCl, with an induction period of 45 min and $R_0 = 2.3 \times 10^{-3}$ $\Delta O.D. min^{-1}$. Metal particles with a very broad and weak plasmon resonance were formed, exhibiting a maximum at 550 nm.

Changes in the plasmon bands were observed when the Au particles were prepared with different (non-neutralized) PEG polymers. Presented in Figure 4 is the dependence of λ_{max} on polymer molar mass. The position of the absorption maximum shifted in a continuous fashion to longer wavelengths with decreasing molar mass when the [PEG] was kept constant at 3 x 10^{-3} M. Broad bands and unstable colloids were formed in solutions of PEG 3350 and PEG 1450. In the case of PEG 8000 an initial red shift of the absorption maximum from 504 to 565 was followed by a blue shift to 532 nm. Experiments with a constant polymer weight (60 g L^{-1}) yielded a smaller shift in λ_{max} with decreasing PEG molar mass, from 518 nm (PEG 20M) to values between 547 to 555 nm for the polymers with lower molar mass. In spite of the higher [polymer] used in these experiments, metal precipitation was noticed in the cases of PEG 3350 and PEG 1450.

Discussion

Polymers are employed frequently as particle stabilizers in the chemical synthesis of colloidal particles (*20*). Interactions between the small metal particles and the macromolecules at the early stages of growth prevent precipitation processes. The appearance of the characteristic surface plasmon resonance of small Au crystallites in the spectra of Figure 1 indicates that Au(III) complexes are reduced to the elemental state by the PEG polymers. Thus, poly(ethylene glycol)s and poly(vinyl alcohol)s can reduce some metal ions to yield metal particles, and also act as stabilizing agents.

Further evidence that the Au particles are formed as a result of interactions between the gold complexes and the polymers is provided by the data of Figure 2. Chloride ions are released faster in solutions containing PEG polymer than through the ligand substitution reactions of the metal complexes with water. Simultaneously, optical signals of Au particles are detected. In addition, it was shown in a previous study that the PEG polymers are fragmented and oxidized after reacting with the Au(III) complexes (14). All these results as well as the conductivity measurements strongly support the assumption that the metal complexes are transformed to the elemental state via reaction 2.

As shown in Figure 3, formation of Au particles is faster with increasing molar mass of the PEG polymers. According to the results of Figure 2 the reaction is faster in the presence of base, and the number of basic groups present in the PEG materials increases with increasing polymer molar mass. However, the accelerating effect induced by increasing the number of oxyethylene groups ($-OCH_2CH_2-$) in the polymer chain is much larger than the acceleration due to the basic groups. For example, the initial rate for the reaction with the polymer containing the fewest basic groups (PEG 1450) was $R_0 = 7.9 \times 10^{-6}$ ΔO.D. min^{-1}, which is 3 orders of magnitude smaller than the value determined with neutralized polymer (2.4×10^{-2} ΔO.D. min^{-1}), or with a similar PEG material (Polysciences, molar mass $= 2 \times 10^4$ g mol^{-1}) that is free of base ($R_0 = 3 \times 10^{-2}$ ΔO.D. min^{-1}) (14). In addition, a plot of log[base] vs PEG molar mass (at constant [PEG]) yielded a straight line that followed the equation: log[base] $= 0.12$(molar mass) $- 5.1$. Thus, a linear increase of $\log(R_0)$ with polymer molar mass is predicted using this relationship. The kinetic data of Figure 3 are not consistent with the prediction because of the large increases in R_0 at the higher molar masses.

It is well known that small cations form crown-ether-like associations with PEG polymers (17). The cations interact with pseudocrown ether structures of the polymers formed by coiling of the macromolecules. Furthermore, it has been shown that $AuCl_4^-$ ions become coordinated to the pseudocrown ether structures via ion-pairs (16). The Au(III) complexes are attracted through electrostatic interactions to cations that are bound to the oxyethylene groups of the polymers. In our systems Na^+ ions can bind to the pseudocrown ether structures, and form ion-pairs with the gold complexes. The bound complexes are reduced via oxidation of the oxyethylene groups by the metal center:

$$(AuCl_4^- \ Na^+)\text{-PEG} \longrightarrow Au(I) + 4Cl^- + 2H^+ + Na^+ + Ox. \text{ Products} \qquad (3)$$

where $(AuCl_4^- \ Na^+)$-PEG represents ion-pairs bound to the pseudocrown ether structures, and Ox. Products represents products generated in the oxidation of the macromolecules. The pseudocrown ether structures are disrupted when oxyethylene groups are oxidized, leaving the Au(I) species free to migrate and react through equilibrium 4

$$3Au(I) \rightleftharpoons 2Au + Au(III) \qquad (4)$$

This reaction is followed by coalescence of metal atoms to generate Au clusters and small metal particles. The metal atoms are probably stabilized by the macromolecules against oxidative reactions induced by O_2 and H^+ ions. Since reaction 3 is faster when pseudocrown ether structures are more abundant, the rate of particle formation is higher with increasing number of these structures. Formation of pseudocrown ether structures is favored for PEG polymers with higher molar mass (17), and with increases in the polymer concentration. The simple mechanism represented by reactions 3 and 4 is consistent with the kinetic results of Figure 3. A slower formation of Au crystallites takes place in polymer solutions that contain large amounts of NaCl. Larger amounts of Na^+ ions bind to the macromolecules under these conditions, increasing to number of sites where the Au(III) complexes can bind to the polymers. However, in this system Cl^- ions are also present at high concentrations, and it is possible that these ions compete with the gold complexes for the pseudocrown ether structures containing bound cations, inhibiting reaction 3. An even slower reduction reaction that occurs at high HCl concentrations. It should be noted that hydrolysis of the PEG materials at low pH values do not appear to occur readily, since the polymers can be synthesized via cationic polymerizations initiated by acids (*17*). Thus, it seems unlikely that degradation of the polymers via hydrolysis reactions can explain the lower reaction rates in acid solutions. On the other hand, it is known that $AuCl_4^-$ ions bound to polymeric pseudocrown ether of low molar mass are very stable in acidic solutions (*16*). We speculate that the slow formation of metal at low pH values is mainly related to the oxidizing effect of protons on the initially formed Au atoms and clusters, which induces corrosion of these species before they can agglomerate to form larger and more stable crystallites. However, protons and Cl^- ions are continuously formed during the reduction step. For this reason particle formation takes place in two steps; the slower step is a result of the inhibiting effect induced by these ions. On the other hand, particle formation is faster in basic solution since the H^+ ions produced via reaction 3 are neutralized by the base.

Red shifts of the broad plasmon bands are observed initially in cases where the particle formation is slow, such as in solutions of polymers with low molar mass. Blue shifts and a narrowing of the plasmon band occur in a subsequent step; smaller blue shifts are noticed when the formation process is fast. Similar effects have been reported before, and have been explained under the assumption that small Au particles agglomerate forming particle networks (*9b,10*). The red shifts and broadening of the plasmon resonance are due to formation of the networks, whereas λ_{max} is shifted to shorter wavelengths as the networks collapse to form large particles with strong plasmon bands (*10*). Results from recent studies support this explanation since agglomerates of small Au particles have been characterized as intermediates in the formation of Au colloids with citrate ions (*7a,8,21*). Hence, the optical results of Figure 4 indicate that networks of small Au particles are the final products in the cases of PEG 3350 and PEG 1450. Larger metal particles with stronger plasmon bands are generated during the faster reactions induced by the heavier PEG polymers.

Acknowledgments

We are thank to Union Carbide for a gift of Carbowax samples, and the U.S. Navy for supporting L. L. through the CIVINS program. This work was supported by the Strategic Defense Initiative Organization's Office of Innovative Science and Technology (SDIO/TNI) through contract N60921-91-C-0078 with the Naval Surface Warfare Center.

Literature Cited

1. Lewis, L. N. *Chem. Rev.* **1993**, *93*, 2693.
2. Henglein, A. *J. Phys. Chem.* **1993**, *97*, 5457.
3. Schmid, G. *Chem. Rev.* **1992**, *92*, 1709.
4. (a) Kreibig, U. *J. Phys. Colloq.* **1977**, *38*, C2-97. (b) Turkevich, J.; Garton, G.; Stevenson, P. C. *J. Colloid Sci. Suppl.* **1954**, *1*, 26.
5. (a) Quinten, M.; Schönauer, D.; Kreibig, U. *Z. Phys. D* **1989**, *12*, 521. (b) Schönauer, D.; Quinten, M.; Kreibig, U. *Z. Phys. D* **1989**, *12*, 527.
6. (a) Mosseri, S.; Henglein, A.; Janata, E. *J. Phys. Chem.* **1989**, *93*, 6791. (b) Meisel, D.; Mulac, W. A.; Matheson, M. S. *J. Phys. Chem.* **1981**, *85*, 179.

7. (a) Biggs, S.; Chow, M. K.; Zukoski, C. F.; Grieser, F. *J. Colloid Interface Sci.* **1993**, *160*, 511. (b) Heard, S. M.; Grieser, F.; Barraclough, C. G.; Sanders, J. V. *J. Colloid Interface Sci.* **1983**, *93*, 545.
8. Chow, M. K.; Zukoski, C. F. *J. Colloid Interface Sci.* **1994**, *165*, 97.
9. (a) Nakao, Y. *J. Chem. Soc. , Chem. Commun.* **1994**, 2067. (b) Ishizuka, H.; Tano, T.; Torigoe, K.; Esumi, K.; Meguro, K. *Colloids Surfaces* **1992**, *63*, 337.
10. Quinn, M.; Mills, G. *J. Phys. Chem*, **1994**, *98*, 9840.
11. (a) Yeung, S. A.; Hobson, R.; Biggs, S.; Grieser, *J. Chem. Soc. , Chem. Commun.* **1993**, 378. (b) Duff, D. G.; Baiker, A.; Edwards, P. P. *J. Chem. Soc., Chem. Commun.* **1993**, 96.
12. Foss, Jr., C. A.; Hornyak, G. L.; Stockert, J. A.; Martin, C. R. *J. Phys. Chem*, **1994**, *98*, 2963.
13. Yi, K. C.; Sanchez Mendieta, V.; Lopez Castanares, R.; Meldrum, F. C.; Wu, C.; Fendler, J. H. *J. Phys. Chem*, **1995**, *99*, 9869.
14. Longenberger, L.; Mills, G. *J. Phys. Chem*, **1995**, *99*, 475.
15. Dunworth, W. P.; Nord, F. F. *Adv. Catal.* **1954**, Vol. VI, p. 125.
16. Warshawsky, A.; Kalir, R.; Deshe, A.; Berkovitz, H.; Patchornik, A. *J. Am. Chem. Soc.* **1979**, *101*, 4249.
17. Bailey, F. E.; Koleske, J. V. *Alkylene Oxides and Their Polymers*; Marcel Dekker: New York, 1991; Chapters 4 and 6.
18. Keim, R. Ed.; *Gmelin Handbook of Inorganic and Organometallic Chemistry, Gold*, 8th Edition; Springer Verlag: Berlin, 1992, Supplement Vol. B1, pp. 96 and 238.
19. Skibsted, L. H. *Adv. Inorg. Bioinorg. Mech.* **1986**, *4*, 137.
20. Hirai, H.; Toshima, N. *Tailored Metal Catalysts*; Iwasawa, Y., Ed.; D. Reidel: Dordrecht, 1986; p. 121.
21. Biggs, S.; Mulvaney, P.; Zukoski, C. F.; Grieser, F. *J. Am. Chem. Soc.* **1994**, *116*, 9150.

RECEIVED November 27, 1995

Chapter 9

Polymer-Protected Platinum Catalysts in the Nanometer Size Range

Andrea B. R. Mayer and James E. Mark

Department of Chemistry and the Polymer Research Center,
University of Cincinnati, Cincinnati, OH 45221−0172

Stable platinum colloids were prepared by reducing dihydrogen hexachloroplatinate H_2PtCl_6 in the presence of protective polymers. In this chapter, we report the results for several nonionic polymers and cationic polyelectrolytes and their ability to stabilize such platinum colloids. The sizes of the platinum particles were investigated by transmission electron microscopy (TEM) and found to be in the nanometer size range. The catalytic activity of these systems was tested by the hydrogenation of cyclohexene, cis-cyclooctene, and 1-hexene. A variety of polymer-protected platinum nanoparticles showed catalytic activity, and conversions of 100 % were obtained in most cases.

Polymeric materials containing finely-divided metal or metal-oxide nanoparticles show unusual properties that could make them especially useful in many technological applications. Transition metal catalysis is certainly a field that can benefit to a great extent by the introduction of these novel materials. The search for synthetic methods to obtain nanoparticles that are stabilized within a polymer matrix, the characterization of these materials, and their application in catalytic reactions are receiving increased attention.

Transition metals are used as catalysts for a wide range of chemical reactions, such as hydrogenations, oxidations, and Fischer-Tropsch reactions. So far, these catalysts have been usually used in the form of metal powders or metal particles supported on inorganic materials, such as silica, alumina, and activated charcoal. Recently, the use of synthetic polymers as protective matrices has become increasingly interesting, since such systems show several advantages in comparison to traditional catalysts (1, 2):

1) Larger surface areas of the metal particles can be obtained due to the possibility of generating very small particles in the nanometer size-range,
2) the protecting polymer can shield the catalysts from deactivation, e.g. by oxygen or water vapor, especially during storage and handling,

0097−6156/96/0622−0137$15.00/0

the polymer can help to obtain stable colloids and prevents the agglomeration of the nanoparticles, and

4) new catalytic properties can be expected due to the small particle sizes and the possibility of employing different types of polymers.

The polymer-protected metal nanoparticles are usually prepared by in-situ reactions, such as chemical reductions, photolyses, and thermal decompositions of metal salt precursors within the polymer matrix (1-8).

The polymers used in these systems must fulfill several requirements:

1) Thermal stability, especially for the temperatures used during the preparation of the materials and the catalytic reactions,

2) solubility in a range of solvents, especially those used in liquid-phase catalysis or needed to dissolve the metal precursors,

3) low permeability, e.g., for oxygen or water vapor, in order to prevent the deactivation of the catalysts during storage and handling, and

4) ability to stabilize the metal colloids and prevent the agglomeration of the metal particles in order to guarantee a high surface area during liquid-phase catalysis (small "gold number" i.e. large protective function) (1).

Water-soluble, nonionic polymers possessing a relatively nonpolar, hydrophobic backbone and hydrophilic side-groups have been found suitable and been employed in the literature thus far (1). These are usually linear addition polymers of high molecular weight (MW). The hydrophobic backbone is directed towards the hydrophobic surfaces of the metal particles, and the hydrophilic side-groups are directed outwards into the dispersion medium. However, polymers bearing carboxylate or amino groups coordinate so strongly to some metals (e.g. palladium and platinum) through chelation that a colloidal dispersion cannot be formed by the mild reduction of the metal precursor.

Only a few polymers have been investigated so far with respect to their usefulness in these novel catalytic systems: Poly(1-vinyl pyrrolidone), poly(vinyl alcohol) and some of their random copolymers, as well as poly(acrylic acid), polyamides, and poly(ethylene imine). In the last three cases complexation and no free metal formation has been reported. A wider variety of polymers thus still needs to be examined (1-8).

The use of different types of polymers could offer an interesting way to combine the protective function of the polymer with the control of the selectivity of various catalytic reactions. In this context, polyelectrolytes or chiral polymers, for instance, are very interesting candidates for investigation.

In this chapter we report some results for several nonionic polymers and cationic polyelectrolytes and their ability to stabilize platinum colloids. Both steric and electrostatic stabilization of the metal colloids can be combined by the use of polyelectrolytes (3). The materials have been examined by transmission electron microscopy (TEM) in order to determine the average particle size, size distribution and particle shape. The catalytic activity of these polymer-protected platinum nanoparticles has been tested by the hydrogenation of cyclohexene, cis-cyclooctene, and 1-hexene.

Experimental

Syntheses. The platinum colloids were prepared according to the method described by H. Hirai, N. Toshima, et al. (*5-8*). Hydrogen hexachloroplatinate H_2PtCl_2 was reduced by refluxing alcoholic solutions containing the dissolved polymers, including polyelectrolytes. Mass ratios of polymer : platinum = 10 : 1 and 25 : 1 were employed. The H_2PtCl_6 and the polymers were obtained from Aldrich and Polysciences, respectively.

Characterization. Transmission electron micrographs were taken with a JEOL-100 CX II transmission electron microscope in order to obtain the particle size, morphology, and particle size distribution of the samples. The TEM samples were prepared by placing a drop of the colloid on a carbon-coated copper grid and letting the solvent evaporate. The particle sizes were measured with a comparator and the average particle sizes and size distributions were determined based on the measurement of at least 100 particles.

Catalytic Hydrogenations. The hydrogenations were carried out with a Parr hydrogenation apparatus at room temperature and hydrogen pressures of about 35 psi. Cyclohexene, cis-cyclooctene, or 1-hexene (0.1 ml) was added to 20 ml methanol. The catalysts were added either as a colloidal dispersion or as a solid after evaporation of the solvent. An amount of catalyst corresponding to 1 wt% platinum (with respect to cyclohexene, cis-cyclooctene, or 1-hexene) was added. The reaction mixtures were analyzed by gas chromatography (SE-30 packed column) with a flame ionization detector, and helium as a carrier gas.

Results and Discussion

Colloid Formation. Several platinum colloids prepared by the alcohol reduction method are listed in Tables I (nonionic polymers) and II (cationic polyelectrolytes). Examples of particles diameters as determined by TEM are given in Table III. In all cases colloids were formed, and most were stable for several weeks, even months. The TEM investigations showed that the particles were in most cases evenly distributed and about 1 - 5 nm in diameter. Depending on the polymer, a range of particle sizes and narrow size distributions were obtained.

Thus, stable platinum colloids can be prepared with a variety of nonionic, water-soluble polymers. Poly(2-ethyl-2-oxazoline), for instance, forms very stable and useful colloids comparable to those obtained with poly(1-vinyl pyrrolidone) or poly(vinyl alcohol) widely used in the literature (see Figures 1 and 2). Further examples for random copolymers, namely poly(1-vinyl pyrrolidone-co-acrylic acid) and poly(1-vinyl pyrrolidone-vinyl acetate) 60/40 are shown in Figures 3 and 4. The requirement for a hydrophobic backbone and hydrophilic side-groups is confirmed. Colloids with somewhat larger particle sizes (20-40 nm, agglomerates) were obtained with polymers, such as poly(2-hydroxypropyl methacrylate) (see Figure 5). This is probably due to the presence of oxygen in the polymer backbone and its restricted hydrophobic character. For non-hydrophobic backbones, such as in poly(ethylene oxide), less-stable colloids are formed. Precipitation of the metal occurs after a comparably short time (1 - 2 days).

Table I. Stabilization of Platinum Nanoparticles with Nonionic Polymers, Selected Examples

Polymer	Color of Colloid
Poly(1-vinyl pyrrolidone), various MW	Black-brown
Poly(1-vinyl pyrrolidone-co-acrylic acid)	Black-brown
Poly(1-vinyl pyrrolidone-vinyl acetate)	Brown
Poly(2-ethyl-2-oxazoline), various MW	Black-brown
Poly(2-hydroxypropyl methacrylate)	Black-brown
Poly(vinyl phosphonic acid)	
Refluxed in alcohol/water	Brown
Refluxed in water without alcohol	Brown

Table II. Stabilization of Platinum Nanoparticles with Cationic Polyelectrolytes

Polyelectrolyte	Color of Colloid
Poly(diallyldimethyl ammonium chloride), low and high MW	Black-brown
Poly(methacrylamidopropyltrimethyl ammonium chloride)	Dark brown
Poly(2-hydroxy-3-methacryloxypropyltrimethyl ammonium chloride)	Dark brown
Poly(3-chloro-2-hydroxypropyl-2-methacryloxyethyl-dimethyl ammonium chloride)	Brown
Hexadimethrine bromide, MW 4000 - 6000	Black

Figure 1. TEM picture showing catalyst particles in poly(2-ethyl-2-oxazoline) av.
MW 200,000. Polymer : Pt = 10 : 1 (Mass)
(Bar = 48 nm).

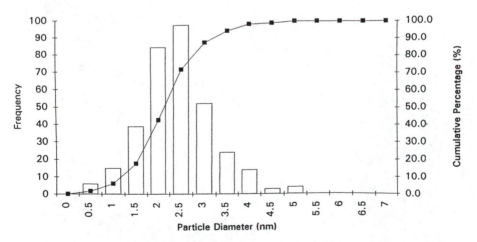

Figure 2. Histogram of particle diameters in poly(2-ethyl-2-oxazoline) av. MW
200,000. Polymer : Pt = 10 : 1 (Mass).

Figure 3. TEM picture showing catalyst particles in poly(1-vinyl pyrrolidone-co-acrylic acid). Polymer : Pt = 10 : 1 (Mass)
(Bar = 54 nm).

Figure 4. TEM picture showing catalyst particles in poly(1-vinyl pyrrolidone-vinyl acetate). Polymer : Pt = 10 : 1 (Mass)
(Bar = 34 nm).

Table III. Average Diameters of the Platinum Nanoparticles by TEM

Polymer	Average Particle Diameter (nm)
Poly(1-vinyl pyrrolidone-co-acrylic acid)	2.4
Poly(1-vinyl pyrrolidone-vinyl acetate)	2.2
Poly(2-ethyl-2-oxazoline)	
av. MW 50,000	2.5
av. MW 200,000	2.6
av. MW 500,000	~ 2-3 [a]
Poly(2-hydroxypropyl methacrylate)	~ 4 [b]
Poly(vinyl phosphonic acid)	
refluxed in alcohol/water mixture	1.6
refluxed in water without alcohol	1.6
Poly(diallyldimethyl ammonium chloride)	1.8
Poly(methacrylamidopropyltrimethyl ammonium chloride)	1.4
Poly(2-hydroxy-3-methacryloxypropyltrimethyl ammonium chloride)	28.5
Poly(3-chloro-2-hydroxypropyl-2-methacryloxyethyl-dimethyl ammonium chloride)	1.7
Hexadimethrine bromide, MW 4000 - 6000	3.9

[a] The particles were organized in loose agglomerates of about 25 nm diameter.
[b] Agglomerates of 20 - 30 nm were present.

In the case of poly(vinyl phosphonic acid), a stable colloid was formed not only when refluxing H_2PtCl_6 in the presence of alcohol as reducing agent. Probably due to the reducing character of the polymer it was sufficient to heat up the combined H_2PtCl_6 and poly(vinyl phosphonic acid) solutions in water. Very small particle sizes and narrow size distributions were obtained in both cases (see Figures 6 and 7). Thus, the investigation of polymers carrying reducing groups seems to be a further interesting option for the preparation of polymer-protected metal nanoparticles. The addition of an additional reducing agent is not necessary in these cases and the polymer can function both as protecting agent and reducing medium.

Cationic polyelectrolytes were found to stabilize platinum colloids very well, and

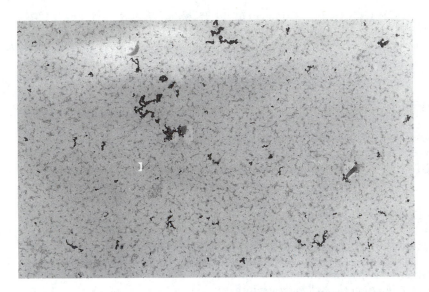

Figure 5.   TEM picture showing catalyst particles in poly(2-hydroxyproyl methacrylate). Polymer : Pt = 10 : 1 (Mass)
(Bar = 157 nm).

Figure 6. TEM picture showing catalyst particles in poly(vinyl phosphonic acid).
Polymer : Pt = 25 : 1 (Mass), refluxed in water without alcohol
(Bar = 22 nm).

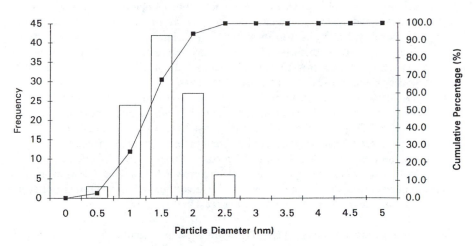

Figure 7. Histogram of particle diameters in poly(vinyl phosphonic acid). Polymer : Pt = 25 : 1 (Mass).

Figure 8. TEM picture showing catalyst particles in poly(3-chloro-2-hydroxy-propyl-2-methacryloxyethyl-dimethyl ammonium chloride). Polymer : Pt = 25 : 1 (Mass)
(Bar = 16 nm).

very small particle sizes (diameters 1 - 5 nm) and narrow size distributions were obtained. An example, namely that for poly(3-chloro-2-hydroxypropyl-2-methacryloxy-ethyldimethyl ammonium chloride) is shown in Figures 8 and 9. An exception is poly(2-hydroxy-3-methacryloxypropyltrimethyl ammonium chloride), where an average particle diameter of 28.5 nm has been found by TEM (see Figures 10 and 11). This colloid also was less stable and partial precipitation of platinum metal occurred after about 1.5 weeks. The reason lies probably in the presence of oxygen in the polymer backbone which leads to larger particle sizes and the destabilization of the metal colloid, as was already observed for nonionic polymers. In comparison, the use of poly(3-chloro-2-hydroxypropyl-2-methacryloxyethyldimethyl ammonium chloride) as stabilizing matrix leads to small particle sizes (average particle size by TEM: 1.7 nm). Even though possessing a similar structure, the presence of chlorine in this polyelectrolyte increases the hydrophobic character which could be a reason for the better stabilization of the platinum colloid. The investigation of additional polymers containing chlorine or fluorine should be of interest in this context.

Catalytic Hydrogenations. Table IV shows some results for the catalytic hydrogenation of cyclohexene, and a selection of results obtained for the hydrogenation of cis-cyclooctene and 1-hexene is given in Table V. The results show that a variety of polymer-protected platinum nanoparticles are catalytically active, and conversions of 100 % were obtained in most cases. The catalysts could be employed either directly as a colloidal dispersion or as a solid after the evaporation of the solvent and the subsequent redissolving in the mixture for liquid-phase hydrogenation. Unlike most catalysts systems, they could be stored in air for several weeks/months and still showed very good catalytic activity.

Conclusions

Both nonionic polymers and cationic polyelectrolytes can be used for the stabilization of platinum colloids and in catalytically active systems for hydrogenations. It can be expected that a wider variety of different types of polymers can be employed in order to function not only as protective matrices but also as active parts of catalytic systems to influence the selectivity of special reactions. In future investigations, other polyelectrolytes (such as phosphonium and sulfonium polyelectrolytes), chiral polymers (such as cyclodextrins), and amphipatic block copolymers will be considered. Polyelectrolytes might enhance the selectivity for catalytic reactions involving charged or polar species, and chiral polymers might provide options for stereoselective reactions. Amphipatic block copolymers are expected to provide very good steric stabilization for colloids. Polymers containing chlorine or fluorine in the backbone could improve the stabilization of metal colloids due to the increased hydrophobic character. Finally, polymers bearing reducing groups can combine both functions as reducing medium for the reduction of the metal precursors, and as protective matrix. We believe that such catalytic systems based on polymer-protected nanoparticles will open new options for the development of catalysts tailored for a variety of reactions.

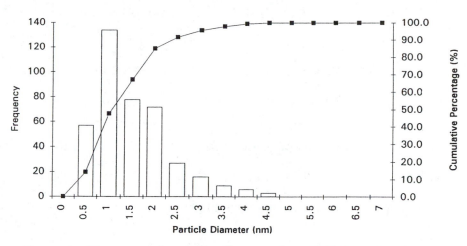

Figure 9. Histogram of particle diameters in poly(3-chloro-2-hydroxy-2-methacryloxyethyl-dimethyl ammonium chloride). Polymer : Pt = 25 : 1 (Mass).

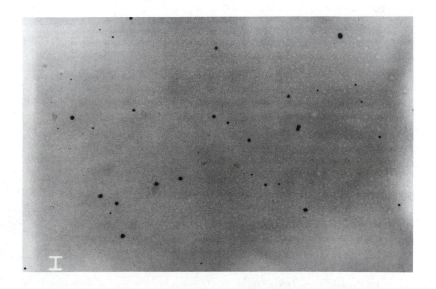

Figure 10.   TEM picture showing catalyst particles in poly(2-hydroxy-3-methacryloxypropyltrimethyl ammonium chloride). Polymer : Pt = 25 : 1 (Mass) (Bar = 230 nm).

Table IV. Hydrogenation of Cyclohexene in MeOH [a]

Polymer, Mass Ratio Polymer : Pt	% Cyclohexane
Poly(1-vinyl pyrrolidone), av. MW 40,000, 10 : 1	100
Poly(1-vinyl pyrrolidone-co-acrylic acid), 10 : 1	100
Poly(1-vinyl pyrrolidone-vinyl acetate) 60/40, 10 : 1	100
Poly(2-ethyl-2-oxazoline), av. MW 50,000, 10 : 1	100
Poly(2-ethyl-2-oxazoline), av. MW 200,000, 10 : 1	95.06
Poly(2-ethyl-2-oxazoline), av. MW 500,000, 10 : 1	100
Poly(2-hydroxypropyl methacrylate), 10 : 1	100
Poly(vinyl phosphonic acid), 25 : 1	100
Poly(diallyldimethyl ammonium chloride), high MW, 10 : 1	100
Poly(methacrylamidopropyltrimethyl ammonium chloride), 25 : 1	100
Poly(3-chloro-2-hydroxypropyl-2-methacryloxyethyl-dimethyl ammonium chloride), 25 : 1	98.94
Poly(2-hydroxy-3-methacryloxypropyltrimethyl ammonium chloride), 25 : 1	100

[a] Hydrogen Pressure ~ 35 psi

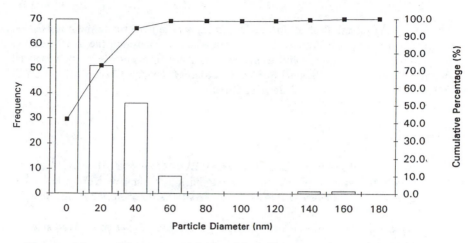

Figure 11.   Histogram of particle diameters in poly(2-hydroxy-3-methacryloxypropyltrimethyl ammonium chloride). Polymer : Pt = 25 : 1 (Mass).

Table V. Catalytic Hydrogenations [a], Selected Examples

Polymer	Starting Material	% Hydrogenated Product
Poly(2-hydroxy-3-methacryloxypropyl- trimethyl ammonium chloride)	1-hexene	100
Poly(1-vinyl pyrrolidone), av. MW 40,000	1-hexene	100
Poly(1-vinyl pyrrolidone-vinyl acetate)	1-hexene	100

Poly(3-chloro-2-hydroxypropyl-2-
    methacryloxyethyldimethyl
    ammonium chloride)
	1-hexene	100

Poly(3-chloro-2-hydroxypropyl-2-
    methacryloxyethyldimethyl
    ammonium chloride)
	cis-cyclooctene	95.8

[a] Hydrogen Pressure ~ 35 psi

Acknowledgments

We would like to thank Prof. R. Morris for his support in performing the transmission electron microscopy and Prof. A. Pinhas for his help in analyzing the hydrogenation mixtures by gas chromatography. It is also a pleasure to acknowledge the financial support provided by the National Science Foundation through Grant DMR-9422223 (Polymers Program, Division of Materials Research).

References

1. Hirai, H. and Toshima, N. In *Catalysis by Metal Complexes, Tailored Metal Catalysts*; Iwasawa, Y., Ed., D. Reidel Publishing Company: Dordrecht, 1986.
2. Bradley, J. S. In *Clusters and Colloids. From Theory to Applications*; Schmid, G., Ed., VCH: Weinheim, 1994.
3. Napper, D. H. *Polymeric Stabilization of Colloidal Dispersions*; Academic Press: London, 1983.
4. Wang, S.; Mark, J. E. *Polym. Bulletin* **1992**, *29*, 343 - 348.
5. Hirai, H.; Nakao, Y.; Toshima, N. *Chem. Lett.* **1978**, 545 - 548.
6. Toshima, N.; Yonezawa, T.; Kushihashi, K. *J. Chem. Soc. Faraday Trans.* **1993**, *89*, 2537 - 2543.
7. Hirai, H.; Nakao, Y.; Toshima, N. *J. Macromol. Sci. - Chem.* **1979**, *A13*, 727 - 750.
8. Hirai, H.; Chawanya, H.; Toshima, N. *Reactive Polymers* **1985**, *3*, 127 - 141.

RECEIVED December 1, 1995

Chapter 10

Polymer Composites of Nanostructured Gold and Their Third-Order Nonlinear Optical Properties

Kenneth E. Gonsalves[1,2], G. Carlson[1], J. Kumar[3], F. Aranda[4], and M. Jose-Yacaman[5]

[1]Polymer Science Program, Institute of Materials Science and
[2]Department of Chemistry, University of Connecticut, Storrs, CT 06269
[3]Center for Advanced Materials, Department of Physics, University of Massachusetts, Lowell, MA 01854
[4]Department of Physics, University of Massachusetts, 100 Morrissey Boulevard, Boston, MA 02125
[5]Instituto de Fisica, Universidad Nacional Autonoma de Mexico, Apartado Postal 20–364, C.P. 01000 Mexico, D.F., Mexico

Using a phase-transfer reaction, 1-10 nm gold particles were synthesized. The particles, functionalized by dodecanethiol, were stable at room temperature over a period of months. The particles were characterized by UV-visible spectroscopy, and transmission electron microscopy (TEM). Composite films were prepared by dispersing the gold in polystyrene, poly(methyl methacrylate), and poly(phenylmethylsilane). Using degenerate four-wave mixing (DFWM), the third-order nonlinear optical susceptibilities of the materials was measured (λ = 532 nm). The nonlinearity of the gold was found to be negative (i.e., opposite sign of CS_2) with a magnitude of 1.0×10^{-10} e.s.u. for a film containing 4.4 mg gold particles per ml of PMMA.

Development of nonlinear optical materials is an important field of research (*1*) because of the emerging technology of photonics, which uses photons for information and image processing. Photonic devices are expected to perform such functions as frequency conversion, optical switching, and data processing (*2*). Polymers and polymer-matrix composites have emerged recently as attractive candidate materials for these devices because they offer a range of properties and processing conditions, which can be optimized for a particular device (*3*). One class of these materials which is of particular interest is those with significant third-order nonlinear optical susceptibility ($\chi^{(3)}$).

Such optical nonlinearity arises in nanostructured metal particles as a consequence of the quantum confinement (*4-6*) of the particle's electron cloud. At

0097–6156/96/0622–0151$15.00/0

sufficiently small size, (below ca. 6 nanometers,) the electronic levels become significantly different from those of bulk materials due to the quantum confinement of carriers. As the number of atoms in the cluster decreases, the electronic energy levels become more discrete. Eventually, the energy levels of very small particles begin to resemble molecular orbitals. This can give rise to enhanced third-order nonlinearity (7-9).

Nanostructured crystallites have a high specific surface area. The resulting high-energy surface must be stabilized to prevent the particles from agglomerating (10). This is accomplished by surface-functionalizing the particles with moieties which introduce steric repulsion between them. When chosen prudently, these groups also render the stable particles soluble in optically clear polymers, which can be formed into the desired shape (such as a thin film) for device applications. Polysilanes are particularly interesting matrix materials because they are optically nonlinear (11) themselves due to sigma conjugation along their backbone.

This chapter describes a process for making stable, nanophase gold and dispersing it into polymer matrices. Characterization of the resulting materials is included, with emphasis on the novel optical properties of the composites.

Experimental Reagents and Instrumentation. Tetrachloroauric acid ($HAuCl_4$,) tetraoctylammonium bromide ($N[C_8H_{17}]_4Br$,) dodecanethiol, sodium borohydride, and sodium metal were used as received from Sigma Chemical Co. Water and toluene were deoxygenated by reflux and distillation under an inert atmosphere. The monomers phenylmethyldichlorosilane (Gelest,) methyl methacrylate, and styrene (Aldrich) were purified by fractional distillation under vacuum. The initiator, azobisisobutylnitrile (AIBN,) was recrystallized from methanol before use.

Ultrasonic treatment was performed using a high-intensity ultrasonic probe (Sonic and Materials VC-600, 1/2" titanium horn, 20 kHz, 100 W/cm^2). Films were cast using a Specialty Coatings Systems P-6204-A spin-coater. Film thickness was measured using a Tencor Alpha-Step 200 profilometer. UV/visible spectra were measured on a Perkin-Elmer Lambda 6 spectrophotometer from 190 to 900 nm using a 1-nm slit at a scan rate of 120 nm/min.

High-resolution TEM samples were prepared from methanol dispersions of the particles and deposited on copper grids with a carbon film. The instrument was a JEOL-4000EX with a point-to-point resolution of approximately 1.7 Å. Transmission electron microscopy was also performed on PMMA/gold composite samples using a Phillips EM 300 instrument. The samples were prepared by microtoming to a thickness of 80 Å.

The nonlinear optical properties were measured using 532-nm, 30-ps pulses from a frequency-doubled, Q-switched neodynium-doped yttrium-aluminum-garnet (Nd:YAG) laser (Quantel). The laser operated at a 10-Hz repetition rate. Average pulse energy was 25 mJ. A neutral density filter was placed in the probe beam to reduce its intensity to about 1% of that of the pump beams. The phase conjugate beam was separated from the signal using a beam splitter. The crossing angle was 6°. Fast silicon photodiodes calibrated against a laser-energy meter were used to monitor the signal, the probe and the pump pulse energies.

Surface-Functionalized Nanostructured Gold

Synthesis. The particles were synthesized using a phase-transfer reaction (*12*). Gold salt (HAuCl$_4$) was dissolved in water (30 ml, 0.030 $\underline{M}$). Normally, the resulting AuCl$_4^-$ anion would be insoluble in organic solvents (*13*). However, when the aqueous solution is stirred for one hour with a toluene solution (80 ml, 0.050 $\underline{M}$) of the phase-transfer catalyst, tetraoctylammonium bromide (N[C$_8$H$_{17}$]$_4$Br,) the gold species is transferred into the toluene.

Next, the surface-functionalizing reagent, dodecanethiol (0.2 ml), was added. Then, sodium borohydride solution (25 ml, 0.40 $\underline{M}$) was dropped in gradually to reduce the gold from Au^{3+} to Au^0. In the absence of the thiol, zerovalent gold would combine rapidly into macroscopic particles of gold. However, the natural affinity of gold for thiol (*14-18*) causes a competing process whereby gold atoms (or clusters of several gold atoms) are coordinated to a thiol group. In the resulting proposed structure (see Figure 1,) gold atoms which combine with each other form the particle core and thiol-functionalized gold forms the surface. The reaction was complete three hours after the borohydride was added.

The product (organic phase) was decanted. Approximately 90% of the toluene was removed by evaporation, then the remaining material was precipitated into ethanol. After filtration, the product was purified by redissolving in toluene and reprecipitating with ethanol. The resulting material was a waxy, purple solid which was stable over a period of months.

Characterization. The solid product was dissolved in toluene and optical absorbance was measured versus a blank containing pure toluene. The spectrum showed a shoulder peak between 520 and 530 nm. This agreement with previous studies (*19*) signifies that the gold is nanostructured.

High-resolution transmission electron microscopy was used to examine the particles directly. Figure 2 shows a HRTEM image of the gold particles. All particles display the characteristic lattice fringes of the common HRTEM images. Some of the gold particles clearly have faceting as indicated by the arrows. The smallest gold particles usually show perfect crystalline structure. However, in some cases, the presence of twin boundaries (Figure 3) are evident. It is also interesting to point out that from these HRTEM images there is no evidence of crystalline defects such as stacking faults or dislocations. The particle size distribution (Figure 4) shows that most of the particle diameters fall in the range of 1.0 - 3.4 nm.

Polymer/Gold Composites

Polysilane Synthesis. The synthesis of poly(phenylmethylsilane) was carried out by the Wurtz coupling of phenylmethyldichlorosilane assisted by ultrasonication (*20*). First, sonication at 40% amplitude for twenty minutes was used to disperse sodium metal in toluene. Then, the silane monomer was added to this dispersion over a thirty-minute period. The reaction continued with sonication at 20% amplitude for one hour. A 50/50 ethanol/water mixture was used to quench the reaction.

The polymer was precipitated into isopropanol, filtered, redissolved in toluene, reprecipitated, filtered again, and vacuum dried. The reaction yielded 17% soluble

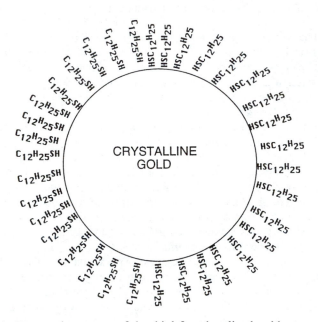

Fig. 1 Proposed structure of the thiol-functionalized gold nanoparticle

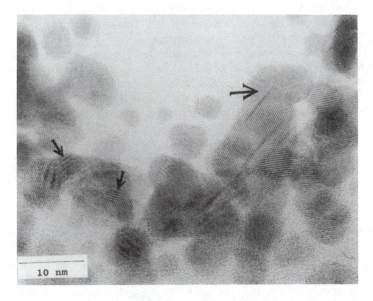

Fig. 2 HRTEM image of gold particles

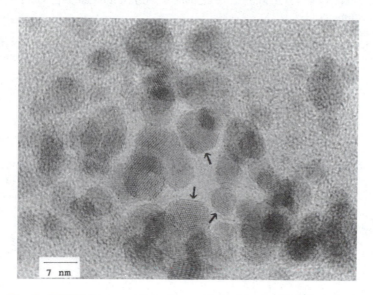

Fig. 3   HRTEM image of smaller gold particles, exhibiting twinning

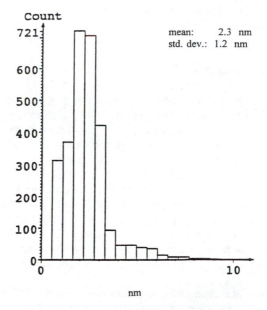

Fig. 4   Particle size distribution

polymer. There was also a significant amount of an insoluble side-product, probably cross-linked material.

Composite Film Preparation. Composites of the gold nanoparticles in polymer were obtained easily by dissolving the particles in a suitable monomer (methyl methacrylate or styrene) and initiating free-radical polymerization using azobisisobutylnitrile under thermal conditions (70 °C). Polymerization proceeded more slowly in the particle-containing reactions than in the control reactions. This may indicate the presence of a chain-terminating side reaction. Polymerization was terminated by cooling when the mixture reached a viscosity suitable for spin-coating.

This method could not be used to form composites with a polysilane matrix because of the harsher polymerization conditions. In this case, the polymer and the nanoparticles were both dissolved in toluene and co-deposited during spin-coating.

Films were cast from solution onto clean quartz substrates. Relatively slow spinning (~250 rpm) was found to produce the best films. At high speeds, centrifugation effects cause the gold concentration to vary radially. After three minutes of spinning, the solvent (toluene or monomer) evaporated, leaving a flat film 5 to 10 microns thick.

Film Characterization. The films were examined by UV-visible spectroscopy (Figs. 5 and 6) and showed similar peak positions to those for the solutions (plain polymer films were used as blanks). The linear absorption coefficient at 532 nm (where the NLO characterization was performed) was approximately 0.6 cm^{-1}. Also, TEM images of thin (8 nm) sections of the composite material (Figure 7) showed that the particles were incorporated into the films without agglomeration.

Nonlinear Optical Measurements

Degenerate Four-Wave Mixing. Third-order nonlinear optical susceptibility of the composite materials was measured using degenerate four-wave mixing (DFWM). In this technique (21), diffraction gratings are formed in the medium due to the interference of incident light waves. The characteristics of the grating, such as diffraction efficiency and decay time, together with the dependence of FWM signal on parameters such as beam intensities, crossing angles, and polarization, provide information about the nonlinear optical properties of the material.

The geometry used for the measurements in this study is the counter-propagating pump geometry. Two strong, counterpropagating pump (or 'write') beams are incident on the medium. These are examined by a weaker 'probe' (or 'read') beam. When the three beams are incident on a nonlinear optical medium, they generate a fourth beam, the phase conjugate, which is counter-propagating to the probe.

The magnitude of the third-order nonlinear susceptibility, $\chi^{(3)}$, was estimated from a measurement of the intensity of the phase-conjugate beam relative to that of a reference sample of CS$_2$ placed in a quartz cell of 2-mm path length. The value of $\chi^{(3)}$ was obtained from (22):

$$\chi^{(3)} / \chi_R^{(3)} = (\eta / \eta_R)^{1/2} \ (n / n_R)^2 \ L/L_R \ (I_{1R} I_{2R} / I_1 I_2)^{1/2}, \tag{1}$$

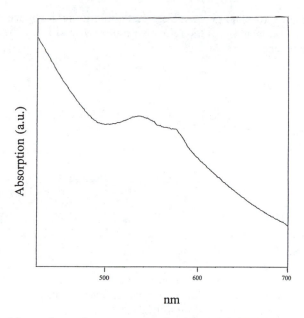

Fig. 5 Linear absorption spectrum of gold particle/PMMA composite

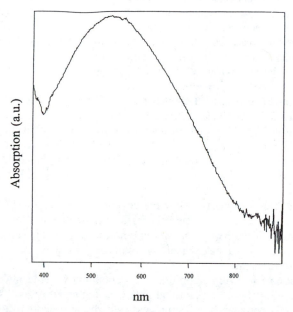

Fig. 6 Linear absorption spectrum of particle/polysilane composite

where the subscript R refers to parameters pertaining to the reference, L is the effective interaction length, n is the refractive index, I_1 and I_2 are the intensities of the pump beams, and

$$\eta = I_s / I_p, \tag{2}$$

is the phase conjugate reflectivity. I_s and I_p are the intensities of the phase conjugate and probe beams, respectively

The measured values of $\chi^{(3)}$ were:

Matrix	Gold Concentration(mg/ml)	$\chi^{(3)}$ (e.s.u.)
PMMA	0	not measurable
PMMA	1.3	1.6×10^{-11}
PMMA	4.4	1.0×10^{-10}
poly(phenylmethylsilane)	0	6.1×10^{-11}
poly(phenylmethylsilane)	5.1	1.5×10^{-10}

Theoretically, $\chi^{(3)}$ is expected to be proportional to particle concentration. However, the proportionality is not exact. Some possible sources of this error are dispersity of particle size and nonuniformity of the films.

Z-Scan. DFWM determines only the magnitude of the third-order susceptibility ($\chi^{(3)}$). To completely characterize $\chi^{(3)}$, its sign (positive or negative) must also be determined. This is done using the Z-scan technique (23). In this technique, a collimated beam is focussed into the sample. The sample is translated along the optical ("z") axis so that it goes from one side of the focus to the other. The change in refractive index due to third-order nonlinearity causes an intensity-dependent lensing effect in the medium which leads to a focussing or defocussing of the beam. This can be measured by an aperture detector.

The behavior of the transmitted intensity is different for materials with positive or negative third-order nonlinearity. By convention, the standard carbon disulfide is assigned a positive sign. The gold particles were found to exhibit the opposite behavior and thus have a negative sign (see Figure 8). This is also known to be the case for polysilanes. Thus, their nonlinearities should add to create a large, negative $\chi^{(3)}$ in the composite.

Time-Resolved Measurements. The time-dependence of the third-order nonlinearity is an important characteristic for device applications and may also indicate the nature of the process causing the nonlinearity (24). The temporal behavior of the phase conjugate signal was monitored by delaying the probe beam with respect to the pump.

Such a study was performed on the gold-containing films using a thirty-picosecond pulse duration. Experimental data (see Figure 9) show that the grating is composed of two components: a fast process (which has a shorter duration than the laser pulse) and a slower one, which decays exponentially. The fast process can be attributed to the electronic response of the medium. The slow process may be the

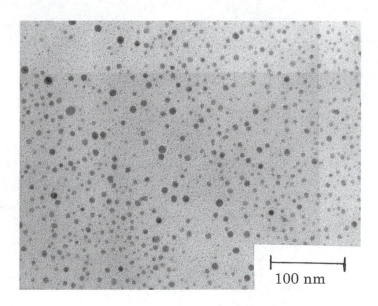

Fig. 7 TEM image of particles dispersed in poly(methyl methacrylate)

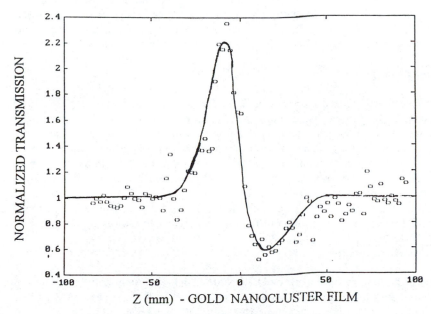

Fig. 8 Z-scan of gold particle composite film

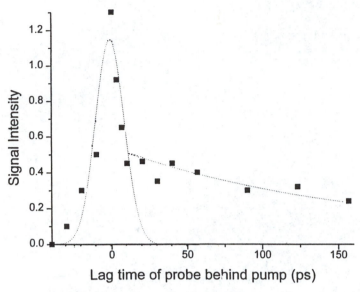

Fig. 9 Time dependence of $\chi^{(3)}$ in gold particle/PMMA composite

result of a physical phenomenon, such as plasmon resonance or lattice heating. This slow process has a longer decay time than in previous studies of nanostructured gold in liquid solution (*25*). This may be evidence of slower diffusion of, or heat dissipation from, the particles when embedded in a polymer matrix.

Conclusions

Surface-functionalized, nanostructured gold was synthesized using a phase-transfer reaction. The particles were approximately 2.2 nm in diameter. They were stable for months at room temperature and soluble in organic solvents and monomers. Polymer composites of the gold were prepared without significant agglomeration.

The third-order nonlinearity ($\chi^{(3)}$) of these composites was measured using DFWM. The nanoparticulate gold was found to have nonlinearities in PMMA and poly(phenylmethylsilane) of -1.0×10^{-10} e.s.u. and -1.5×10^{-10} e.s.u., respectively. In the future, this work will be extended to different types of surface-functionalized nanoparticles and modification of the matrix polymer to increase $\chi^{(3)}$.

Acknowledgements

Partial funding for this work by ONR N0014-94-1-0833 is gratefully acknowledged. Also, thanks to Dr. Ramiro Perez, P. Santiago and L. Rendon at UNAM and L. Khairallah at UConn for technical assistance with the TEM and to Dr. P. Rangarajan for guidance with the sonochemical synthesis.

Literature Cited

1. Hinton, H.S. *IEEE Spectrum* **Feb. 1992**, 42.
2. Stamatoff, J., et al. *Angew. Makr. Chem.* **1990,** *183*, 151.
3. Prasad, P. N. in *Materials For Nonlinear Optics;* Marder, S.R. et al., Eds.; ACS Sym. Ser. no. 455; ACS: Washington, D.C., 1991. pp.50-66.
4. Brus, L. E. *J. Chem. Phys.* **1984,** *80,* 4403.
5. Brus, L. E. *J. Phys. Chem.* **1986,** *90,* 2555.
6. Brus, L. E.*IEEE J. Quantum Electron.* **1986,** *QE-22*, 1909.
7. Wang, Y.; Mahler, W. *Opt. Commun.* **1987,** *61,* 233.
8. Hilinski, E. et al. *J. Chem. Phys.* **1988,** *89,* 3435.
9. Wang,Y. et al. *J. Chem. Phys.* **1990,** *92,* 6927.
10. Andres, R. P.; Averback, R. S.; Brown, W. L.; Brus, L. E.; Goddard III, W. A.; Kaldor, A.; Louie, S. G.; Moscovits, M.; Peercy, P. S.; Riley, S. J.; Siegel, R. W.; Spaepen, F. Y.; Wang, Y. *J. Mater. Res.* **May/June 1989**, *4, # 3*, Panel Report.
11. Mark, J. E. et al. *Inorganic Polymers*; Prentice-Hall: Englewood Cliffs, NJ, **1992**.
12. Brust, M.; Walker, M.; Bethell, D.; Schiffrin, D. J.; Whyman, R. *J. Chem. Soc., Chem. Commun.* **1994**, 801.
13. Dehmlow, E. V. *Phase Transfer Catalysis;* Weinheim: New York, NY, **1993.**
14. Nuzzo, R. G.; Allara, D. L. *J. Am. Chem. Soc.* **1983,** *105*, 4481.
15. Nuzzo, R.G.; Fusco, F. A.; Allara, D. L. *J. Am. Chem. Soc.* **1987,** *109*, 2358.
16. Li, T. T.-T.; Weaver, M. J. *J. Am. Chem. Soc.* **1984,** *106*, 6107.
17. Finklea, H. O.; Avery, S.; Lynch, M.; Furtsch, T. *Langmuir* **1987,** *3,* 409.
18. Porter, M. D. *J. Am. Chem. Soc.* **1987,** *109*, 3559.
19. Olsen, A. W., and Kafafi, Z. H., *J. Am. Chem. Soc.* **1991,** *113,* 7758.
20. Kim, H. K.; Matyjasewski, K. *J. Am. Chem. Soc.* **1988,** *110,* 3323.
21. Prasad, P. N. ; Williams, D. J. *Introduction to Non-linear Optical Effects in Molecules and Polymers;* Wiley & Sons, New York, NY, **1992**.
22. Bogdan, A. R.; Prior, Y.; Bloembergen, N. *Opt. Lett.* **1981,** *6,* 82.
23. Cheung, Y. M.; Gayen, S. K. *J. Opt. Soc. Am. B* **1994,** *11,* 636.
24. Dutton, T.; VanWonterghem, B.; Saltiel, S.; Chestnoy, N. V.; Rentzepis, P. M.; Shen, T. P.; Rogovin, D. *J. Phys. Chem.***1990,** *94,* 1100.
25. Bloemer, M. J.; Haus, J. W.; Ashley, P. R. *J. Opt. Soc. Am. B* **1990,** *7,* 790.

RECEIVED December 1, 1995

Chapter 11

Nanoparticle Synthesis at Membrane Interfaces

Michael A. Markowitz, Gan-Moog Chow, and Alok Singh

Laboratory for Molecular Interfacial Interactions,
Center for Bio/Molecular Science and Engineering,
Naval Research Laboratory, Code 6930, 4555 Overlook Avenue, S.W.,
Washington, DC 20375–5348

A comparison of the synthesis of metal nanoparticles at the surfaces of polymerized and unpolymerized phospholipid vesicles has been undertaken. Vesicles were prepared from mixtures of charge-neutral phospholipid, 1,2-*bis*(tricosa-10,12-diynoyl)-*sn*-glycero-3-phosphocholine, and palladium ion-bound negatively charged phospholipid, 1,2-*bis*(tricosa-10,12-diynoyl)-*sn*-glycero-3-phosphohydroxyethanol. The divalent palladium ion bound to the negatively charged phospholipids in the vesicle membrane initiated electroless metallization of gold or cobalt on the vesicle surface leading to the formation of metallic nanoparticles. By using the internal volume of polymerized vesicles as the reaction vials, two types of unagglomerated nanoparticles, Au (4-10 nm) and $Co/Co(OH)_x$ (30-100 nm), have been synthesized. Metallization of unpolymerized vesicles resulted in the formation of size controlled, unagglomerated Au nanoparticles (20-28 nm). Dry powders of the Au particles were redispersable in water. TEM results suggest that composites of lipid and metal were formed. For both the polymerized and unpolymerized vesicles, HRTEM studies have demonstrated that particle nucleation and growth occurred at the vesicle membrane surfaces and that particle growth could be initiated by either single or multiple nucleation sites.

Self-assembled membranes constructed from phospholipids and other surfactants have been extensively investigated to understand their formation, encapsulation and release, and templating properties (*1-23*). Lipids and surfactants are amphiphilic molecules with hydrophilic, polar headgroups and nonpolar tails. As a result of the hydrogen bonding and electrostatic interactions of the hydrophilic headgroups and the van der Waals interactions between the hydrophobic tails, amphiphiles form organized membrane microstructures when dispersed in water or oil. When dispersed, the surfactant membrane organizes to form the minimum energy structures segregating the nonpolar tails from an aqueous phase or the polar headgroups from an oil phase. Depending on the structure of the amphiphile and the nature of the dispersing conditions, a variety of monolayer and bilayer microstructures can be formed.

Micelles (1-20 nm) are constructed from flexible, surfactant monolayers comprised of single chain or short double chain amphiphiles (*3,4*). In dilute solutions, micelles are generally spherical but as the surfactant concentration increases, significant polymorphism is observed. Oil in water or water in oil surfactant emulsions are colloidal but disordered systems (*4*). Vesicles are closed spherical structures with an aqueous core. Generally, vesicle membranes are comprised of single or multibilayers of phospholipids and surfactants (*1-5*). The size ranges of small unilamellar vesicles, large unilamellar vesicles and multilamellar vesicles are 20-100 nm, 100-300 nm, and 100-800 nm, respectively. However, the formation of vesicles constructed from bipolar lipid monolayers also has been reported (*6-9*). In addition to forming vesicles, diacetylenic phospholipids can self-assemble into hollow, open-ended microcylinders formed from single or multiple bilayer membranes (*10-12*). The diameters of lipid tubules range from 0.1 to 1 μm and the length can extend up to 100 μm. The encapsulation and release of a number of chemical and biological agents from vesicles and lipid tubules has been extensively studied (*2,17,18,22,23,24*). Furthermore, bilayer vesicles and lipid tubules can be stabilized towards thermal, mechanical and chemical perturbations by polymerization of the membrane lipids (*17,25*). Typically, polymerizable lipids and surfactants contain photo- or chemically reactive methacrylate, diacetylene, vinyl, styrene, sorbyl, thiol, or isocyano groups.

Micelles, vesicles and lipid tubules have been used as reaction cages or templates in the construction of more complex materials. Enzymes and antibodies have been tethered covalently and ionically to headgroups on the external surfaces of vesicles in order to fabricate biosensors (*19-21*). Metallized lipid tubules have been used to fabricate microcathode arrays (*26*). More recently, micelles, emulsions and vesicles have been utilized in the synthesis of nanoscale particles (1-100 nm) (*14-16*). As the size of the particle is reduced in size to the nanoscale range, an increasing proportion of the particle's atoms are located on the particle surface resulting in an increasing contribution of surface effects to the properties of the particles (*27-30*). Consequently, it is expected that materials fabricated from metal and ceramic nanoparticles will have enhanced electrical, optical, transport, mechanical and processing characteristics resulting from the large surface energy and the large number of interfaces of the packed particles. In order for these property enhancements to be realized, nanoparticles must be highly pure, size controlled, redispersable in a variety of solvents, produced in high enough yield for particular applications, and must remain unagglomerated during synthesis and processing. Since agglomeration arises from the thermodynamic driving forces to lower the surface energy of the particles, the synthesis of a sufficient yield of unagglomerated nanoparticles necessary for quantitative analysis and testing in applications has been difficult to achieve. Reverse micelles (headgroups segregated from an oil phase) have been utilized to stabilize nanoscale semiconductor crystallites to produce quantum dot materials with enhanced electrical and optical

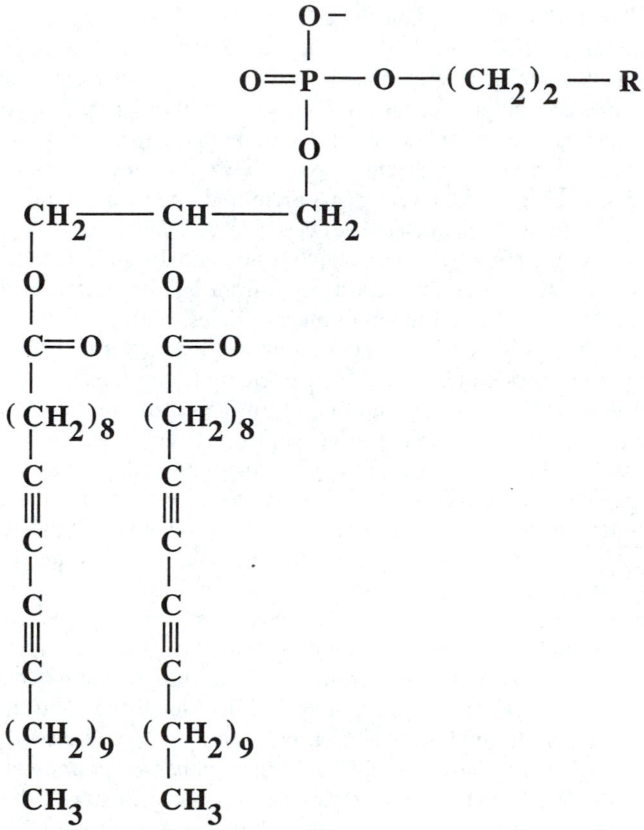

$$O=P \begin{matrix} O^- \\ | \\ \\ | \\ O \end{matrix} O-(CH_2)_2-R$$

$$CH_2-CH-CH_2$$
$$| \qquad |$$
$$O \qquad O$$
$$| \qquad |$$
$$C=O \quad C=O$$
$$| \qquad |$$
$$(CH_2)_8 \ (CH_2)_8$$
$$| \qquad |$$
$$C \qquad C$$
$$||| \qquad |||$$
$$C \qquad C$$
$$| \qquad |$$
$$C \qquad C$$
$$||| \qquad |||$$
$$C \qquad C$$
$$| \qquad |$$
$$(CH_2)_9 \ (CH_2)_9$$
$$| \qquad |$$
$$CH_3 \qquad CH_3$$

1: $R = OH$

2: $R = \overset{+}{N}(CH_3)_3$

Figure 1. Structure of diacetylenic phospholipids.

properties (*31-33*). The segregated water pools of water in oil microemulsions and vesicles have been used as reaction vials for the precipitation of metal and ceramic oxide and hydroxide precursors and products (*34-40*). In these instances, the membranes play passive roles in the synthetic process and serve only as physical barriers to unwanted electronic interactions or agglomeration. Generally, little control over the final product can be realized due to limitations in membrane diffusivity and stability. The chemical and physical characteristics of the precipitated particle are determined by the rate at which the reactants can diffuse through the membrane while the range of possible reactions is limited because only some potential reactants, ie. anions but not cations, can pass readily through unpolymerized membranes (*41*). The lack of mechanical, chemical, and thermal stability of the membranes limits the types of reactions which can be performed, the reaction conditions which can be used, and the ability to form redispersable powders of the nanoparticles. If reaction sites could be incorporated into the membrane surface, the process of nucleation and particle growth possibly could be more finely tuned. By tailoring the membrane surface to predictably alter the number and density of nucleation sites, the rate and growth of particle formation and potentially chemical composition of the particle could be controlled. Also, since metal cations pass through polymerized membranes, the use of polymerized vesicles as the reaction vials should enable a greater variety of reactions to be performed. Recently, the positive effects of incorporating reaction sites in the membrane and using polymerized vesicles as reaction cages for the synthesis of metal nanoparticles have been demonstrated (*42,43*). Polymerized and unpolymerized vesicles formed from mixtures of charge-neutral and palladium ion-bound negatively charged phospholipids have been utilized as templates for the synthesis of unagglomerated, Au and $Co/Co(OH)_x$ nanoparticles. In this paper, we will compare the results of particle formation from metallization of the polymerized and unpolymerized membranes. In addition, the effects of metal ion concentration on the crystallinity of intravesicularly precipitated metal hydroxide as well as the effect of incorporating the membrane in the nanoparticle on the formation and dispersity of dry metal nanoparticle powders will be discussed.

Materials and Methods

1,2-*bis*(tricosa-10,12-diynoyl)-*sn*-glycero-3-phosphocholine (**1**) and 1,2-*Bis*(tricosa-10,12-diynoyl)-*sn*-glycero-3-phosphohydroxyethanol (**2**) were synthesized according to literature procedures (Figure 1) (*44-46*). $Pd(NH_3)_4Cl_2$ catalyst solution was prepared as previously described (*47*). The hydrodynamic radius of the particles formed was determined by light scattering (Coulter Submicron Particle Analyzer, N4MD) and the extent of monomer reaction after exposure to UV radiation was determined by phosphate analysis of the thin layer chromatographed vesicles using established literature procedures (*48*).

Samples, mounted on carbon or formvar coated copper grids, were examined by transmission electron microscopy (TEM) using a JEOL JEM200CX microscope operated at 200 KV. High resolution transmission electron microscopy (HRTEM) was performed by using a Hitachi 9000 UHR microscope operated at 300 kV. The sample grids were prepared by direct application of the aqueous particle suspensions to the grids. Crystallographic information was obtained by using selected area electron diffraction technique as well as HRTEM lattice imaging. X-ray diffraction patterns of metal particles were recorded in a Phillips X-ray diffractometer in the theta-two theta mode using Cu-K_α radiation. The average crystallite size of the gold particles formed using nonpolymerized vesicles was determined from line broadening of the gold (111) peak.

Vesicle Preparation. A thin film of the lipid mixture was hydrated at 60-70 °C in water 30 min in the presence $Pd(NH_3)_4Cl_2$. The total concentration of lipid in each sample ranged from 2 mg/mL to 20 mg/mL. The concentration of $Pd(NH_3)_4Cl_2$ was equimolar to **1**. The mixture was vortexed and then sonicated (Branson sonifier, Model 450) at 60-70 °C for 20 min at which point the dispersions were transluscent. Polymerization of the vesicles was accomplished by exposure to UV radiation (254 nm, Rayonet Photochemical Reactor, model RPR-100) at 20 °C for 15 min. For unpolymerized vesicles, the dispersions were dialyzed overnight against water to remove excess $Pd(NH_3)_4Cl_2$. For polymerized vesicles in which metal particles were to be formed internally an amount of metal ion chelating EDTA (tetrasodium salt) equimolar to the amount of $Pd(NH_3)_4Cl_2$ present was added and then the vesicles were then immediately gel filtered (Sephadex G-50) through centifuge minicolumns (5 mL) within a centrifuge at 2,000 rpm for 2 min.

Metal Particle Formation. Metal particle deposition or formation of vesicles was accomplished with a Au or Co plating bath. The gold plating bath was prepared by mixing equal volumes of 20 mM $AuCl_2$ and a second solution consisting of equal volumes of 10 mM sodium hypophosphite, 0.2 mM dibasic sodium phosphate, and 0.02 mM sodium cyanide. The cobalt plating bath was prepared as previously described (*47*). The vesicle dispersions were diluted with an equal volume of the plating bath. Preparation of Co or $Co(OH)_x$ particles was carried out in a bath sonicator (Ney 300). Plating of unpolymerized vesicles was allowed to continue for 1.5 to 3 hrs. The dispersions were then dialyzed against water overnight to remove excess plating bath. In the case of metal particle formation inside polymerized vesicles, plating was allowed to continue for 6 hrs.

Results and Discussion

The first approach in using vesicle membranes in nanoparticle synthesis was to polymerize vesicles formed from 9:1 and 1:1 mixtures of **1** and **2** in the presence of $Pd(NH_3)_4Cl_2$. Palladium ions bound to negatively charged phospholipids in bilayer membranes had been previously demonstrated to serve as catalytic sites for electroless metallization (*47*). Light scattering revealed that bimodal populations of vesicles were formed both from the 9:1 (31 $\pm$ 15 nm and 114 $\pm$ 40 nm) and 1:1 (38 $\pm$ 13 nm and 109 $\pm$ 41 nm) mixtures. Exposure of the vesicles to UV radiation (254 nm) resulted in the reaction of 33 % $\pm$ 6% of the lipid monomer when 10 % of **2** was present and 50% $\pm$ 4% for vesicles containing 50% **2**. The differences in the amount of reaction of monomer are probably due to both ion and pH effects. Non-cross linked polymerized vesicle membranes consist of many individual polymer chains and contain breaks in the polymeric network which are the transport paths through the membrane (*49*). The density of these channels depends on vesicle diameter and polymer size (*50-52*). Because diacetylenes react topotactically, the polymerization of diacetylenic phosphocholine vesicles produces a high number of short chain polymers (*53*). Therefore it is reasonable to suggest that the partially polymerized vesicles composed of mixtures of diacetylenic phospholipids **1** and **2** are stabilized by a number of short chain polymers resulting in a high number of polymer boundaries within the membrane.

In order to synthesize the metal particles inside the vesicles, the nucleating palladium ions on the external membrane surfaces must be removed and the metal ions must pass through the polymerized vesicle membranes. Without the removal of the externally bound ions, the external membranes have been selectively metallized when gold or nickel plating baths were added (*42,54*). When the vesicles were pretreated with metal chelating EDTA (tetrasodium salt) prior to the addition of gold or cobalt plating baths, crystalline Au or $Co/Co(OH)_x$ nanoparticles formed only on the internal membrane vesicle surfaces demonstrating that the EDTA treatment removed the palladium ions from the external vesicle membrane and that metal cations diffused through the membrane walls of polymerized lipid vesicles (Figure 2). In figure 2a, transmission electron micrographs of gold particles (4-15 nm) formed inside polymerized vesicles containing 50% **2** are shown. HRTEM of the same sample showed the lattice fringes which correspond to Au (111). A comparison of the electron diffraction patterns of gold particles formed inside polymerized vesicles containing 10% and 50% **2** revealed that a mixture of gold and impurities problably due to unreacted precursors were present in the vesicles containing 10% **2** but only gold was present in the vesicles containing 50% **2**. Also, HRTEM revealed that nucleation of gold occurred both from single nucleation (resulting in the formation of single crystals) and multiple nucleation sites for different vesicles.

Figure 2b reveals transmission electron micrographs of a mixture of cobalt and cobalt hydroxide polycrystalline particles (30-100 nm) produced inside vesicles, as determined by electron diffraction, containing 50% of **1** (w/w) after bath sonication for 2.5 hrs (temperature increased from 23 °C to 50 °C during this time period). Extensive TEM analysis indicated that the vesicles remained intact during bath sonication but the possibility that some vesicles did shatter can not be ruled out. The deposited metal particle appears to have grown into the vesicle membrane as well as out from it. As was the case with the formation of gold nanoparticles, selected area electron diffraction performed on individual vesicles showed many vesicles contained single crystals while others had disordered fringes surrounding the whole inner surface of the membrane of some vesicles suggesting the presence of multiple nucleation sites. HRTEM lattice imaging on these crystals showed some vesicles with materials deposited on the inner surface of the membrane, while the center remained hollow (Figure 3a). HRTEM of the gold particles formed inside polymerized vesicles also demonstrated that particle nucleation and growth proceeded out from the inner membrane (Figure 3b). A possible suggestion of the observed particle size distribution is that nucleation of these particles inside the membrane did not occur at the same time.

The formation of Co and $Co(OH)_x$ inside the vesicles indicates that two

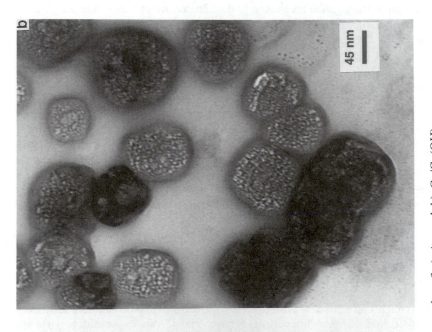

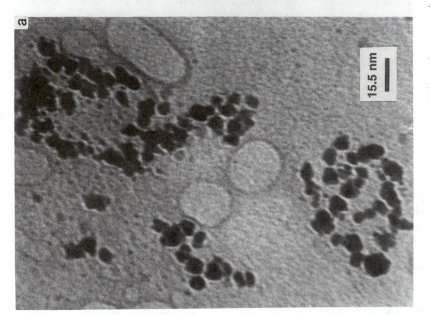

Figure 2. Transmission electron micrographs of a) Au and b) $Co/Co(OH)_x$ nanoparticles formed in polymerized vesicles.

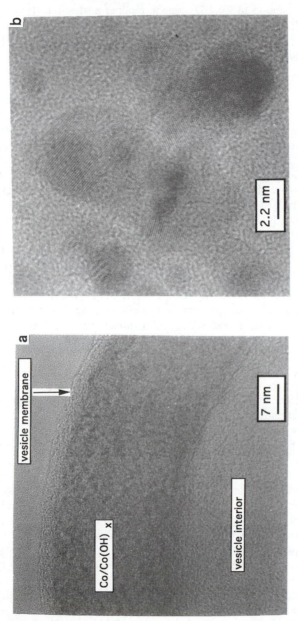

Figure 3. High resolution transmission electron micrographs of a) $Co/Co(OH)_x$ and b) Au nanoparticles.

competing processes are occurring. One is the nucleation and growth of Co particles on the internal membrane surface and the other process is the intravesicular precipitation of $Co(OH)_x$ due to random nucleation. It is interesting to note that $Co(OH)_x$ particle formation occurred only inside the vesicles and not in the surrounding aqueous environment. This suggests that the particles are forming as a result of a combination of the alkaline reaction conditions and supersaturation of $CoCl_2$ inside the vesicles. Because of the relative ease of formation of $Co(OH)_x$ inside the vesicles, this seemed like a good system to use to investigate the effect of $CoCl_2$ concentration on the crystallinity of the intravesicularly precipitated $Co(OH)_x$. The experiments were carried out as described for the formation of $Co/Co(OH)_x$ except no reducing agent was added. TEM, HRTEM and electron diffraction of $Co(OH)_x$ particles formed inside polymerized vesicles formed from 1:1 mixtures of **1** and **2** revealed that there was indeed a concentration effect on particle formation and crystallinity. $Co(OH)_x$ particles formed inside the vesicles when the $[CoCl_2] = 100$ and 50 mM but no particles were formed when the $CoCl_2$ was lowered to 25 mM (Figure 4). A comparison of the HRTEM and electron diffraction patterns revealed that the particles formed when $[CoCl_2] = 50$ mM were larger, more well formed and more crystalline with longer range order than the particles formed when $[CoCl_2] = 100$ mM. At the higher $CoCl_2$, smaller, more randomly ordered particles were formed.

The second approach to the synthesis of unagglomerated metal nanoparticles involved the metallization of unpolymerized vesicles prepared from 9:1 and 1:1 mixtures of **1** and **2** (*43*). In this approach, the hope was that unpolymerized **1** would be able to reorganize and stabilize the metal particles formed from the nucleation of Pd ion bound **2** in the external vesicle membranes. As the synthesis progressed, the vesicles would break up and all of the Pd ion bound **2** would become available for nucleation. $Pd(NH_3)_4Cl_2$ was added to the bath during vesicle formation. The vesicles were dialyzed to remove excess palladium ions and a gold plating bath was added. TEM of initial experiments revealed that unagglomerated polycrystalline gold particles were formed (40-70 nm) and suggested that particle stabilization was achieved through reorganization of unreacted lipid into vesicles which absorbed to the surface of the gold particles to provide steric barriers to agglomeration. More detailed experiments in which the reaction conditions, ie. temperature, were more carefully controlled demonstrated that size controlled, unagglomerated, redispersable gold particles could be formed using this approach (*55*). Vesicles were formed from a 9:1 mixture of **1** and **2** in the presence of $Pd(NH_3)_4Cl_2$. After dialysis against water, the gold plating bath was added and the mixture was incubated at 25 °C for 3 hrs. After dialysis against water overnight to remove excess plating bath, the mixture was freeze-dried. The resulting powder was examined by X-ray diffraction, which revealed the formation of polycrystalline gold. TEM of the powder redispersed in water revealed that a size controlled population of unagglomerated gold particles (20-28 nm, average crystallite size = 21 nm) was formed (Figure 5a). Examination of the particles at higher magnification revealed that discontinuities or gaps in the individual particles were present (Figure 5b). These discontinuities are problably due to the presence of nonmetallic material which suggests that the particles are composites of phospholipid and gold. If this were so, then there could be lipid material extending

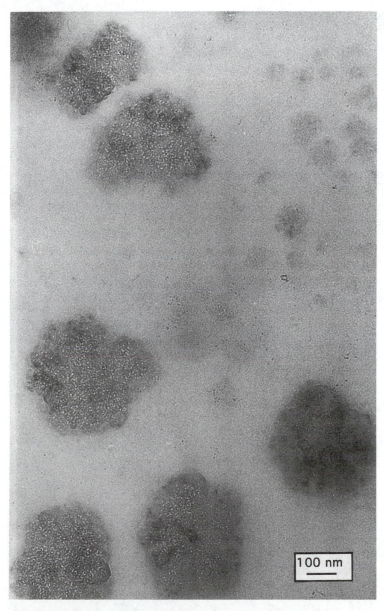

Figure 4. Transmission electron micrographs of Co(OH)$_x$ particles formed by intravesicular precipation at [CoCl$_2$] = 100 mM.

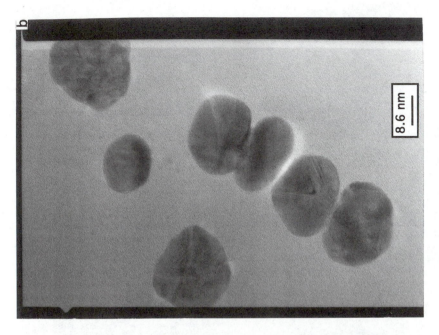

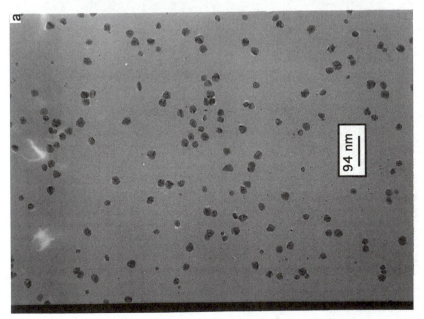

Figure 5. High resolution transmission electron micrographs of gold-phospholipid composite particles at a) low and b) high magnification.

from the surface of the particle providing steric stabilization when the particles are dispersed in water. Additional experiments have demonstrated that reaction temperature plays a key role in determining particle size distribution and extent of agglomeration. Particles formed at 40 °C were unagglomerated but had a wider size distribution while particles formed at 10 °C were agglomerated and did not redisperse. The exact nature of this temperature effect is not clear although it is probably due to changes in the stability, fluidity, and morphology of the vesicles.

Conclusion

Vesicles formed from mixtures of charge-neutral and palladium ion-bound negatively charged phospholipids have been successfully utilized in the synthesis of unagglomerated metal nanoparticles. The nucleation process is initiated at the membrane and single or multiple nucleation events may occur on a vesicle membrane surface. Nanoparticles formed using unpolymerized vesicle surfaces as templates remained unagglomerated after being freeze-dried and redispersed in water. The particles had a narrow size distribution and appeared to be composites of lipid and metal. The formation of gold and cobalt/cobalt hydroxide particles inside polymerized vesicles demonstrated that metal cations diffuse through the polymerized membrane affording the ability to use vesicles as reaction vials for a wider range of reactions than can be accomplished inside unpolymerized vesicles.

Acknowledgments

This research was funded by ONR through NRL programs on synthetic membranes and nanocomposites.

Literature Cited

1.	Bangham, A. D. *Progess Biophys. Mol. Biol.* **1968**, *18*, 31-95.
2.	Bangham, A. D.; Hill, M. W.; Miller, N. G. A. in *Methods in Membrane Biology, Vol. 1*; Korn, E. D., Ed.; Plenum Press: New York, 1985, 1.
3.	*Physics of Amphiphiles: Micelles, Vesicles and Microemulsions*; Degiorgio, V.; Corti, M., Eds.; North Holland: Amsterdam, 1985.
4.	*Micelles, Membranes, Microemulsions, and Monolayers*; Gelbart, W. M.; Avinoam, B-.S.; Didier, R., Eds.;Springer-Verlag:New York, 1994.
5.	*Phospholipid Bilayers*; Cevc, G.; Marsh, D.; John Wiley and Sons: New York, 1987.
6.	Okahata, Y; Kunitake, T. *J. Am. Chem. Soc.* **1979**, *101*, 5231.
7.	Kushwaha, S. C.; Kates, M.; Sprott, G. D.; Smith, I. C. P. *Science* **1981**, *211*, 1163.
8.	Langworthy, T. A. *Curr. Top. Membr. Transp.* **1982**, *17*, 45.
9.	Furhop, J. -H.; Fritsch, D. *Acc. Chem. Res.* **1986**, *19*, 130.
10.	Yager, P.; Schoen, P. E. *Mol. Cryst. Liq. Cryst.* **1984**, *106*, 371.
11.	M. Markowitz and A. Singh *Langmuir*, **1991**, *7*, 16-18.

12.   Schnur, J. M., *Science*, **1993**, *262*, 1669.
13.   J. M. Seddon *Biochim. Biophys. Acta*, **1990**, *1031*, 1.
14.   Chow, G. M.; Markowitz, M. A.; Singh, A. *J. Min. Met. and Mater.* **1992**, *45*, 62.
15.   Heuer, A.; Fink, D. J.; Laraia, V. J.; Arias, J. L.; Calvert, P. D.; Kendall, K.; Messing, G. L.; Blackwell, J; Rieke, P. C.; Thompson, D. H.; Wheeler, A. P.; Veis, A.; Caplan, A. I. *Science*, **1992**, *255*, 1098.
16.   Alper, M. *MRS Bulletin*, **1992**, *17(11)*, 53.
17.   Singh, A.; Schnur, J. M. in *Phospholipids Handbook*; Cevc, G. Ed.; Marcel Dekker: New York, 1993, pp 233-291.
18.   *Biotechnological Applications of Lipid Microstructures*, Gaber, B. P.; Schnur, J. M.; Chapman, D., Eds.; Plenum Press: New York, 1988.
19.   Bhatia, S. K.; Cooney, M. J., Shriver-Lake, L. C.; Fare, T. L.; Ligler, F. L. *Sensors and Actuators* **1991**, *B 3*, 311.
20.   Hua, J. D.; Liu, L.; Qiu, Y-M. *J. Macromol. Sci - Chem* **1989**, *A26*, 45.
21.   Singh, A.; Markowitz, M. A.; Tsao, L.-I.; Deschamps, J. *Polymeric Materials in Diagnostics and Biosensors*, Usmani, A. Ed.; American Chemical Society: Washington D. C., 1994; p. 252.
22.   Farmer, M. C.; Gaber, B. P. *Methods Enzymol.* **1987**, *14*, 184.
23.   Rudolph, A. S. *Cryobiology* **1988**, *25*, 277.
24.   Schnur, J. M.; Price, R.; Rudolph, A. S. *J. Controlled Release* **1994**, *28*, 3-13.
25.   Singh, A.; Markowitz, M. A. *New J. Chem.* **1994**, *18*, 377.
26.   Chow, G. M.; Stockton, W. B.; Price, R.; Baral, S.; Ting, A. C.; Ratna, B. R.; Schoen, P. E.; Schnur, J. M.; Bergeron, G. L.; Czarnaski, M. A.; Hickman, J. J. ; Kirkpatrick, D. A. *Materials Science and Engineering* **12**, *A158*, 1.
27.   Hayashi, C. *J. Vac. Sci. Technol. A*, **1987**, *5*, 1375.
28.   *Research Opportunities for Materials with Ultrafine Microstructures*, **1989**, National Academy Press: Washington, DC.
29.   Ichinose, N.; Ozaki, Y.; Kashu, S. *Superfine Particle Technology*, **1992**, Springer-Verlag: London.
30.   Gleiter, H. *Nanostructured Materials*, **1992**, *1*, 1.
31.   Kortan, A. R.; Hull, R.; Opila, R. L.; Bawendi, M. G.; Steigerwald, M. L.; Carroll, P. J.; Brus, L. E. *J. Am. Chem. Soc.* **1990**, *112*, 1327.
32.   Gobe, M.; Kon-No, K.; Kandori, K.; Kitahara, A. *J. Colloid Interface. Sci.* **1983**, *93*, 293.
33.   Barnickel, P.; Wokaun, A.; Sager, W.; Eicke, H. F. *J. Colloid Interface Sci.* **1992**, *148*, 80.
34.   Mann, S.; Williams, R. J. P. *J. Chem. Soc. Dalton Trans.* **1983**, 311.
35.   Mann, S.; Hannington, J. P.; Williams, R. J. P. *Nature* **1986**, *324*, 565.
36.   Bhandarkar, S.; Yaacob, I.; Bose, A. *J. Colloid Interface. Sci.* **1990**, *135*, 531.
37.   Bhandarkar, S.; Yaacob, I.; Bose, A. *Mat. Res. Soc. Symp. Proc.* **1990**, *180*, 637.
38.   Liu, H.; Graff, G. L.; Hyde, H.; Sarikaya, M.; Aksay, I. A. *Mat. Res. Soc. Symp. Proc.* **1991**, *218*, 115.

39.   Heywood, B. R.; Fendler, J. H.; Mann, S. *J. Colloid Interface. Sci.* **1990**, *138*, 295.
40.   Meldrum, F. C.; Heywood, B. R.; Mann, S. *J. Colloid Interface. Sci.* **1993**, *161*, 66.
41.   Michaelson, D. M.; Horwitz, A. F.; Klein, M. P. *Biochemistry*, **1973**, *12*, 2637.
42.   Markowitz, M. A.; Chow, G. M.; Singh, A. *Langmuir*, **1994**, *10(11)*, 4095-4102.
43.   Singh, A.; Markowitz, M. A.; Chow, G. M. *Nanostructured Materials* **1995**, *5(2)*, 141.
44.   Singh, A. *J. Lipid Res.*, **1990**, *31*, 1522.
45.   Singh, A.; Schnur, J. M.; *Syn. Commun.*, **1986**, *16*, 847
46.   Singh, A.; Markowitz, M. A.; Tsao, Li-I *Syn. Commun.*, **1992**, *22*, 2293.
47.   Markowitz, M. A.; Baral, S.; Brandow, S.; Singh, A. *Thin Solid Films* **1993**, *224*,   242.
48.   Regen, S. L.; Kirszenstejn, P.; Singh, A. *Macromolecules*, **1983**, *16*, 335.
49.   Stefely, J.; Markowitz, M. A.; Regen, S. L. *J. Am. Chem. Soc.* **1988**, *110*, 7463.
50.   Samuel, N. K. P.; Singh, M.; Yamaguchi, K.; Regen, S. L. *J. Am. Chem. Soc.* **1985**, *107*, 42.
51.   Dorn, K.; Patton, E. V.; Klingbiel, R. T.; O'Brien, D. F. *Makromol. Chem., Rapid Commun.* **1983**, *4*, 513.
52.   Bolikal, D.; Regen, S. L. *Macromolecules* **1984**, *17*, 1287.
53.   Patel, G. N.; Chance, R. R.; Witt, J. D. *J. Chem. Phys.* **1979**, *70*, 4387.
54.   Markowitz, M. A.; Chow, G. M.; Baral, S; Singh, A. in *Metallized Plastics IV*, Mittal, K. L. Ed.; Plenum Press: New York, in press.
55.   Chow, G. M.; Markowitz, M. A.; Singh, A., in preparation.

RECEIVED December 13, 1995

SEMICONDUCTORS, METALS, AND NANOCOMPOSITES

Chapter 12

Synthesis, Characterization, and Immobilization of Nanocrystalline Binary and Ternary III–V (13–15) Compound Semiconductors

L. I. Halaoui, S. S. Kher, M. S. Lube, S. R. Aubuchon, C. R. S. Hagan, R. L. Wells[1], and L. A. Coury, Jr.[1]

Department of Chemistry, Duke University, Box 90346, Durham, NC 27708–0346

Two synthetic routes to nanocrystalline III-V (13-15) materials are discussed. The first employs dehalosilylation reactions between Group III trihalides and $E(SiMe_3)_3$ (E = P, As) in hydrocarbon solvents affording nanocrystalline III-V semiconductors or their precursors. The second involves reactions of MX_3 (M = Ga, X = Cl, I; M = In, X = Cl, I) in glymes with *in situ* synthesized $(Na/K)_3E$ (E = P, As, Sb) in aromatic solvents, yielding nanocrystalline GaP, GaAs, GaSb, InP, InAs and InSb after refluxing reaction mixtures. Materials are characterized by TEM, XRD, Elemental Analysis, NMR, UV-vis, and STM. STM images of InAs give particle size distributions and confirm sample conductivity. Scanning tunneling spectroscopy shows a larger bandgap for nanocrystalline InAs than for InAs wafers, consistent with quantum confinement.

Much of the research interest in nanomaterials is attributable to the remarkably different properties displayed by these fascinating materials. For example, the prediction of size-dependent bandgaps for nanocrystalline semiconductors has excited speculation about their exploitation in novel optoelectronic and photoelectrochemical applications. Due to the relative ease with which they may be synthesized, most work to date has focused on metal nanoparticles and nanocrystalline II-VI (12-16) semiconductors. Despite the tremendous potential of unique properties and applications, however, nanocrystalline III-V (13-15) semiconductors remain largely unexplored. To this end, the work reported here details two different routes for the synthesis of nanocrystalline III-V materials, and discusses the characterization of these materials. In particular, it will be shown that stable, conductive, nanocrystalline materials with a reasonably narrow size distribution can be prepared which have a markedly different bandgap than commercial wafers of the bulk material.

---

[1]Corresponding authors

Dehalosilylation as a Route to III-V (13-15) Compound Semiconductors

Introduction. Dehalosilylation (or silyl halide elimination) has come to the fore as a viable synthetic technique in main group chemistry. In 1986, we first reported the use of dehalosilylation as a means to the formation of Ga-As bonds (*1*). Since then, researchers in our laboratory (*2-8*) as well as numerous other investigators (*9-15*) have applied this method, or adaptations thereof, in the preparation of compounds containing bonds between elements of Group III and Group V, as well as III-V semiconductor materials. Buhro and coworkers have applied dehalosilylation in the formation of ternary II-IV-V materials, (*16*) and Cowley and coworkers have recently synthesized BiP through a dehalosilylation route (*17*). Research in our laboratory has focused on the preparation of single-source precursors to binary and ternary III-V materials utilizing primarily the dehalosilylation method. However, Alivisatos and co-workers (*15*) observed that nanocrystalline GaAs was obtainable from the 1:1 mole ratio reaction of $GaCl_3$ and $As(SiMe_3)_3$ in solution, a reaction originally reported from our laboratories (*18-19*). Therefore, closer examination of the materials derived from the thermolyses of our III-V precursors was warranted.

Binary III-V Investigations: Precursors and Nanocrystalline Materials. We have reported the syntheses and characterization of a new class of dimeric species containing four-membered ring cores of alternating gallium and phosphorus atoms, with all exocyclic ligands on the metal centers being halogens; *viz.*, $[X_2GaP(SiMe_3)_2]_2$ (X = Cl, Br, I) (*7-8*). These dimers were prepared from the 1:1 mole ratio reaction of GaX_3 with $P(SiMe_3)_3$, resulting in the elimination of one molar equivalent of Me_3SiX to yield the dimeric complex. Subsequent thermolysis of these dimers at 400 °C resulted in the elimination of the remaining Me_3SiX, yielding powders containing nanocrystalline GaP of 3 nm average domain size (*8*).

Also reported were novel precursors of formula $(Ga_2ECl_3)_n$ (E = P, As) which result from the separate 2:1 mole ratio reactions of $GaCl_3$ with either $P(SiMe_3)_3$ (*7*) or $As(SiMe_3)_3$ (*20*). These powders undergo $GaCl_3$ elimination at temperatures > 300 °C to produce nanocrystalline GaP (*7*) or GaAs (*21*) (domain size *ca.* 3 nm). Structural data on the precursors are unavailable, however, as both have been found to be highly insoluble in hydrocarbon solvents.

Initial research on the reactions of $InCl_3$ with $As(SiMe_3)_3$ showed the 1:1 mole ratio reaction to proceed directly to crystalline InAs, however neither the size of the crystallites nor the effect of using a different indium(III) halide was determined (*18-19*). Subsequent research to investigate these points indicated that all 1:1 mole ratio reactions of InX_3 (X = Cl, Br, I) produce a black or brown-black powder, which upon annealing at 400 °C gives nanocrystalline InAs (characterized by XRD, XPS, TEM, and elemental analysis) with domain sizes ranging from 9-12 nm (*22*). When $InCl_3$ was allowed to react with $As(SiMe_3)_3$ in a 2:1 mole ratio reaction in an attempt to isolate a compound similar to the aforementioned $(Ga_2ECl_3)_n$, a red-brown powder resulted. This powder was not found to have the expected 3:2:1 ratio of Cl:In:As, however it did eliminate a yellow powder upon thermolysis (presumably $InCl_3$) at 400 °C to yield nanocrystalline InAs of 16 nm domain size (*22*).

Barron and coworkers originally investigated the 1:1 mole ratio reactions of InX_3 (X = Cl, Br, I) with $P(SiMe_3)_3$, yielding insoluble powders which were identified by elemental analysis to be oligomeric species of formula $[X_2InP(SiMe_3)_2]_n$ (14,23). Upon thermolysis, these powders were found to decompose to crystalline InP, however particle sizes of these samples were not reported. Further investigation of these reactions in our laboratories revealed that these powders progressed to nanocrystalline InP upon thermolysis at 400 °C, with domain sizes around 3 nm (22). Also, the 1:1 mole ratio reaction of InI_3 with $P(SiMe_3)_3$ was found to yield the 1:1 Lewis acid-base adduct $I_3In \cdot P(SiMe_3)_3$. This compound was found by TGA to eliminate three molar equivalents of Me_3SiI to yield nanocrystalline InP (identified by XRD, XPS, TEM, and elemental analysis), of domain size 2 nm (22).

Ternary III-V Investigations. The focus in our laboratories has recently been expanded to include investigations into ternary III-V compounds and materials. Dehalosilylation has proven an effective pathway to both precursors and materials with ternary formulations, whether the target compound contains two different group III elements and a pnicogen, or one group III element and two pnicogens.

As mentioned earlier, the 2:1 reaction of $GaCl_3$ with $E(SiMe_3)_3$ produced an oligomeric precursor compound $(Ga_2ECl_3)_n$ which upon thermolysis eliminated $GaCl_3$ to produce nanocrystalline GaE (E = P, As) (8, 20-21). Since either pnicogen can be used to synthesize a compound of formula $(Ga_2ECl_3)_n$, it seemed possible that a mixture of $P(SiMe_3)_3$ and $As(SiMe_3)_3$ could react with $GaCl_3$ in a 2:1 metal:pnicogen ratio to produce a similar mixed-pnicogen precursor of formula $[Ga_2(P/As)Cl_3]_n$. An off-white, insoluble powder was isolated from such a reaction and identified by elemental analyses as this mixed-pnicogen oligomer. Thermolysis of $[Ga_2(P/As)Cl_3]_n$ at 400 °C resulted in elimination of $GaCl_3$ and subsequent formation of a dark brown powder. This powder was confirmed by XRD, XPS, and elemental analyses to be the ternary III-V semiconductor $GaAs_xP_y$ ($0.6 \leq x,y \leq 0.9$) (21). Furthermore, X-ray powder diffraction studies of the $GaAs_xP_y$ showed the powder to be nanocrystalline, with domain size of *ca.* 3 nm. The reflections observed in this pattern fall between those expected for GaAs (24) and GaP (25), which would be expected according to Vegard's Law (26-27), further confirming the identity of this mixed-pnicogen semiconductor.

Based on the success of the 2:1 mole ratio metal:pnicogen mixed-pnicogen reaction described above, several 1:1 mole ratio metal:pnicogen preparations were investigated in order to synthesize mixed-pnicogen ring complexes or to develop routes to mixed-metal or mixed-pnicogen ternary materials. A Ga-As-Ga-P ring compound,

$\overline{I_2GaAs(SiMe_3)_2Ga(I)_2}P(SiMe_3)_2$, had previously been synthesized in our laboratories through equilibration of its constituent dimeric complexes $[I_2GaE(SiMe_3)_2]_2$ (E = P, As) (28) and seemed to be a good candidate for synthesis by a more direct method. To this end, GaI_3, $As(SiMe_3)_3$, and $P(SiMe_3)_3$ were allowed to react in solution in a 2:1:1 mole ratio to produce a yellow powder which was fully characterized as being

$\overline{I_2GaAs(SiMe_3)_2Ga(I)_2}P(SiMe_3)_2$ (29). This compound was thermolyzed at 400 °C, and observed both in bulk decomposition and TGA studies to eliminate four molar equivalents of Me_3SiI to produce $GaAs_xP_y$ as a brown-black powder. XRD studies of this powder (Figure 1A) displayed the (111) peak between the expected values for GaAs (*24*) and GaP (*25*) (Table I), which by Vegard's Law (*26-27*) is evidence for the presence of $GaAs_xP_y$ in the sample. However, due to the small particle size of the crystallites (domain size *ca.* 1 nm) the (220) and (311) peaks were broadened such that they could not be easily identified. Thus, an additional sample of the cyclic precursor was heated at 450 °C for 12 hours. The resulting brown powder was shown to be $GaAs_xP_y$ of 2.4 nm particle size, with the three major peaks in the XRD being readily identifiable (see Table I). Elemental analysis of the $GaAs_xP_y$ powder obtained at 400 °C showed x = 0.65 and y = 0.52, with significant contamination by C, H, and I.

Table I. Comparison of Prepared Ternary III-V Materials with JCPDS Files for Binary III-V Materials (d-spacings in angstroms).

	(111)	(220)	(311)
GaAsP			
GaAs standard (*24*)	3.26	2.00	1.70
GaAsP sample (400 °C)	3.24	N/A[a]	N/A[a]
GaAsP sample (450 °C)	3.21	1.97	1.70
GaP standard (*25*)	3.14	1.92	1.64
GaInP			
GaP standard (*25*)	3.14	1.92	1.64
GaInP sample	3.18	(1.98)[b]	(1.68)[b]
InP standard (*30*)	3.39	2.08	1.77
InAsP			
InP standard (*30*)	3.39	2.08	1.77
InAsP sample	3.46	2.12	1.81
InAs standard (*31*)	3.50	2.14	1.83

[a]   Data inconclusive: line-broadening due to small particle size obscured these peaks.

[b]   Line broadening due to small particle size results in poor signal-to-noise ratio; values obtained from a compressed spectrum.

Similar direct preparations using different metal/pnicogen combinations have resulted in insoluble powders which were found to decompose to a ternary material. The reaction of a solution-phase mixture of $GaCl_3$ and $InCl_3$ with two molar equivalents of $P(SiMe_3)_3$ resulted in a light yellow powder with a Ga:In:P ratio of 1.04:1.00:1.05, however no crystalline sample suitable for single-crystal X-ray analysis could be obtained, nor could a compound be identified from these data.

Thermolysis of this powder at 400 °C yielded a brown powder with a Ga:In:P ratio of 2.69:1.00:4.16, and significant C, H, and Cl contamination. An XRD pattern of this sample (Figure 1B) showed it to be nanocrystalline (domain size *ca.* 1 nm), with (111), (220) and (311) reflections located between those expected for GaP (*25*) and InP (*30*) (Table I), indicative of the presence of a ternary GaInP mixed-metal semiconductor in the powder sample.

A similar one-pot synthesis was also conducted in an attempt to form the ternary mixed-pnicogen semiconductor InAsP. Two molar equivalents of $InCl_3$ were allowed to react in solution with a mixture of one molar equivalent each of $As(SiMe_3)_3$ and $P(SiMe_3)_3$, yielding a brown powder with an In:As:P ratio of 3.71:1.85:1.00. Once again, no crystalline sample could be obtained from this powder. Subsequent thermolysis of this sample at 400 °C yielded a lustrous black powder with a In:As:P ratio of 2.38:1.89:1.00. The XRD pattern of this sample (Figure 1C) also showed it to be nanocrystalline (domain size *ca.* 9 nm), with (111), (220) and (311) reflections located between those expected for InP (*30*) and InAs (*31*) (Table I), again indicative of the presence of a ternary semiconductor, InAsP. A high-resolution TEM image of this sample (Figure 2) shows lattice planes for several nanocrystalline InAsP particles ranging in size from 6 to 15 nm.

Although the current results from preparations of ternary materials utilizing the silyl cleavage method are preliminary, the aforementioned data has encouraged further investigation into synthesizing precursors to III-V ternary and quaternary materials using this versatile reaction pathway.

Synthesis of III-V Semiconductor Nanocrystals by Solution Phase Metathesis

Introduction.   We have recently published a straightforward new method for preparing nanocrystalline III-V semiconductors (*32-34*). This method utilizes *in situ* reactions of Group III halides in chelating solvents with $(Na/K)_3E$ (E = P, As, Sb) in aromatic solvents. Semiconductor nanocrystallites with average particle size of 4-35 nm can be prepared using this method and GaP (diameter = 11 nm), GaAs (10 nm), GaSb (35 nm), InP (4 nm), InAs (11 nm) and InSb (26 nm) have each been obtained. The particle sizes of the semiconductors depend on the nature of Group III halide, nature of the solvent, concentration and chain length of the glyme solvents used. When $GaCl_3$ was dissolved in various solvents and subsequently reacted with $(Na/K)_3As$, synthesized *in situ* in refluxing toluene, different average particle sizes of GaAs were obtained: toluene (36 nm), dioxane (36 nm), monoglyme (17 nm) and diglyme (10 nm). The chelating nature of multi-dentate glyme solvents seem to play a crucial role in limiting the growth of GaAs crystallites beyond a certain size. It was also observed that dimeric Group III halides ($GaCl_3$, GaI_3 and InI_3) gave final products with much smaller particle sizes. Oligomeric $InBr_3$ and $InCl_3$, on the other hand, gave nanocrystallites with larger particle size.

Characterization and Surface Chemistry.   Thus obtained quantum crystallites have been characterized by various techniques. Figure 3 shows a high resolution transmission electron micrograph (HRTEM) of GaP nanocrystallites. Numerous lattice fringes

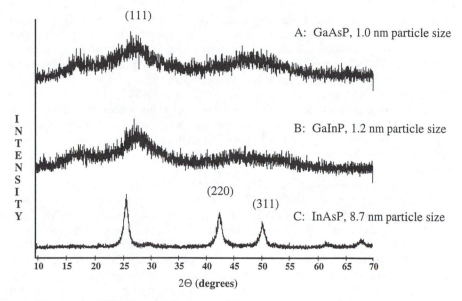

Figure 1. XRD Powder Patterns of Ternary III-V Materials Prepared by Dehalosilylation.

Figure 2. High-Resolution TEM of InAsP Nanocrystals.

originating from 3-12 nm crystallites are observed in the figure. The XRD pattern indicated that the average particle size of this GaP sample was 11 nm (*32*). Our earlier reports (*32-34*) dealt with particles from which excess Group V element was sublimed away. Currently we have focused on nanocrystallites in the as-prepared state, *i.e.*, obtained by simply refluxing the reaction mixture. The as-prepared GaAs nanocrystallites have been investigated in great detail (*35*). These materials are quite interesting since they are capped and can form remarkably stable colloidal suspensions without requiring any surfactants. For example, upon repeated extractions of as-synthesized GaAs with methanol, grey colloidal suspensions are formed which have been stable for more than 16 months despite repeated exposure to atmosphere and light.

These capped GaAs nanoclusters present in the colloid have been characterized by XRD, multi-nuclear NMR, HRTEM, XPS, FT-IR photoacoustic spectroscopy (PAS), Elemental Analysis, UV-Vis and atomic force microscopy (AFM) (*35*). FT-IR PAS, NMR and XPS analysis showed that the GaAs particles were capped by methanol used during the extractions, and no other impurities were detected. FT-IR PAS and NMR indicated that methanol and the residual water in the methanol were molecularly bound to the nanocrystal surface, and features assignable to dissociative binding of methanol and water were not observed. The hydrogen bonding between surface-bound and free solvent molecules in the colloid is a likely cause of the remarkable stability of these GaAs suspensions.

The average crystallite size of the particles obtained by evaporating methanol from the colloid was 5 nm. HRTEM of the solids from the grey colloid showed lattice planes due to 3-11 nm particles, although majority of the particles were 4-8 nm as evident from the fringe patterns. Figure 4 shows HRTEM images of GaAs quantum dots in the grey colloid. The micrograph shows several crystallites clustered together due to solvent evaporation from the colloid; however, due to methanol capping, the nanocrystals exist in the colloid in an isolated state as observed in AFM studies (*36*). Centrifugation of the grey colloid at 1315 G force for 30 min resulted in settling of larger crystallites and a reddish-orange colloid was obtained. HRTEM of the solids in the reddish-orange colloid showed that it mostly contained crystallites smaller than 2 nm, and larger crystallites such as those seen in Fig. 4 were not present in this colloid.

The XRD pattern of the particles in the reddish-orange colloid was inconclusive as it showed two broad humps. Crystallites smaller than ~3 nm do not yield conclusive diffraction pattern and appear to be "XRD amorphous" (*37-38*). Previously we have reported lattice fringe patterns from crystallites as small as 1 nm (*32*). The UV-Vis spectrum of the reddish-orange colloid showed rise in absorption at ~510 nm. The ^{71}Ga NMR of this colloid showed that it contained GaAs (*39*). When several drops of reddish-orange colloid were placed on a glass slide and the solvent was allowed to evaporate, orange GaAs particles were obtained. The as-prepared grey colloidal suspensions contain GaAs nanocrystals with a wide particle size distribution. Fischer, *et al.*, have recently used size exclusion and hydrodynamic chromatography techniques to separate nanocrystalline particles by size (*40*) and these techniques will be explored in future studies to obtain monodisperse crystallites and probe their properties.

Figure 3. High-Resolution TEM of GaP Nanocrystals. The bar in the figure indicates a scale of 6 nm.

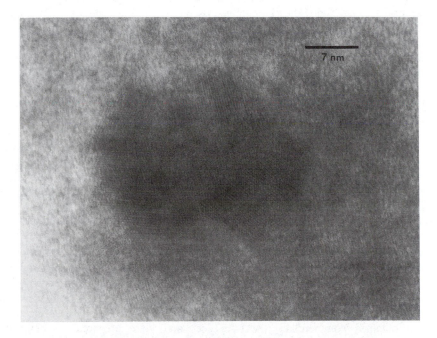

Figure 4. High-Resolution TEM of GaAs quantum dots obtained from evaporating solvent from the grey colloidal suspension. The image shows several crystallites clustered together due to solvent evaporation. The bar indicates a scale of 7 nm.

Characterization of Nanocrystalline InAs Prepared by Solution Phase Metathesis and Dehalosilylation Reactions Using Scanning Tunneling Microscopy (STM) and Spectroscopy (STS)

Background Information. Nanocrystalline semiconductor materials are increasingly being suggested as possible components for new electro-optical devices (*41*). The number of experimental reports in which electrical characterization has been attempted on these systems, however, is quite limited (*42*). Aside from estimates of band gaps from absorbance measurements (*15,43,44*), very little data have appeared to date for materials other than II-VI compounds (*45*). Because of the relative purity and stability of our III-V materials (*8,18-22*), however, such investigations are now feasible. Preliminary data from scanning tunneling microscopy (STM) and scanning tunneling spectroscopy (STS) experiments are thus reported below for nanocrystalline InAs samples.

STM and STS are two members of a family of characterization techniques capable of providing information about conductive samples with nanometer resolution (*46*). In STM, an extremely sharp electrode called a "tip" is positioned within a few nanometers of a conductive sample. A small dc-potential (50 - 100 mV) called the "bias voltage" is applied between the tip and surface, which induces a tunneling current. An image may then be obtained by moving the tip in the x-y plane above the sample by plotting the fluctuations in tunneling current as a function of position ("constant height mode"). Alternatively, the value of the tunneling current can be fixed via electronic feedback, causing the tip to move in the z-direction at each (x,y) position to re-establish the desired current. This method is referred to as the "constant current mode," and was employed for the studies reported below. The images shown thus represent the tip displacement (z-coordinate) for each location in the x-y plane.

The STS technique is useful for probing the electronic properties of a sample (*45*). In the implementation used here, the tip is held over a single (or possibly several) semiconductor particle(s) while the bias voltage is swept over a range of potentials. The tunneling current is then monitored as a function of the bias voltage value to give information about the location of the semiconductor band edges. For example, the region along the potential axis where very little tunneling current is monitored can be used to estimate the band gap of the sample.

Characterization Methods. Measurements were made with a Digital Instruments NanoScope II using Pt-Ir or electrochemically-etched W tips (*42*). Diglyme-capped, nanocrystalline InAs particles prepared using the solution metathesis reaction, as well as InAs from the dehalosilylation synthetic route, were deposited from sonicated, methanol suspensions onto polycrystalline Pt or Au electrodes. These electrodes were obtained commercially (AAI-AbTech) and were prepared by magnetron sputtering onto borosilicate glass. STM images were obtained in air using the constant current mode of the instrument, typically with bias voltages in the range 50-100 mV at a set-point current of 1.2 nA with 400 samples per scan.

STS measurements were made on InAs samples using silicone oil as a bathing fluid. Various other liquids were examined as alternatives (*e.g.,* methanol, diglyme) but only silicone oil allowed for STS scans without dislodging the particles. The typical voltage range used was ± 1.1 to 1.4 V to maintain tunneling currents during scanning to less than ± 50 nA. The experiment was implemented by positioning the tip using a 0.4 V bias at a set-point current of 1.5 nA. Comparison data were obtained for freshly etched, single crystal Zn-doped *p*-InAs wafers, which had been passivated by treatment with a sodium sulfide solution. Evaporated films of Au (1350 Å) over Zn (153 Å) provided ohmic back contact to the wafers for these investigations.

Representative Data. Figure 5 shows an STM image of InAs particles deposited onto a Pt electrode. (For comparison, an image of the featureless, bare Pt surface is shown in Figure 6.) The particles appear fairly uniform in size, and are surprisingly evenly dispersed across the surface. The lack of aggregation seen may, in fact, be due to the diglyme capping agents, which also allow for the particles to form methanol suspensions which are stable for months.

The fact that an image was obtainable for nanocrystalline InAs suggests that it is sufficiently conductive to support a substantial tunneling current in the as-prepared state (*i.e.,* without deliberate doping). We have routinely been able to obtain STM images for the nanocrystalline III-V materials we have prepared by both the solution metathesis and dehalosilylation synthetic routes (*vide supra*). This includes data for nanocrystalline GaAs, a material with a substantially larger bandgap and lower intrinsic conductivity (*47*). However, in the case of nanocrystalline GaAs, a much larger bias voltage is needed to obtain an image (*circa* 2.0 V compared to 52 mV for this image of InAs), consistent with the different electronic properties of GaAs.

By taking sequential cross-sections of images such as the one shown in Figure 5, it is possible to obtain a particle size distribution for the nanocrystals. Figure 7 shows one such distribution, in this case revealing an arithmetic mean diameter of 14 nm with a sample standard deviation of 5 nm for a sample comprised of 124 particles. The best-fit gaussian function for the histogram shown was:

$$y = (-0.933) + (37.1)e^{\left(-0.5\left(\frac{x-15.2}{5.72}\right)^2\right)}$$

giving a coefficient of determination (R^2) of 0.981, and a centroid of 15.2 nm. Since the exciton diameter for InAs is estimated to be 62.5 nm (*22*), this particular sample should exhibit quantum confinement effects (*e.g.*, a larger band gap than bulk InAs).

STS experiments were subsequently performed to investigate this possibility. Representative results are shown in Figure 8, comparing data for nanocrystalline InAs with that for a wafer of single crystal *p*-InAs. As is evident, the "zero-current" region is noticeably larger for the InAs particles, demonstrating a larger band gap. It is important to note that this measurement probes the band gap of one (or a few) particle(s). This is in contrast to techniques such as photoluminescence,

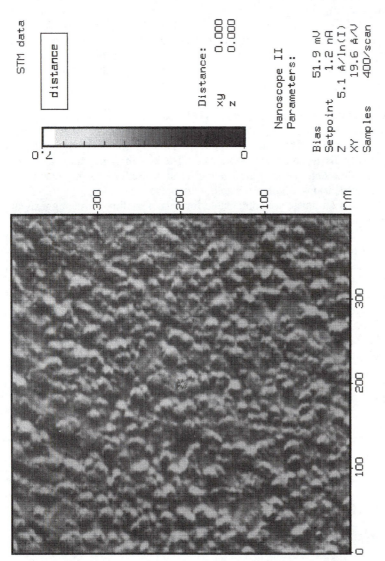

STM data

distance

7.0

0

Distance:
xy   0.000
z   0.000

Nanoscope II
Parameters:

Bias   51.9 mV
Setpoint   1.2 nA
Z   5.1 Å/ln(I)
XY   19.6 Å/V
Samples 400/scan

300

200

100

nm

300

200

100

0

Data taken Wed Jun 08 14:34:27 1994
Buffer 2(PTINAS.11(F)), Rotated 0°, XY axes [nm], Z axis [nm]

Figure 5. STM of InAs Nanocrystals deposited on Pt substrate. *Panel A:* top view.

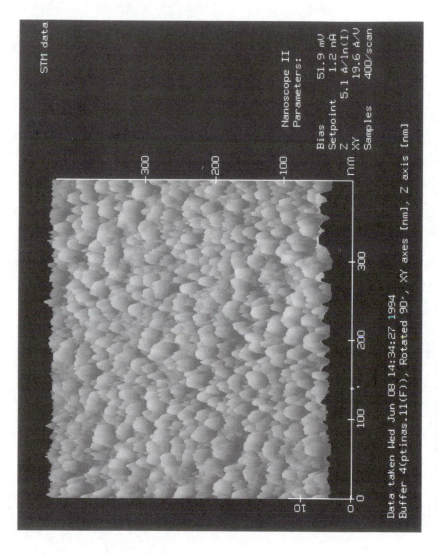

Figure 5. STM of InAs Nanocrystals deposited on Pt substrate. *Panel B*: side view.

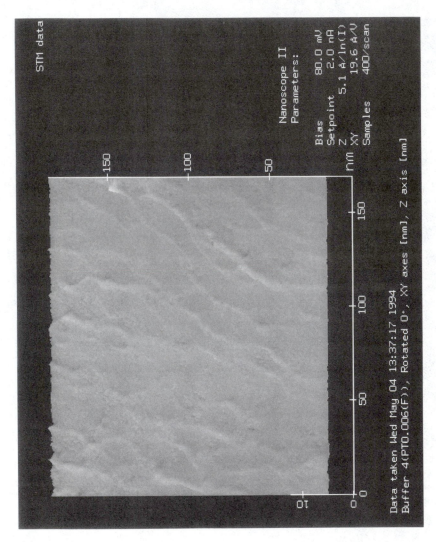

Figure 6. STM of bare Pt substrate electrode.

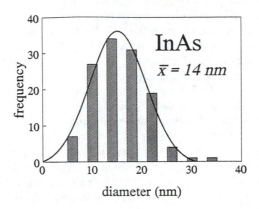

Figure 7. Particle size distribution obtained from cross-sections of STM image.

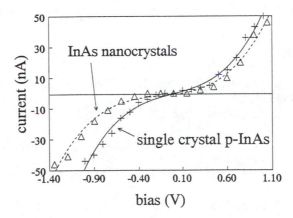

Figure 8. STS plots comparing data for nanocrystalline InAs (Δ) with that for a wafer of single-crystal *p*-InAs (+).

XRD, BET or absorption spectroscopy which yield a composite (weighted-average) value. Furthermore, STS is not subject to the effects of photon scattering (42), since it is a "dark" or ground-state measurement.

To assess the precision of the STS experiments, replicate measurements were made for both the nanocrystalline material and the sulfide-passivated single crystal wafer. The mean band gap taken from 38 experiments (with each experiment representing the average of 40 current measurements at each voltage) on a sulfide-passivated p-InAs wafer was found to be 0.41 eV with a sample standard deviation of 0.08 eV. This agrees (within experimental error) with the literature value of 0.36 eV for non-passivated InAs. By contrast, 21 measurements on the nanocrystalline InAs material yielded a mean band gap value of 0.837 eV with a sample standard deviation of 0.126 eV. The greater amount of scatter in the latter data set is entirely expected, since a distribution in particle sizes (and hence band gaps) exists for the sample (cf., Figure 7). We have previously published the UV/vis absorption spectrum for a methanol suspension of nanocrystalline InAs prepared by the dehalosilylation method (22), and that spectrum showed an absorption edge which was severely blue-shifted ($\lambda = 322$ nm) relative to that expected for bulk InAs (3444 nm). Thus, these STS data are consistent with the spectroscopic results, and confirm the presence of quantum confinement effects in nanocrystalline InAs.

Acknowledgments

We are grateful for the generous support of the Air Force Office of Scientific Research, The Office of Naval Research, and The Lord Foundation of North Carolina.

Literature Cited

(1)   Pitt, C. G.; Purdy, A. P.; Higa, K. T.; Wells, R. L. *Organometallics* **1986**, *5*, 1266.

(2)   Wells, R. L. *Coord. Chem. Rev.* **1992**, *112*, 273, and references therein.

(3)   Wells, R. L.; Jones, L. J.; McPhail, A. T.; Alvanipour, A. *Organometallics* **1991**, *10*, 2345.

(4)   Wells, R. L.; McPhail, A. T.; Speer, T. M. *Organometallics* **1992**, *11*, 960.

(5)   Wells, R. L.; McPhail, A. T.; Jones, L. J.; Self, M. F. *Polyhedron* **1993**, *12*, 141.

(6)   Jones, L. J.; McPhail, A. T.; Wells, R. L. *Organometallics* **1994**, *13*, 3634.

(7)   Wells, R. L.; Self, M. F.; McPhail, A.T.; Aubuchon, S. R.; Woudenberg, R. C.; Jasinski, J. D. *Organometallics* **1993**, *12*, 2832.

(8)   Aubuchon, S. R.; McPhail, A. T.; Wells, R. L.; Giambra, J. A.; Bowser, J. B. *Chem. Mater.* **1994**, *6*, 82.

(9)   Uchida, H.; Matsunga, T.; Yoneyama, H.; Sakata, T.; Mori, H.;
        Sasaki, T. *Chem. Mater.* **1993**, *5*, 716.
(10)   Stuczynski, S. M.; Opila, R. L.; Marsh, P.; Brennan, J. G.;
        Steigerwald, M. L. *Chem. Mater.* **1991**, *3*, 379.
(11)   Douglas, T.; Theopold, K. H. *Inorg. Chem.* **1991**, *30*, 594
(12)   Butler, L.; Redmond, G.; Fitzmaurice, D. *J. Phys. Chem.* **1993**,
        97, 10750.
(13)   Mićić, O. I.; Curtis, C. J.; Jones, K. M.; Sprague, J. R.; Nozik,
        A. J. *J. Phys. Chem.* **1994**, *98*, 4966.
(14)   Healy, M. D.; Laibinis, P. E.; Stupik, P. D.; Barron, A. R. *J.
        Chem. Soc. Chem. Commun.* **1989**, 359.
(15)   Olshavsky, M. A.; Goldstein, A. N.; Alivisatos, A. P. *J. Am.
        Chem. Soc.* **1990**, *112*, 9438.
(16)   Goel, S. C.; Buhro, W. E.; Adolphi, N. L.; Conradi, M. S. *J.
        Organomet. Chem.* **1993**, *449*, 9.
(17)   Carmalt, C. J.; Cowley, A. H.; Hector, A. L.; Norman, N. C.;
        Parkin, I. V. *J. Chem. Soc. Chem. Commun.* **1994**, 1987.
(18)   Wells, R. L.; Pitt, C. G.; McPhail, A. T.; Purdy, A. P.; Shafie-
        ezad, S.; Hallock, R. B. *Chem. Mater.* **1989**, *1*, 4.
(19)   Wells, R. L.; Pitt, C. G.; McPhail, A. T.; Purdy, A. P.; Shafie-
        ezad, S.; Hallock, R. B. *Mater. Res. Soc. Symp. Proc.* **1989**, *131*,
        45.
(20)   Wells, R. L.; Hallock, R. B.; McPhail, A. T.; Pitt, C. G.; Johan-
        sen, J. D. *Chem. Mater.* **1991**, *3*, 381.
(21)   Aubuchon, S. R.; Lube, M. S.; Wells, R. L. *Chem. Vap. Deposi-
        tion* **1995**, *1*, 28.
(22)   Wells, R. L.; Aubuchon, S. R.; Kher, S. S.; Lube, M. S.; White,
        P. S. *Chem. Mater.* **1995**, *7*, 793.
(23)   Healy, M. D.; Laibinis, P. E.; Stupik, P. D.; Barron, A. R.
        Mater. Res. Soc. Symp. Proc. **1989**, *131*, 83.
(24)   *Joint Committee on Powder Diffraction Standards (JCPDS)*, File No.
        14-450, GaAs.
(25)   *JCPDS*, File No. 12-191, GaP.
(26)   Cullity, B. D. *Elements of X-Ray Diffraction, 2nd edn.*; Addison-
        Wesley: Reading, MA, 1978; pp 375-377.
(27)   Wiley, J. B.; Kaner, R. B. *Science* **1992**, *255*, 1093.
(28)   Wells, R. L.; McPhail, A. T.; White, P. S.; Lube, M. S.; Jones,
        L. J. *Phosphorus, Sulfur, and Silicon* **1994**, *93-94*, 329.
(29)   Wells, R. L.; Aubuchon, S. R.; Lube, M. S. *Main Group Chem-
        istry* **1995**, *1*, 81.
(30)   *JCPDS*, File No. 13-232, InP.
(31)   *JCPDS*, File No. 15-869, InAs.
(32)   Kher, S. S.; Wells, R. L. *Chem. Mater.* **1994**, *6*, 2056.
(33)   Kher, S. S.; Wells, R. L. *Mater. Res. Soc. Symp. Proc.* **1994**, *351*,
        293.

(34)   Kher, S. S.; Wells, R. L. *U. S. Patent* Application No.
        08/189,232.
(35)   Kher, S. S.; Wells, R. L. *to be published.*
(36)   Falvo, M.; Superfine, R.; Kher, S. S.; Wells, R. L. *to be published.*
(37)   Herron, N.; Wang, Y.; Eckert, H. *J. Am. Chem. Soc.* **1990**, *112*,
        1322.
(38)   Murray, C. B.; Norris, D. J.; Bawendi, M. G. *J. Am. Chem. Soc.* **1993**,
        115, 8706.
(39)   Potter, L.; Wu, Y.; Kher, S. S.; Wells, R. L. *to be published.*
(40)   Fischer, Ch.-H.; Giersig, M. *J. Chromatogr. A* **1994**, *688*, 97, and
        references therein.
(41)   Meyer, G. J.; Searson, P. C. *Electrochem. Soc. Interface* **1993**, *2*,
        23.
(42)   Hagan, C. R. S.; Kher, S. S.; Halaoui, L. I.; Wells, R. L.; Coury,
        L. A., Jr. *Anal. Chem.* **1995**, *67*, 528.
(43)   Uchida, H.; Curtis, C. J.; Nozik, A. J. *J. Phys. Chem.* **1991**, *95*, 5382.
(44)   Mićić, O. I.; Sprague, J. R.; Curtis, C. J.; Jones, K. M.; Machol, J. L.;
        Nozik, A. J.; Giessen, H.; Fluegel, B.; Mohs, G.; Peyghambarian, N. *J. Phys. Chem.* **1995**, *99*, 7754.
(45)   Ogawa, S.; Fan, F.-R. F.; Bard, A. J. *J. Phys. Chem.* **1995**, *99*,
        11182.
(46)   Christmann, K. *Introduction to Surface Physical Chemistry*;
        Springer-Verlag: New York, 1991; pp 102-107.
(47)   Solymar, L.; Walsh, D. *Lectures on the Electrical Properties of
        Materials, 4th edn.*; Oxford: New York, 1990.

RECEIVED November 27, 1995

Chapter 13

Nanocomposites Containing Nanoclusters of Selected First-Row Transition Metal Phosphides

C. M. Lukehart[1], Stephen B. Milne[1], S. R. Stock[2], R. D. Shull[3], and James E. Wittig[4]

[1]Department of Chemistry, Vanderbilt University, Nashville, TN 37235
[2]School of Materials Science and Engineering, Georgia Institute of Technology, Atlanta, GA 30332
[3]Materials Science and Engineering Laboratory, National Institute of Standards and Technology, Gaithersburg, MD 20899
[4]Department of Applied and Engineering Sciences, Vanderbilt University, Nashville, TN 37235

Addition of the metal phosphine complexes, $Fe(CO)_4L$, $Co_2(CO)_6L'_2$, NiL_4, or trans-NiL'_2Cl_2, where L = $PPh_2(CH_2)_2Si(OMe)_3$ and L' = $PPh_2(CH_2)_2Si(OEt)_3$, to conventional silica sol-gel formulations gives apparent covalent incorporation of these complexes into the resulting silica xerogel matrix. Subsequent thermal treatment under a reducing atmosphere affords nanocomposite materials containing crystalline nanoclusters of Fe_2P, Co_2P, or Ni_2P, respectively, highly dispersed throughout the xerogel matrix. These results indicate that metal phosphine complexes can serve as single-source molecular precursors for nanoparticulate metal phosphide formation. The synthesis and characterization of these nanocomposites is described.

Metal phosphides are a well known class of compounds that are technologically important as semiconductors and as phosphorescent or electronic materials (1, 2). Conventional syntheses of metal phosphides has entailed direct reaction of the elements for prolonged periods at high temperature, reaction of phosphine with metals or metal oxides under similar conditions, reduction of metal phosphates with carbon at high temperature, electrolysis of molten phosphates, chemical vapor deposition, or by solid-state metathesis reactions (3). The metal phosphides, Cd_3P_2, $ZnGeP_2$, and $CdGeP_2$, have been prepared as powders from suitable molecular precursors (4). Frequently, such syntheses of transition metal phosphides as bulk materials using traditional methods give products of variable purity. Little is known about the synthesis and properties of transition metal phosphides as nano-scale particulates.

We have discovered a synthetic route for the preparation of nanocomposite materials in which nanoclusters of transition metal phosphides are formed highly dispersed throughout a silica xerogel host matrix. In this method, a precursor molecule containing (1) the desired metal, (2) at least one phosphine ligand, and (3) a bifunctional

ligand is synthesized. The bifunctional ligand should contain a -Si(OR)$_3$ group.
Addition of such precursors to a conventional silica sol-gel formulation affords a silica
xerogel host matrix into which the precursor molecules have been apparently covalently
incorporated. Subsequent thermal treatment of these molecularly doped xerogels under
appropriate conditions gives silica xerogels containing nanoparticulate metal phosphides.
    The preparation of metal phosphide nanocomposites of selected first-row transition
metals has been achieved. Using the iron, cobalt, or nickel precursors, Fe(CO)$_4$L,
Co$_2$(CO)$_6$L'$_2$, NiL'$_4$, or trans-NiL'$_2$Cl$_2$ [where L is PPh$_2$(CH$_2$)$_2$Si(OMe)$_3$ and L' is
PPh$_2$(CH$_2$)$_2$Si(OEt)$_3$], one can prepare nanocomposites containing, respectively,
nanoclusters of Fe$_2$P (syn-barringerite), Co$_2$P, or Ni$_2$P. The synthesis and
characterization of these nanocomposite materials are presented below.

Experimental

Synthesis of the molecular precursor complexes, Fe(CO)$_4$L, NiL'$_4$, trans-NiL'$_2$Cl$_2$, or
Co$_2$(CO)$_6$L'$_2$, and the associated bifunctional phosphine ligands, L or L', was
accomplished using published procedures or procedures reported for the preparation of
analogous complexes (5-9).
    Conventional sol-gel formulations and procedures were used in the syntheses of the
molecularly doped silica xerogels (10). These xerogels were formed at room
temperature using aqueous ammonia as a catalyst for hydrolysis of the host matrix
precursor, Si(OMe)$_4$, TMOS, and for the silyl groups of the bifunctional ligands.
Stoichiometries for the molar ratio of TMOS to molecular dopant complex were typically
10:1. Product xerogels were washed several times with alcohol prior to air drying.
Representative chemical microanalytical data for the xerogel containing the iron
precursor indicated elemental weight percentages of 4.20 for Fe and 27.38 for Si giving
a Si/Fe atomic ratio of ca. 13 : 1. Corresponding data for the xerogel doped with the
NiL'$_4$ precursor indicated elemental weight percentages of 2.58 for Ni and 19.94 for Si
giving a Si/Ni atomic ratio of ca. 16:1.
    Conversion of the molecularly doped xerogels to nanocomposite materials was
accomplished by placing the powdered xerogel into an alumina boat. Such samples
were then introduced into a quartz tube placed inside of a tube furnace. A hydrogen
atmosphere was established within the tube by passing a continuous flow of hydrogen
through the tube. Temperatures at the sample location were measured using an internal
thermocouple. The thermal conditions used for nanocomposite formation were the
following: 0.5 h at 900 C for the Fe-doped xerogel; 2 h at 800 C for the Co-doped
xerogel; and, 3 h at 750 C for the Ni-doped xerogel. These heating temperatures were
those determined to give nanoclusters of sufficient size and crystallinity to exhibit
electron and X-ray diffraction.
    Representative chemical microanalytical data for the resulting nanocomposite
obtained from the iron precursor indicated elemental weight percentages of 5.82 for Fe
and 41.33 for Si giving a Si/Fe atomic ratio of ca. 14 : 1. Corresponding data for the
nanocomposite obtained from the xerogel doped with the NiL'$_4$ precursor indicated
elemental weight percentages of 5.10 for Ni and 35.61 for Si giving a Si/Ni atomic ratio
of ca. 15:1. Characterization of these nanocomposites revealed nanocrystals of Fe$_2$P or
crystalline nanoclusters of Co$_2$P or of Ni$_2$P dispersed uniformly throughout a silica
xerogel host matrix for these respective products.
    Techniques used to characterize these nanocomposites included transmission
electron microscopy (TEM) using a Philips CM20T TEM operating at 200 kV and
equipped for selected area electron diffraction and energy dispersive spectroscopy
(EDS), X-ray diffraction (XRD) using a Philips PW1800 $\theta/2\theta$ automated powder
diffractometer equipped with a Cu target and a post-sample monchromator, and
magnetization measurements recorded as a function of field strength and temperature
using a vibrating sample magnetometer.
    Bulk chemical microanalyses were performed by Galbraith Laboratories, Inc.,
Knoxville, TN.

Results and Discussion

A previously reported synthetic method for incorporating inorganic complexes into silica xerogel matrices has now been extended to include metal phosphine complexes of Fe, Co, or Ni (*10*). Metal complexes containing bifunctional ligands which possess silyl ester groups, such as $Si(OR)_3$, undergo hydrolysis and condensation reactions in conventional sol-gel formulations to give apparent covalent incorporation of the dopant complex into the silica xerogel matrix. Thermal treatment of silica xerogels doped with selected metal phosphine complexes under reducing conditions leads to the formation of crystalline, metal phosphide nanoparticulates highly dispersed throughout the xerogel matrix. Detailed procedures for the preparation of silica xerogels covalently doped with the molecular precursors, $Fe(CO)_4L$, $Co_2(CO)_6L'_2$, NiL'_4, or trans-NiL'_2Cl_2 [where L is $PPh_2(CH_2)_2Si(OMe)_3$ and L' is $PPh_2(CH_2)_2Si(OEt)_3$], and for the conversion of these doped xerogels to nanocomposites containing nanoclusters of Fe_2P, Co_2P, or Ni_2P, respectively, are provided in Scheme I.

Thermal treatment of silica xerogel doped with $Fe(CO)_4L$, where L is $PPh_2(CH_2)_2Si(OMe)_3$, gives a nanocomposite containing nanocrystals of Fe_2P. A TEM micrograph of this nanocomposite (Figure 1) reveals particulate features having sharp edges and corners with some particles oriented properly to give a hexagonal projection. A histogram of particle sizes, shown in Figure 2, indicates a monomodal distribution of particle diameters ranging in value from 2 to 8 nm with an average particle diameter of 4.7 nm.

On-particle EDS spectra indicate a Fe:P ratio of 1.98:1 which is consistent with the formula, Fe_2P. Selected area diffraction ring patterns obtained from this nanocomposite reveal seven rings having *d*-spacings which match well with those reported for standard samples of hexagonal Fe_2P, known as syn-barringerite. A spot pattern obtained by electron diffraction from one nanocrystal of Fe_2P was successfully interpreted consistent with the reported cell parameters for this phase. An XRD scan ($32\,° < 2\theta < 52\,°$) of this nanocomposite shows peaks that match well in 2θ values and in relative intensities with the corresponding 111, 201, and 210 XRD peaks of barringerite (Figure 3). Measurement of the full width at half maximum of each of these three peaks and application of Scherrer's equation (*11*) gives a volume-averaged mean particle diameter of 10 nm. This nanocomposite apparently contains a small fraction of Fe_2P particles having diameters significantly larger than 4.7 nm.

Preliminary magnetization measurements for the Fe_2P nanocomposite have been recorded. Magnetization versus temperature data are shown in Figure 4 during field cooling and during zero-field cooling. These results indicate that this nanocomposite is weakly superparamagnetic at room temperature and does not become a bulk ferromagnetic upon cooling to low temperature. A very high blocking temperature near 250 K is observed; normally, blocking temperatures are near to or less than 150 K as observed with nanoparticulate γ-Fe_2O_3 (*12*). Magnetization versus applied field measurements at four temperatures are shown in Figure 5. These curves show zero hysteresis at temperatures above 250 K and the appearance of hysteresis at temperatures below 250 K as expected for very small ferromagnetic particles of Fe_2P separated from one another by a silica xerogel host matrix. At temperatures below 50 K, hysteresis data reveal displaced hysteresis loops after field cooling and indicate that this nanocomposite exhibits magnetic spin-glass behavior. Further study of this latter phenomenon is underway.

Fujii and coworkers report that bulk Fe_2P is paramagnetic and exhibits ferromagnetism at temperatures below 217 K (*13*). Particles of Fe_2P have been used by Seeger and coworkers as an ink in the fabrication of integrated circuits (*14*). In this application, the mean diameter of the Fe_2P particulates ranges from 3 - 22 microns.

Thermal treatment of silica xerogel doped with $Co_2(CO)_6L'_2$, where L' is $PPh_2(CH_2)_2Si(OEt)_3$, gives a nanocomposite containing nanoclusters of Co_2P. A TEM

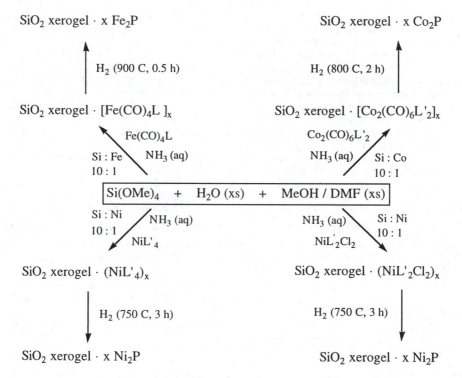

Scheme I. Syntheses of silica xerogel nanocomposites containing nanoclusters of Fe₂P, Co₂P, or Ni₂P.

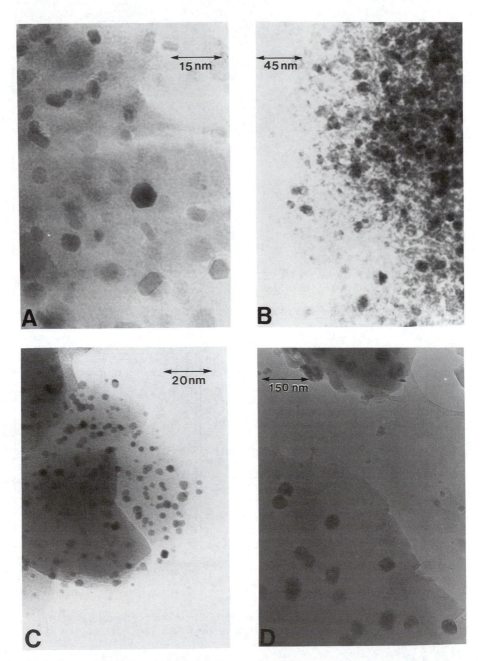

Figure 1. TEM micrographs of silica xerogel nanocomposites containing nano-clusters of Fe_2P (A), Co_2P (B), or Ni_2P (C - NiL'_4 precursor; D - NiL'_2Cl_2 precursor).

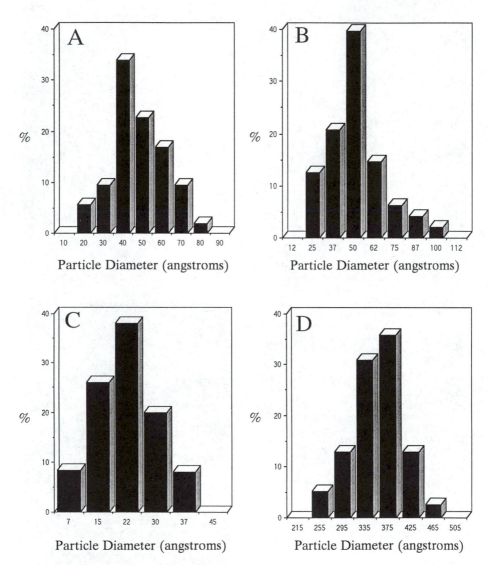

Figure 2.   Histograms of nanocluster particle sizes in silica xerogel
nanocomposites containing nanoclusters of Fe_2P (A), Co_2P (B), or Ni_2P (C -
NiL'_4 precursor; D - NiL'_2Cl_2 precursor).

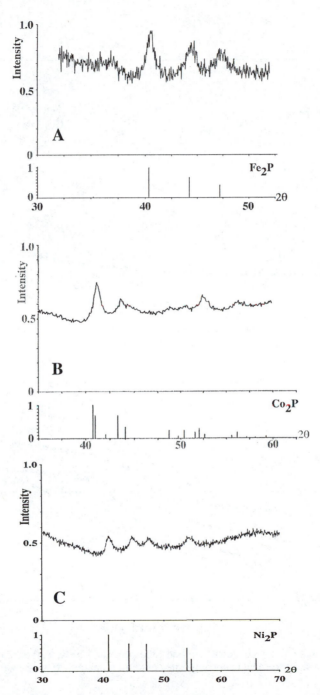

Figure 3. XRD scans of silica xerogel nanocomposites containing nanoclusters of Fe$_2$P (A), Co$_2$P (B), or Ni$_2$P (C - NiL'$_4$ precursor) along with the XRD scans of the corresponding pure substances.

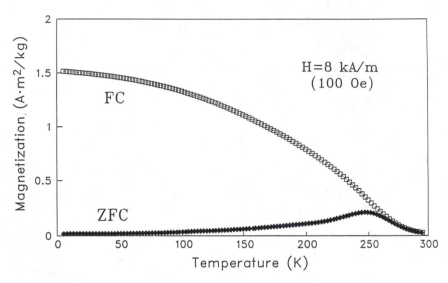

Figure 4. Magnetization versus temperature plots for a silica xerogel nano-composite containing nanoclusters of Fe_2P.

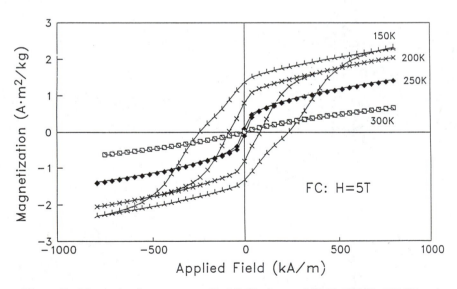

Figure 5. Magnetization versus applied field plots at 150 K, 200 K, 250 K, and 300 K for a silica xerogel nanocomposite containing nanoclusters of Fe_2P.

micrograph of this nanocomposite (Figure 1) reveals crystalline particulate features. A histogram of particle sizes, shown in Figure 2, indicates a monomodal distribution of particle diameters ranging in value from 2.5 to 10 nm with an average particle diameter of 5 nm.

EDS spectra indicate the presence of both Co and P, and *d*-spacings derived from the observed selected area electron diffraction ring patterns (six rings) for this nanocomposite are consistent with the known orthorhombic Co_2P crystalline phase. On-particle EDS spectra indicate a Co:P ratio of 2.09:1 which is consistent with the formula, Co_2P. An XRD scan of this nanocomposite ($35 °< 2\theta < 60 °$) confirms the presence of crystalline Co_2P with no other crystalline phases being evident (Figure 3).

Thermal treatment of silica xerogel doped with NiL'_4, or trans-NiL'_2Cl_2, where L' is $PPh_2(CH_2)_2Si(OEt)_3$, gives a nanocomposite containing crystalline nanoclusters of Ni_2P. A TEM micrograph of the nanocomposite (Figure 1) prepared from the NiL'_4 precursor reveals nearly spheroidal particulate features . A histogram of Ni_2P particle sizes obtained from this precursor, shown in Figure 2, indicates a monomodal distribution of particle diameters ranging in value from 0.7 to 3.7 nm with an average particle diameter of 2.6 nm. Larger particles of Ni_2P are obtained using trans-NiL'_2Cl_2 as precursor under the same synthetic and thermal treatment conditions (see Figures 1 and 2). These nanoclusters have diameters ranging from 25.5 to 46.5 nm with an average particle diameter of 35.5 nm. On-particle EDS spectra of this nanocomposite indicate a Ni:P ratio of 2.06:1 which is consistent with the formula, Ni_2P. Investigation of the apparent dependence of Ni_2P particle size on the identity of Ni molecular precursor is in progress.

EDS spectra obtained from the Ni_2P nanocomposite prepared from the NiL'_4 precursor confirm the presence of both Ni and P in this product xerogel. Selected area diffraction ring patterns obtained from this nanocomposite reveal five rings having *d*-spacings which match well with those reported for standard samples of hexagonal Ni_2P. An XRD scan ($20 °< 2\theta < 70 °$) of this nanocomposite shows peaks that match well in 2θ values and in relative intensities with the corresponding 111, 201, and 120 XRD peaks of Ni_2P (Figure 3). Measurement of the full width at half maximum of each of these three peaks and application of Scherrer's equation (*11*) gives a volume-averaged mean particle diameter of ca. 9 nm. This nanocomposite apparently contains a small fraction of Ni_2P particles having diameters significantly larger than 2.6 nm.

Ni_2P is commercially available as a chemical reagent. Sharon and coworkers report that hexagonal Ni_2P has a calculated band gap of 1.0 eV and is, therefore, a semiconductor (*15*).

Conclusions

Metal phosphine complexes of iron, cobalt, or nickel act as single-source molecular precursors in the formation of the known metal phosphides, Fe_2P, Co_2P, or Ni_2P, respectively. When such precursors are incorporated into a silica xerogel matrix and are subjected to reducing thermal treatments, the corresponding metal phosphide phases are formed as crystalline nanoparticulates to give a nanocomposite. This synthetic strategy provides a method for the synthesis of quantum dots of selected metal phosphide phases. While the M:P core stoichiometry of the molecular precursor does not control the M:P stoichiometry of the resulting nanoparticulate material, the synthetic conditions employed are selective for the formation of only one crystalline metal phosphide composition. This composition is the congruently melting phase having the highest phosphorus content. Optical characterization of those metal phosphides which are expected to exhibit semiconductor properties is anticipated. Magnetic measurements reveal that the nanocomposite containing Fe_2P nanocrystals is superparamagnetic at room temperature. The general scope of using this synthetic method as a route to nanocomposite materials containing nanoclusters of metal phosphides is being explored.

Acknowledgments

C. M. L. thanks the donors of The Petroleum Research Fund, administered by the American Chemical Society, for partial support of this research and to support by, or in part by, the U. S. Army Research Office.

Literature Cited

1. von Schnering, H. G.; Honle, W. *Encyclopedia of Inorganic Chemistry*; John Wiley & Sons: Chichester, U.K., 1994, Vol. 6, pp. 3106.
2. Arinsson, B.; Landstrom, T.; Rundquist, S. *Borides, Silicides and Phosphides*; Wiley: New York, NY, 1965.
3. Hector, A. L.; Parkin, I. P. *J. Mater. Chem.*, **1994**, *4*, 279.
4. Goel, S. C.; Chiang, M. Y.; Buhro, W. E. *J. Am. Chem. Soc.*, **1990**, *112*, 5636.
5. Therien, M. J.; Trogler, W. C. *Inorg. Synth.*, **1990**, *28*, 173.
6. Cundy, C. S. *J. Organomet. Chem.*, **1974**, *69*, 305.
7. Bemi, L.; Clark, H. C.; Davis, J. A.; Fyfe, C. A.; Wasylishen, R. E. *J. Am. Chem. Soc.*, **1982**, *104*, 438.
8. Schubert, U.; Rose, K.; Schmidt, H. *J. Non-Cryst. Solids*, **1988**, *105*, 165.
9. Niebergall, H. *Makromol. Chem.*, **1962**, *59*, 218.
10. Lukehart, C. M.; Carpenter, J. P.; Milne, S. B.; Burnam, K. J. *Chemtech*, **1993**, *23*, No. 8, 29.
11. Klug, H. P.; Alexander, L. E. *X-Ray Diffraction Procedures for Polycrystalline and Amorphous Materials*; John Wiley & Sons: New York, NY, Second Edition, 1974.
12. Vassiliou, J. K.; Mehrotra, V.; Russell, M. W.; McMichael, R. D.; Shull, R. D.; Ziolo, R. F. *J. Appl. Phys.*, **1993**, *73*, 5109.
13. Fujii, H.; Uwatoko, Y.; Motoya, K.; Okamoto, T. *J. Phys. Soc. Jpn*, **1988**, *57*, 2143.
14. Seeger, R. E., Jr.; Morgan, N. H.; Landry, J. R., Jr. U. S. Patent No 4 759 970 (26 July 1988).
15. Sharon, M.; Tamizhmani, G.; Levy-Clement, C.; Rioux, J. *Sol. Cell.*, **1989**, *26*, 303.

RECEIVED December 11, 1995

Chapter 14

Preparation of Copper Sulfide and Gold Nanoparticles Dispersed in Hydroxypropylcellulose–Silica Film with Gas Diffusion Method

W. Yang, H. Inoue, Y. Nakazono, H. Samura, and T. Saegusa

Advanced Materials Laboratory, Kansai Research Institute, Kyoto Research Park 17, Chudoji Minami-machi, Shimogyo-Ku, Kyoto 600, Japan

A new method for preparation of silica films containing dispersed nano-particles of CuS and Au is reported. These particles were generated through reaction with H_2S gas with their precursors which were uniformly dissolved in silica films as promoted by a compatibilizer, hydroxypropyl cellulose(HPC). All films with characteristic colors were transparent. The particle size, size distribution and composite structure of the films depended on the precursor concentration and matrix composition. Electric conductivity was observed for the CuS embedded film with the value of $2.8 \times 10^2 \ \Omega/\square$. Au particles dispersed homogeneously in HPC-silica film with narrow radius distribution were prepared from $HAuCl_4$ by H_2S gas treatment. Au particles with diameter of as small as 3 nm were prepared with this method.

Nano-composite materials with fine semiconductor particles dispersed in the matrix have attracted considerable interest because the properties of the particles are much different from their bulks when the diameters are less than the Bohr exciton radius. Such particles, which are generally named as nano-particles, are characterized by non-stoichiometric surface structure and quantum size effect[1,2]. These properties would lead to new phenomena, new theoretical insights, and new materials and devices.

Due to the surface of nano-particle is very active, it is difficult to prepare stable nano-particles without aggregation. To clear this problem, techniques such as addition of surface active agents, or modification of the particle surface, have been developed. With these methods, stable nano-particle colloids of CdS, PbS, ZnS, Fe_2O_3, ZnO, CdSe, etc. were obtained[3~5].

Another approach to prepare nano-particles dispersed materials is to synthesize nano-particles in solid matrices. In this case, precursors are dissolved in the matrices and converted to nano-particles by chemical reaction. The formation of particles in solid is dominated by the solid matrix network sterically. According to this, it is expected that very small particles with narrow radius distribution could be prepared. The distribution of the particles can also be designed by choosing different matrices. For example, a thin film of silica with layer structure was prepared by converting

0097–6156/96/0622–0205$15.00/0

tetraethoxysilane (TEOS) to silica in LB membrane[6], and hybrid materials with clay as matrix were also reported[7,8]. Using gas diffusion method, CdS particles dispersed in organic polymers such as poly(acrylonitrile-styrene) were synthesized from its precursor. Their nonlinear optical properties were studied[9].

Homogeneous inorganic matrices such as silica have the advantage of durability and transparency, which are important in the application of electronics or optical devices. To achieve this, sol-gel process is an effective method, in which silica sols containing nano-particle precursors are used as starting materials to form films. Using this method, silica films containing semiconductor nano-particles such as chalcogenides could be fabricated. However, in our preliminary study aimed at preparation of chalcogenide nano-particles[10], it was found that when metal salts were used as precursors, they precipitated in the silica films after drying. As a result, chalcogenide particles formed by H_2S gas treatment were characterized with large particle size and broad size distribution. The particle size broadening is mainly due to the inhomogenous distribution of the precursors in the silica matrices. One method to narrowing the particle size distribution is to choose a matrix which can uniformly distribute the precursors. Our basic consideration to solve this problem, is to add a third component to compatibilize the precursor with silica matrix. It was found that organic polymers with hydroxyl group such as HPC were effective in preparation of homogenous organic polymer-silica films containing precursors such as $CuCl_2$, $Pb(CH_3COO)_2$, $Cd(CH_3COO)_2$, etc.. Upon treating with H_2S gas, these precursors could be converted to the corresponding chalcogenide particles in diameter less than 10 nm .

In this paper, we report the preparation, structure characterization and properties of CuS and Au nano-particles dispersed in HPC-silica films with gas diffusion method.

Experiment

The procedure for preparation of HPC-silica films containing CuS and Au nano-particles is shown in Figure 1.

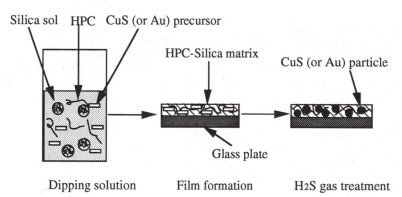

Figure 1 Procedure for preparation of HPC-silica film containing
CuS and Au nano-particles

Generally, a dipping solution for preparation of CuS precursor films was prepared by dissolving $CuCl_2$, $AlCl_3$ and silica sol in 100 ml of methanol containing

0.5 g of HPC by the amount of corresponding to the composition listed in Table 1 and 2. Dipping solution for Au precursor film was prepared by dissolving 0.5 g of $HAuCl_4 \cdot 4H_2O$ and 0.29 g of silica sol (nominal SiO_2 weight) in 100 ml of methanol containing 0.5 g of HPC.

The silica sol was prepared from hydrolysis of tetraethoxysilane (TEOS) with a procedure described as following: To 10 g of TEOS, 2.1 g of water, 1.5 g of methanol and 100 mg of hydrochloric acid(35%) were added. The mixture was stirred at room temperature. After a few minutes the mixture became homogenous, the solution was stirred for further 1 hr to give the silica sol solution.

Precursor films were formed on glass plates by dipping method. The films were dried immediately by air flow at room temperature. The number of dip was one and the film thicknesses were about 0.1 μm. The CuS nano-particles dispersed films were obtained by exposing the precursor films to H_2S gas for 1 to 2 seconds. $CuCl_2$ was converted to CuS and the color of the films changed from light yellow to blue. The films were heated at 150 °C under vacuum for 15 to 30 minutes before characterization. The Au nano-particles dispersed films were prepared by both H_2S gas treatment and thermal decomposition. In the case of H_2S gas treatment, the precursor film was exposed to H_2S gas for 1 to 2 seconds and heated at 150 °C under vacuum for 1 hr. The thermal decomposed one was prepared by heating the precursor film at 150 °C under vacuum for 1 hr.

Self supporting films for transmission electron microscope(TEM) observation were prepared as following: The nano-particles dispersed films were generated on the glass plates coated with thin polycarbonate (PC) film. The PC film was removed with chloroform to give the self supporting films suspended in chloroform.

The particle size, size distribution and composite structure of the films were characterized by TEM (H-7100). Surface electric conductivity of the CuS dispersed films was measured using a MCP-T350 four probes surface resistance meter. UV-Visible absorption spectra were recorded with a Ubest-50 UV-Visible spectrophotometer.

Results and Discussion

CuS nano-particles. CuS can generally be synthesized by passing H_2S gas through a solution of copper(II) salt. On the other hand, if H_2S gas is allowed to diffuse into a solid matrix containing copper(II) ions, precipitated CuS crystal particles yield. The formation of particles can be generally divided into two steps: core formation step and particle growth step. Since the diffusion of copper(II) ions which govern the particle growth is suppressed by the matrix network, this method is expected to be useful for preparing CuS crystals with small radius. The particle size distribution is considered to be mainly attributed to the inhomogeneous distribution of copper(II) ions in the matrix. In order to narrow the particle size distribution, it is necessary to choose a matrix which can uniformly distribute copper(II) salt. In this study, we found that HPC is one of the effective matrices which dissolves CuS precursors such as $CuCl_2$ in solid state. HPC is the derivative of cellulose with chemical structure shown below:

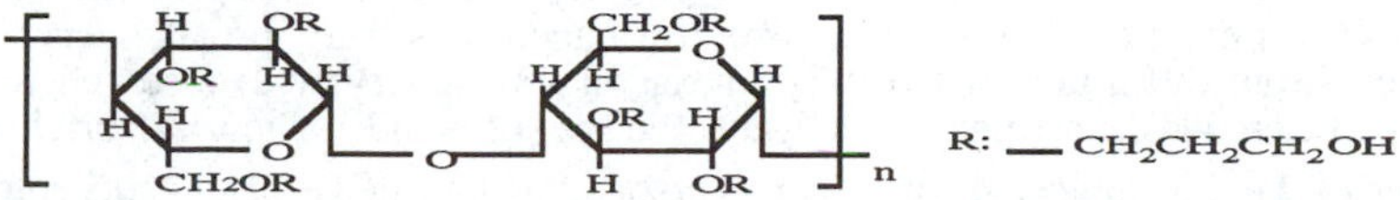

Figure 2   Chemical structure of HPC

Owing to the hydroxyl and ether groups in the polymer, HPC is also compatible with silica through hydrogen bonds[11~13] to form molecular hybrid. This means that HPC can work as a compatibilizer to disperse $CuCl_2$ in silica matrix of molecular level, which is the key step in preparation of CuS nano-particles dispersed films containing silica. It is emphasizing that, like other organic-inorganic hybrid materials, HPC-silica hybrid is a multifunctional material offering a wide range of interesting properties from its organic and inorganic components, which can be easily controlled by changing the components composition.

The effect of HPC to increase the solubility of $CuCl_2$ was examined with polarized microscope observation of the resulting precursor films. Upon reaching the solubility limit, $CuCl_2$ crystals precipitated inside the films were observed. The results are summarized in Table 1.

Table 1   Solubility of $CuCl_2$ in HPC, silica films

No	Film composition			Solubility limit
	HPC g	SiO_2 g	$AlCl_3$ g	$CuCl_2 \cdot 2H_2O$ g
1	0	1.0	0	<0.1
2	0.256	0.744	0	0.3
3	0.633	0.376	0	0.6
4	1.0	0	0	1.0
5	0.505	0.293	0.202	0.7

As can be seen in Table 1, the solubility of $CuCl_2 \cdot 2H_2O$ in silica was only 0.1 g/gram of SiO_2. Above this, the film became opaque and crystals appeared. However, the solubility in HPC was as high as 1.0 g/gram of HPC, which is 10 times higher than that in silica. After HPC was added to silica, the solubility limit of $CuCl_2 \cdot 2H_2O$ increased with increasing the amount of HPC used. Another interesting observation is that addition of $AlCl_3$ also increased the solubility of $CuCl_2$. Homogeneous precursor film with high content of $CuCl_2$ could be obtained by addition of $AlCl_3$ due to unknown reason.

When the films were exposed to H_2S gas, the colored and transparent conducting films were formed. Energy dispersive X-ray spectroscopy(EDX) examination showed that the molar ratio of Cu/S was from 1.1 to 1.3, indicating the formation of CuS. The particles precipitated in the films were characterized with TEM observation. The particle size, size distribution and composite structure of the films were influenced by the precursor film composition. Figure 3 shows the TEM images from the CuS particles dispersed HPC films without addition of silica sol and $AlCl_3$.

The concentration of $CuCl_2 \cdot 2H_2O$ in the precursor films were 16.7%, 33.6% and 50%, respectively. It is clear that the size and size distribution of the resulting CuS particles increased with increasing concentration of $CuCl_2$. In the bright filed image for $CuCl_2 \cdot 2H_2O$ concentration of 50%, aggregates of CuS particles with diameter of 40 to 100 nm were observed. The size distribution was very broad. Very sharp electron beam diffraction pattern with hexagonal symmetry was observed for the CuS particles, indicating high crystallinity. Pictures of a and b show the bright filed images of the films prepared from the precursor films of $CuCl_2 \cdot 2H_2O$ concentration of 16.7% and 33.6%, respectively. The isolated CuS particles can be seen embedded

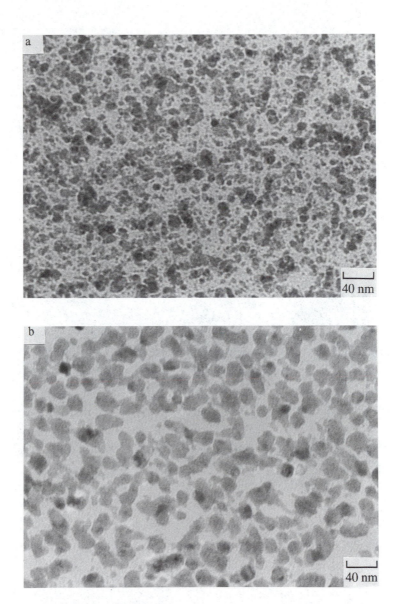

Figure 3 TEM images of CuS dispersed films with precursor film composition of a, $CuCl_2 \cdot 2H_2O/HPC=0.1/0.5(g/g)$; b, $CuCl_2 \cdot 2H_2O/HPC=0.253/0.5(g/g)$; c, $CuCl_2 \cdot 2H_2O/HPC=0.5/0.5(g/g)$.

Continued on next page

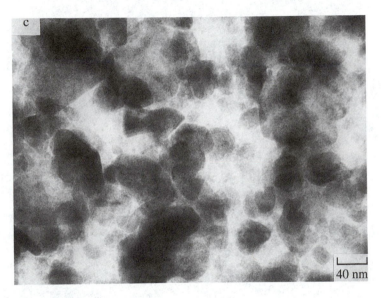

Figure 3. *Continued*

in the HPC matrices homogeneously. The diameter of the CuS particles was about 10 nm for $CuCl_2 \cdot 2H_2O$ concentration of 16.7%, and about 20 nm for $CuCl_2 \cdot 2H_2O$ concentration of 33.6%. The particle surfaces were normally clean and abrupt, as is more clear in the TEM image of $CuCl_2 \cdot 2H_2O$ concentration of 33.6%. The particles were clearly seen to be faceted. As a result, individual grains were polyhedral.

Figure 4 shows the TEM images of the CuS dispersed films after silica sol and $AlCl_3$ were added to HPC. For convenience of comparison, the same concentration of $CuCl_2 \cdot 2H_2O$ in the precursor films(33.6%) was used for the samples. Addition of silica sol caused little change in the particle size, but the grain boundary became less abrupt. It is even more announced that when silica sol and $AlCl_3$ were added, almost no isolated particle was observed in the TEM image. At the same time, diffraction dots in the electron beam diffraction pattern became continuous rings. It means that CuS formed a continuous network interpenetrated with the matrix.

Along with the morphology change of the CuS particles in the presence of silica and $AlCl_3$, the electric conductivity increased. Table 2 shows the surface resistance of the resulting films. As shown in Table 2, the surface resistance of the film added with silica and $AlCl_3$ was 2.8×10^2 $\Omega/\square$, which is more than 20 times higher than that prepared with only HPC as matrix. This increase of conductivity is believed to be attributed to the highly developed CuS network. Another important finding is that the mechanical strength of the film was also improved by addition of silica sol and $AlCl_3$. The film could be washed with water or organic solvents under ultrasonic irradiation without any damage. The films were transparent with blue color. Figure 5 shows the transparency of the films listed in Table 2. Addition of $AlCl_3$ and silica sol caused a little decrease in transmittance.

Table 2   Relation of surface resistance with film composition

Precursor film composition				Surface resistance
HPC	$CuCl_2 \cdot 2H_2O$/HPC	SiO_2 sol	$AlCl_3$	
g	g	g	g	$\Omega/\square$
0.5	0.25			8.0×10^3
0.5	0.4	0.29		4.7×10^3
0.5	0.5	0.29	0.2	2.8×10^2

Figure 6 shows the temperature dependence of the conductivity. From -140 to -20 °C, the film resistance remained almost constant, above -20 °C, the resistance increased. This behavior indicates that the conductivity is resulted from electric conduction. Figure 7 shows the durability of the conducting film. After heating at 80 °C in air for about 300 hrs, the film resistance increased only by about twice. It is well-known that CuS is usually unstable under atmospheric condition, especially for small particles with large surface area. Considering the diameter of the CuS particles in the film, the stability must be due to the hybrid effect.

Au nano-particles. It is well-known that $HAuCl_4$ is a useful source to prepare Au particles by thermal decomposition. The diameter of the particles formed with this method is usually more than 10 to 20 nm because the melting point of Au is very low[14]. In this study, it was found that Au particles with diameter of about 3 nm could be obtained when $HAuCl_4$ was used as precursor, and followed by treatment with H_2S gas.

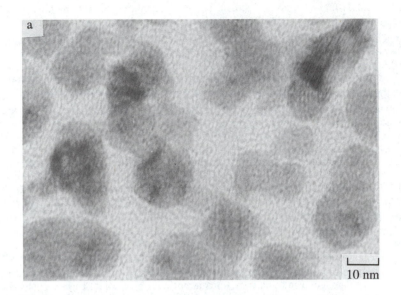

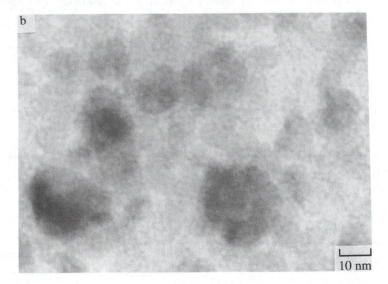

Figure 4   TEM images of CuS dispersed films with precursor film composition of
            a, $CuCl_2\cdot2H_2O/HPC=0.253/0.5(g/g)$; b,$CuCl_2\cdot2H_2O/HPC/SiO_2$
            sol=0.5/0.4/0.29(g/g); c, $CuCl_2\cdot2H_2O/HPC/SiO_2$
            sol/$AlCl_3$=0.5/0.5/0.29/0.2(g/g).

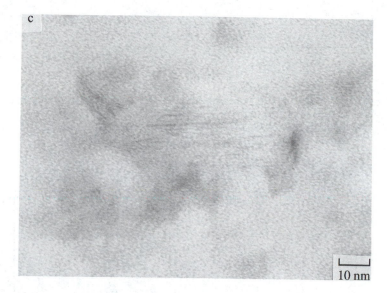

Figure 4. *Continued*

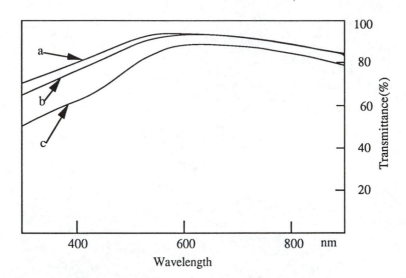

Figure 5   UV-Visible transmittance spectrum of CuS dispersed films with precursor
film composition of a, $CuCl_2 \cdot 2H_2O/HPC$=0.253/0.5(g/g);
b,$CuCl_2 \cdot 2H_2O/HPC/SiO_2$ sol=0.5/0.4/0.29(g/g); c,
$CuCl_2 \cdot 2H_2O/HPC/SiO_2$ sol/$AlCl_3$=0.5/0.5/0.29/0.2(g/g).

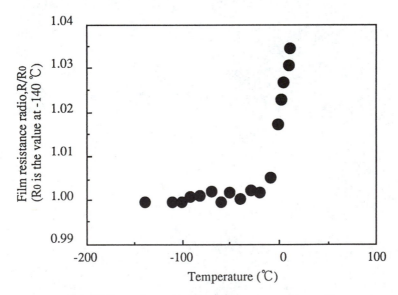

Figure 6   Relation of film resistance with temperature: precursor film
composition, $CuCl_2 \cdot 2H_2O/HPC/SiO_2$ sol/$AlCl_3$=0.5/0.5/0.29/0.2(g/g).

Thermal decomposition was carried out at 150°C for 1 hr under vacuum in a vacuum oven. The resulting film was wine-colored. The H_2S gas treatment was conducted with the same method as the preparation of CuS, followed by heating at 150°C for 1 hr under vacuum. Different from that of the film of thermal decomposition, the color of the H_2S treated one was brown. Figure 8 shows the TEM images of the Au particles embedded HPC-silica films prepared by thermal decomposition and H_2S gas treatment. The size of the Au particles generated by thermal decomposition was uneven, ranging from 10 to 20 nm in diameter. In contrast with this, the particle diameter of the H_2S treated particles was about 3 nm. By comparison of the electron beam diffraction pattern of the film treated with H_2S gas with that of Au particles, it was found that the particles formed were Au instead of Au_2S_3. Since H_2S is a reductant, $HAuCl_4$ was probably reduced to Au metal upon exposing to H_2S gas. Another reasonable explanation for the Au particle formation is the decomposition of Au_2S_3 during the heating process. Investigation of the Au particle formation mechanism upon H_2S gas treatment is now in progress. It is noteworthy that the size distribution of the Au particles prepared with H_2S treatment was extremely narrow. Another remarkable fact is that the particles dispersed in the HPC-silica film very homogeneously. These characteristics are believed to be related to the Au particle formation mechanism.

Figure 9 shows the UV-Visible absorption spectra of the Au particles dispersed films. In the spectrum of the film prepared by thermal decomposition, plasmon resonant peak around 550 nm was obviously observed. On the other hand, only a weak shoulder was observed for the H_2S treated film. However, when the film was heated at 200 °C, an increase in the peak intensity and blue shift in the peak position from 550 nm to 525 nm were observed with extending heating time as shown in Figure 10. Generally, heat treatment will cause enlargement of particle size. As a result, a red shift of the plasmon resonant absorption is expected. At the present time, the reason for the blue shift observed in this study is unknown. One possible explanation is that the Au particles generated by H_2S treatment are disordered crystals with structure defects. Therefore, disorder-order transformation occurs further during the heating with volume reduction. Further studies of the relationship between the blue shift and thermal treatment are now in progress.

Conclusion

Silica films containing CuS and Au nano-particles were fabricated from their precursor films through reaction with H_2S gas. It was found that HPC was an effective compatibilizer to dissolve the corresponding precursors in silica matrix. The particle size, size distribution and composite structure of the films depended on the precursor concentration and matrix composition.

CuS dispersed HPC-silica film showed high electric conductivity. Surface resistance of 2.8×10^2 $\Omega/\square$ was observed for the film prepared from the precursor film with composition of $CuCl_2 \cdot 2H_2O$ / HPC / SiO_2 sol / $AlCl_3$ = 0.5/0.5/0.29/0.2(g/g). The conduction mechanism was found to be electronic conduction. The durability of the conducting film was evaluated and the surface resistance increased only by about twice after heating at 80°C in air for about 300 hrs.

Au particles generated with H_2S gas showed very narrow particle radius distribution, which was dispersed in the film homogeneously. The particle diameter was about 3 nm, much smaller than that prepared by thermal decomposition. Only a very weak plasmon resonant absorption was observed for the H_2S treated Au

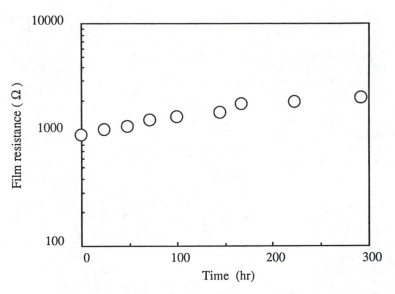

Figure 7   Film resistance change at 80 °C under air atmosphere: precursor film
composition, $CuCl_2 \cdot 2H_2O/HPC/SiO_2$ sol/$AlCl_3$=0.5/0.5/0.29/0.2(g/g).

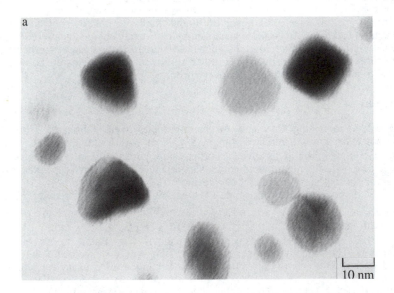

Figure 8   TEM images of Au particles dispersed films:
a, thermal decomposed at 150 °C for 1 hr under vacuum; b and c, H_2S gas
treatment followed by heating at 150 °C for 1 hr under vacuum; precursor
film composition, $HAuCl_4 \cdot 3H_2O/HPC/SiO_2$ sol=0.5/0.5/0.29(g/g).

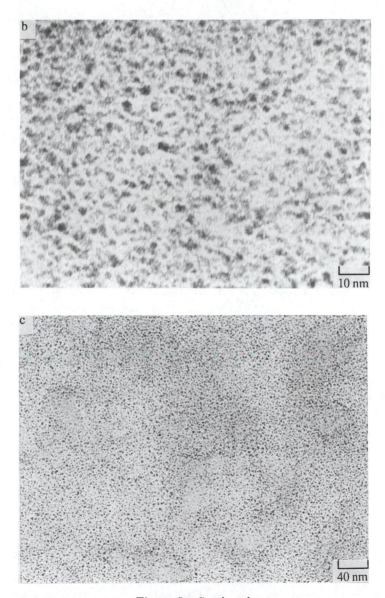

Figure 8. *Continued*

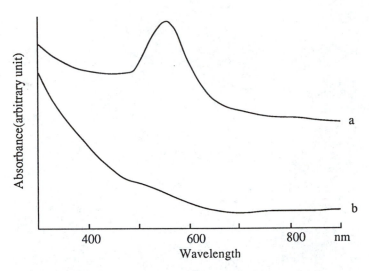

Figure 9 UV-Visible absorption spectra of Au particles dispersed films:
a, thermal decomposed at 150 °C for 1 hr under vacuum; b, H_2S gas
treatment followed by heating at 150 °C for 1 hr under vacuum; precursor
film composition, $HAuCl_4 \cdot 3H_2O/HPC/SiO_2$ sol=0.5/0.5/0.29(g/g).

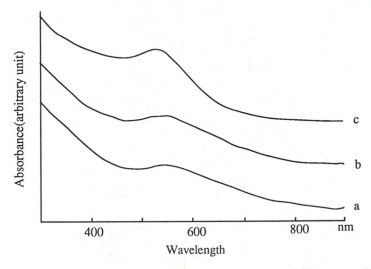

Figure 10 UV-visible spectra change of Au particles dispersed film prepared
with H_2S treatment heated at 200 °C for different time: a, 4 hr; b, 6 hr;
c, 20 hr; precursor film composition, $HAuCl_4 \cdot 3H_2O/HPC/SiO_2$
sol=0.5/0.5/0.29(g/g).

particles dispersed film. However, when treated at 200 °C, an increase in the peak intensity and blue shift in the peak position from 550 nm to 525 nm were observed with increasing heating time.

Literature Cited

1. Tain R. K.; Lin R. C. *J. Opt. Soc. Am.* **1983**, *73*, 647.
2. Henglein A. *Chem. Rev.* 1989, *89*,1861.
3. Steigerwald M. L.; Brus L. E. *Acc. Chem. Rev.* **1990**, *23*, 183.
4. Wang Y.; Herron N. *J. Phys. Chem.* 1991, *95*, 525.
5. Uchida H.; Curtis C. J.; Kamat P. V.; Jones K. M.; Kim M.; Nozik A. J. *J. Phys. Chem.* **1992**, *96*, 1156.
6. Kunitake T.; Nakashima N.; Shimomura M.; Kajiyama T.; Harada A.; Okuyama K.; Takayanagi M. *Thin Solid Films* **1984**, *121*, L89.
7. Halsey T. C. *Science* **1992**, *258*, 761.
8. Kojima Y.; Usuki A.; Kawasmi M.; Okada A.; Kurauchi T.; Kamigaito O. *J. Polym. Sci., Part A* **1993**, *31*, 175.
9. Misawa K.; Yao H.; Hayashi T.; Kobayashi T. *Chem. Phys. Lett.* **1991**, *183*, 113.
10. Yang W.; Inoue H.; Ohnaka T.; Samura H.; Saegusa T. *Polym. Prepr. Japan* **1995**, IIIPd070.
11. Toki M.; Takeuchi T.; Miyasita S.; Kanbe S. *J. Materials Sci.* **1992**, *27*, 2857.
12. Toki M.; Chow T. Y.; Ohnaka T.; Samura H.; Saegusa T. *Polym. Bull.* **1992**, *29*, 653.
13. Toki M. *J. Sol-Gel Science and Technology* **1994**, *2*, 97.
14. Ohkubo K.; Kawata M.; Orito T.; Ishida H. *J. Chem. Soc., Perkin Trans.* **1989**, *1*, 666.

RECEIVED December 14, 1995

Chapter 15

Preparation and Characterization of Nanophase Iron and Ferrous Alloys

Kenneth E. Gonsalves[1], S. P. Rangarajan[1], C. C. Law[2], C. R. Feng[3], Gan-Moog Chow[4], and A. Garcia-Ruiz[1,5]

[1]Institute of Materials Science and Department of Chemistry, University of Connecticut, Storrs, CT 06269
[2]Pratt & Whitney, United Technologies Corporation, East Hartford, CT 06108
[3]Division of Materials Science and Engineering and [4]Laboratory for Molecular Interfacial Interactions, Center for Bio/Molecular Science and Engineering, Naval Research Laboratory, Code 6930, 4555 Overlook Avenue, S.W., Washington, DC 20375–5348

Two approaches for the synthesis of nanostructured M50 type steel (composed of 4.0% Cr, 4.5% Mo, 1.0% V, 0.8% C and balance Fe) powders and their consolidation are reported in this chapter. One approach involved the sonochemical decomposition of organometallic precursors; and the other involved the reduction of the metal halides with lithium triethyl borohydride followed by vaccum sublimation of the powders to remove lithium chloride. The as-synthesized powders are amorphous by X-ray diffraction (XRD) but the peaks corresponding to bcc α-Fe are observed in the compacts. The morphology and composition of the powders synthesized by both techniques, as well as the compacts, were examined by scanning electron microscopy (SEM) and transmission electron microscopy (TEM). Hardness, density, particle size and impurity contents were also determined for the compacts. In addition, pure nanosized iron particles obtained by the ultrasound decompositon of iron pentacarbonyl were consolidated and the properties of the latter were studied.

Nanophase materials in which the average grain size, phases or crystallites are in the nanometer regime have recently been the focus of intense research effort (1-3). This interest has developed due to their superior properties compared to conventional materials which have particle sizes on the order of a micron (4-5). Nanostructured materials have traditionally been prepared by a variety of techniques which include physical methods such as gas-phase condensation, metal evaporation, spray pyrolysis, laser ablation and plasma synthesis (6-12). Chemical methods to synthesize such materials have frequently been used due to the better control of the stoichiometry in the end-product, the molecular level mixing of the constituent phases and the feasibility of low cost bulk production of these materials. Various chemical

[5]Permanent address: UPIICSA-COFAA, IPN. Té 950 Esq. Resina. 08400 México, D.F. and from Instituto de Fisica, Universidad Nacional Autonoma de Mexico, Apartado Postal 20–364, C.P. 01000 México, D.F., México

0097–6156/96/0622–0220$15.00/0
© 1996 American Chemical Society

methods commonly used for the preparation of nanostructured materials include sol-gel synthesis, chemical vapor deposition, chemical reduction and laser pyrolysis (13-18). Nanocrystalline materials are being studied for applications in structural materials due to their enhanced hardness and wear resistance compared to conventional structural materials composed of micron sized particles (19-20). Iron-based alloys are industrially important materials and M50 steel (composed of 4.0% Cr, 4.5% Mo, 1.0% V, 0.8% C and balance Fe (in weight %)) due to its good resistance to wear and rolling contact fatigue is extensively used in the aircraft industry as main-shaft bearings in gas-turbine engines (21-22). Conventional hardened M50 consists of a body centered tetragonal martensite phase and a dispersion of carbide particles (23-25), some of which are several microns in diameter. These large particles often act as fatigue crack initiation sites and are detrimental to the end-product. In contrast, nanostructured steel would not have such a coarse microstructure and would be expected to exhibit superior mechanical properties.

In this chapter are presented, two different liquid phase chemical synthesis methods for the preparation of M50 steel and iron nano-powders. Also, the compaction of these powders into a consolidated sample and a structural and morphological study of the powders and compacts is described here. One synthetic methodology involves the ultrasound assisted decomposition (26-28) of organometallic precursors and the other involves the use of a reducing agent to produce the fine particles from the elemental halides. In addition, the ultrasound-assisted synthesis of pure iron nanopowders, its compaction and the physical properties are reported.

Experimental

Chemicals and Instrumentation. All manipulations for the preparation the powders were performed in the dry-box or by Schlenck line techniques. For the sonochemical reaction, the reactants, iron pentacarbonyl, $Fe(CO)_5$; bis(ethylbenzene)chromium, $Cr(Et_xC_6H_{6-x})_2$ (where x = 0-4) and cyclopentadienyl molybdenum tricarbonyl, $CpMo(CO)_3$; were purchased from Strem Chemicals, Inc. In addition, a surfactant (polyoxyethylene sorbitan trioleate) that was added to the sonochemical reaction was purchased from Sigma Chemical Co. For the chemical reduction method, $FeCl_3$, $MoCl_3$, VCl_3 $CrCl_3$ and lithium triethylborohydride were purchased from Aldrich Chemicals. The solvents decalin, tetrahydrofuran (THF) and methanol were distilled over CaH_2 and degassed by bubbling argon through them for at least 5h.

Ultrasonic treatment was done by means of a high intensity ultrasonic probe (Sonic and Materials, model VC-600, 0.5 in Ti horn, 20 kHz, 100 Wcm^{-2}). X-ray powder diffraction data for the as-synthesised powders and the consolidated sample was collected on a Norelco/Phillips diffractometer using CuK$_\alpha$ radiation (λ = 1.5418 Å). Scanning electron micrographs (SEM) were taken on a Cambridge (Mark 250) electron microscope fitted with an x-ray analyzer. The hardness of the sample was measured on a Clark microhardness tester (DMH-2) interfaced with a Compaq computer and software. Transmission electron micrographs (TEM) were taken on a JEOL 200CX electron microscope with an accelerating voltage of 200 KV.

Powder Synthesis. Two synthetic approaches to produce M50 type steel nanopowders are outlined in Scheme 1. Due to the repeated segregation of vanadium in the M50 powders produced in earlier experiments by using either $V(CO)_6$ or $CpV(CO)_4$ as a vanadium precursor during ultrasound decomposition, vanadium has been excluded in the following reported sonochemical synthesis of M50 type steel, however efforts are currently in progress to remedy this segregation.

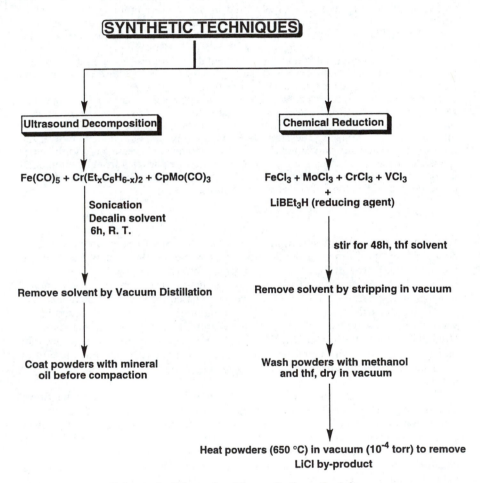

Scheme 1. Schematic of the synthetic methodology.

Sonochemical Synthesis of M50 Type Steel Nanopowders. A dispersion of 15g (0.0765 mol) of $Fe(CO)_5$, 0.66g of $Cr(Et_xC_6H_{6-x})_2$, 0.75g (0.0015 mol) of $CpMo(CO)_3$ and 1.0 g of polyoxyethylene sorbitan trioleate (surfactant) in dry decalin was sonicated at 50% amplitude for 7h at room temperature in a sonochemical reactor fitted with a condenser and gas inlet and outlet tubes connected to a mercury bubbler. The color of the solution turned dark and then black within a few minutes and this reaction mixture was sonicated until the formation of shiny metallic particles was observed on the walls of the reaction vessel. The sonication was then stopped and the decalin solvent was removed from the reaction flask via vacuum distillation. Fine black powder (Yield: 4.448g) remained at the bottom of the reactor, which was then isolated, transferred to a vial and coated with mineral oil before the compaction.

Chemical Reduction Method for the Synthesis of M50 Type Steel Nanopowders. To a suspension of 10g $FeCl_3$, 0.5g $MoCl_3$, 0.445g $CrCl_3$ and 0.111g VCl_3 in 100 mL thf, 202.98 mL of 1.0M lithium triethyl borohydride in THF was added slowly while stirring at room temperature by a liquid addition funnel in the dry-box. Slow effervescence was observed for a few minutes. After the reaction was stirred in the glove box for 48h at room temperature, a black suspension was formed. The solvent (THF) was removed from the reaction flask by vacuum distillation and the black powders were washed with ≈ 100 mL of distilled degassed methanol till no further bubbling was observed. The fine black solid was washed again with 50 mL of thf and dried under vacuum.

The lithium chloride by-product was removed from the above solid by vacuum sublimation in a tube furnace at 650 °C/10^{-4} torr. The yield of the powders after the sublimation was 3.382g.

Sonochemical Synthesis of Iron Nanopowders. A dispersion of 15g (0.076 mol) of $Fe(CO)_5$ in dry decalin was sonicated at 50% amplitude for 6h at room temperature in a sonochemical reactor as described previously. The color of the solution turned dark and then black within a few minutes and this reaction mixture was sonicated till the formation of shiny metallic particles was observed on the walls of the reaction vessel. The sonication was then stopped and the decalin solvent was removed from the reaction flask via vacuum distillation. The black powders (Yield: 3.881g) at the bottom of the reactor was then isolated, transferred to a vial and coated with mineral oil before the compaction.

Consolidation. The powders were transferred to a cylindrical steel die with a 17.7 mm diameter cavity in the dry-box (see Scheme II) and the die was filled with powder to a height of 32 mm after manually tapping and pressing in the dry-box. The filled die was then wrapped in a nickel foil and transferred to a vacuum hot press for consolidation. Prior to compaction, the powders were treated in a hydrogen retort (to maintain the carbon content in the powder to less than 1 weight %) for 2h at 420 °C with a hydrogen flow rate of 20 standard cubic foot per hour (SCFH). Consolidation of the powders was done by using a vacuum hot press (VHP) custom designed by Centorr as per Pratt & Whitney specifications. The powders were compressed at 275 MPa at 700 °C for 2h.

Results and Discussion

Precursor Characteristics. The precursors used in the sonochemical synthesis were chosen due to their ability to readily decompose to produce metallic clusters, as well as their cost and commercial availability. For example, $Fe(CO)_5$ is a liquid that can readily decompose at its boiling point to Fe clusters (29-30). The chromium and molybdenum precursors were chosen for similar reasons. In the case of the chem

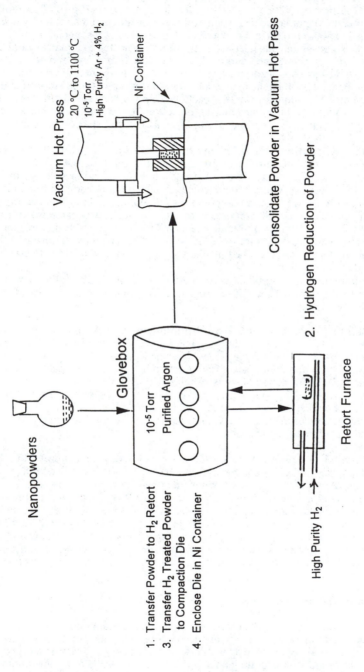

Scheme 2. Schematic of the consolidation process.

-ical reduction technique, the halides are the most attractive precursors due to their low cost and commercial availability.

Sonochemically Produced M50 Type Steel. The morphology and microstructure of the M50 steel powders produced by the ultrasound assisted decomposition of the metal-organic precursors were examined by SEM and TEM microscopy. The SEM micrograph of the as-synthesized powders at 4400X magnification shown in Figure 1 indicates that the powders have a porous coral like microstructure usually observed for nano-powders produced by ultrasound decomposition (31-32). In the energy dispersive analysis by X-ray (EDAX) spectrum of the as-synthesized powder (Figure 2), the peaks corresponding to Fe, Cr and Mo are observed. The TEM (Figure 3) at 118,500X magnification showed that the particles were agglomerated and were composed of smaller particles. The XRD analysis indicated that the as-synthesised powders are amorphous as shown in Figure 4. However the consolidated sample displayed sharp crystalline peaks (Figure 5) which were assigned to α-Fe as confirmed by refinements of the diffraction patterns using the Rietveld method (33). The average particle size calculated from X-ray line broadening analysis was 25 nm.

Figure 6 shows a SEM micrograph of the compacted M50 steel specimen. By comparing the Figure 6 (SEM micrograph) with the corresponding EDAX spectra (Figures 7(a)–(d)) the compact shows a uniform microstructure with scattered Cr-rich precipitations. The specimen is 100% dense. The carbon and oxygen content of the consolidated specimen was found to be 0.54% and 4.1% respectively.

In Figure 8(a) and (b) is shown a bright field and dark field TEM micrograph respectively of the matrix in the consolidated sample. It appears that there is a distribution in grain size, and the range varies between 5 to 70 nm. TEM micrographs obtained by tilting the sample (+5° to -5°) in the same region confirmed that the microstructural contrast observed in Figure 8(a) is mainly due to orientation effects of α-Fe crystallites, altough phase separation between the matrix and the precipitates could also contribute to this effect. The electron diffraction (Figure 8(c)) of the matrix showed a spotty diffraction pattern corresponding to the α-Fe bcc phase. Figures 8(d) and 8(e) show the bright field and dark field TEM micrographs of the precipitate, respectively. They indicate a very fine precipitate with diameters averaging $\approx$10 nm. The electron diffraction pattern of the precipitate (Figure 8(f)) indicated that it had a CF8 structure with the best fit corresponding to the Mo_2C phase.

M50 Steel Produced by Chemical Reduction. Nanopowders of M50 steel produced by the chemical reduction technique were characterized by SEM/EDAX. The SEM of the powders showed a similar porous coral-like morphology as shown in Figure 1 for powders synthesized by the sonochemical method. In Figures 9(a) and (b) is shown the EDAX spectrum of the powders before and after the vacuum heat treatment at 650 °C. It indicates that the lithium chloride by-product has been successfully removed by vacuum sublimation. The average particle size, calculated from X-ray line broadening analysis was 34 nm. Current efforts are under way to compact these powders and to study their structural and physical properties.

Sonochemically Prepared Nanostructured Iron. The consolidated iron pellet had a homogenous microstructure as confirmed by SEM taken at 100X magnification (Figure 10) and it had a density of 100%. The carbon and oxygen contents were determined to be 0.05% and 1.1% respectively. In the XRD spectra the major peaks were assigned to the α-Fe phase and line broadening analysis revealed the average crystallite size in the consolidated specimen to be $\approx$40 nm.

Figure 1. SEM micrograph of the sonochemically synthesized M50 steel powders.

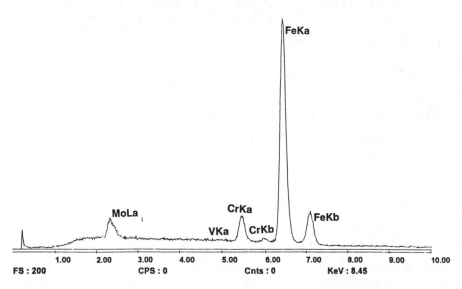

Figure 2. EDAX spectrum of the sonochemically synthesized M50 steel powders.

Figure 3. TEM micrograph of the sonochemically synthesized M50 steel powders.

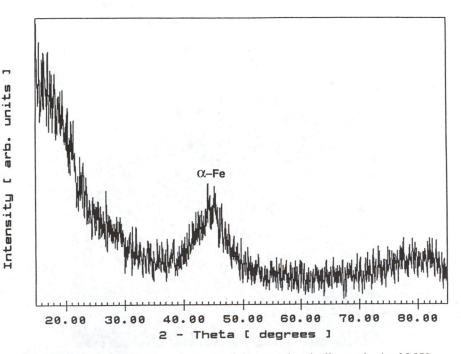

Figure 4. X-ray diffraction spectrum of the sonochemically synthesized M50 steel powders.

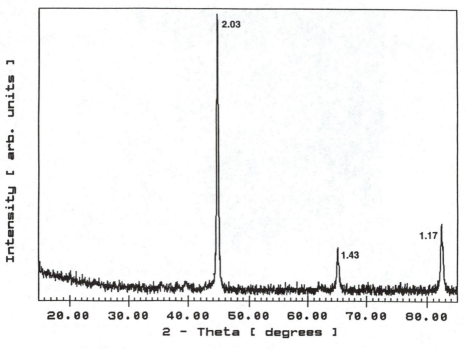

Figure 5. X-ray diffraction spectrum of the M50 type consolidated steel specimen showing the major peaks due to α-Fe.

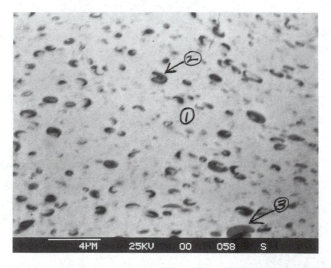

Figure 6. SEM micrograph of consolidated M50 steel sample.

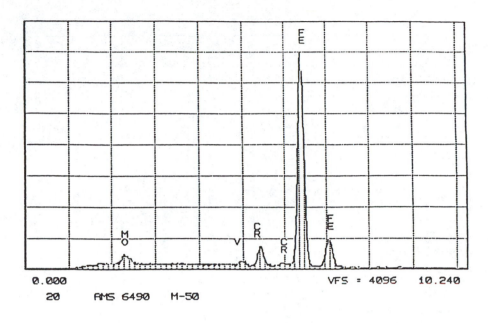

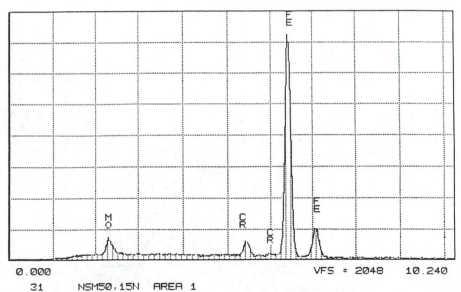

Figure 7. EDAX spectrum of (a) commercial M50 steel standard (b) Spot 1 in Figure 6 (c) Spot 2 in Figure 6 and (d) Spot 3 in Figure 6.

Continued on next page

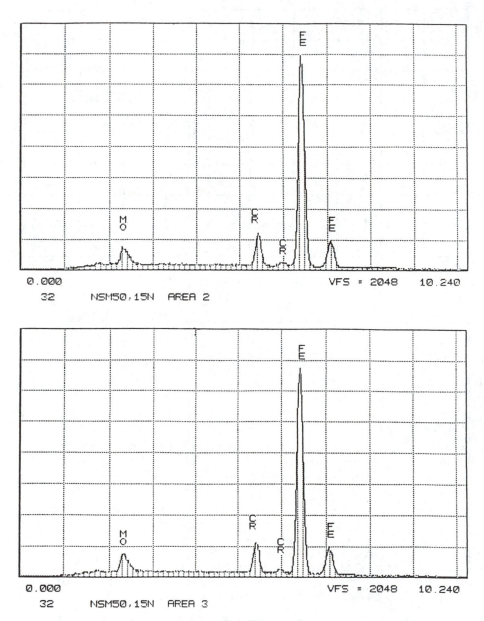

Figure 7. *Continued*

a

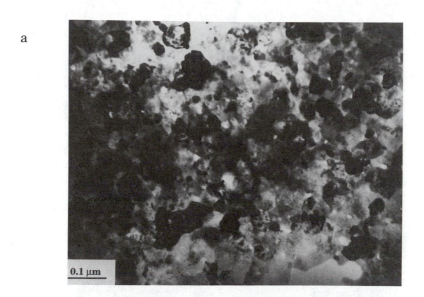

b

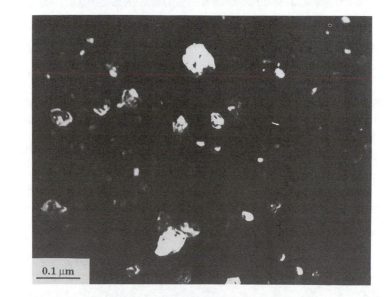

Figure 8. TEM micrographs of the consolidated M50 steel specimen: (a) Bright field image of the matrix (b) Dark field image of the matrix (c) Diffraction pattern of the matrix (d) Bright field image of the precipitate (e) Dark field image of the precipitate and (f) Diffraction pattern of the precipitate and the matrix

Continued on next page

c

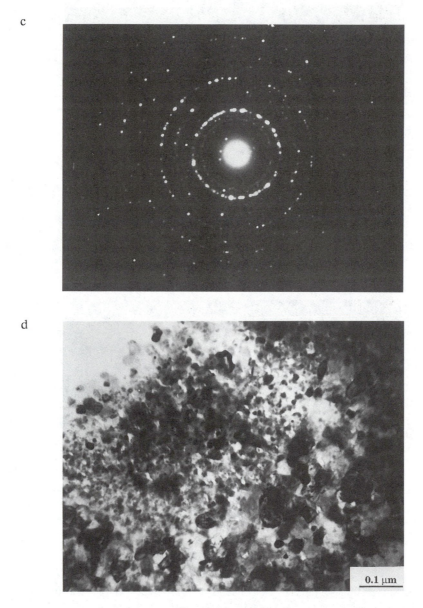

d

0.1 μm

Figure 8. *Continued*

e

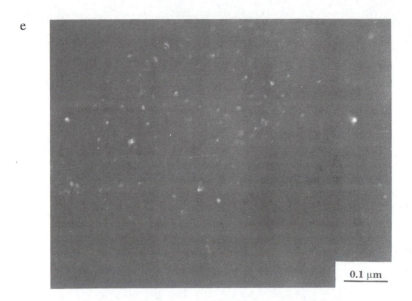

0.1 μm

f

Figure 8. *Continued*

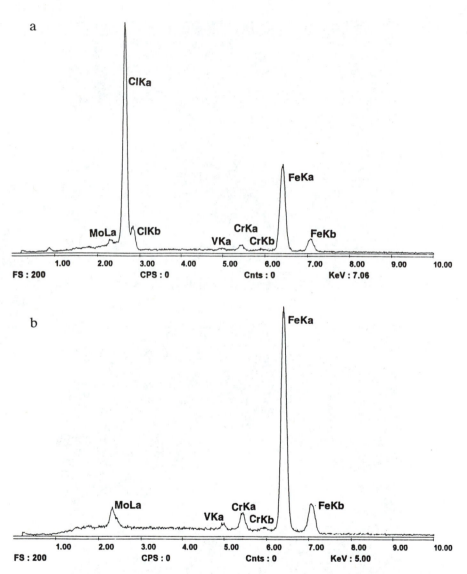

Figure 9. EDAX spectrum of (a) M50 steel powders produced by chemical reduction method before the vacuum sublimation and (b) After the vacuum sublimation at 650 °C/10^{-4} torr.

Figure 10. SEM micrograph of the pure consolidated iron sample.

Hardness. The nanostructured M50 compact had a hardness of 69 Rockwell C (RC) as compared to a hardness of 58-62 RC for conventional, commercial M50 steel. The hardness of the consolidated iron sample was 37 RC as compared to that of conventional micron sized iron compacts (4-5 RC). Since the nanostructured iron has a hardness approximately seven times that of the conventional, iron, it is therefore reasonable to argue that the marginally higher hardness in the case of the M50 compact results from the iron matrix rather than from the precipitate.

Conclusions

Two new techniques have been developed for the synthesis of pure nanostructured iron and a M50 type nanophase steel specimen. TEM analysis of the compacted specimen indicated the crystallite size to be between 5-70 nm with random orientation of the Fe crystallites. The iron compact had a highly homogenous microstructure and low carbon and oxygen contents. It also exhibited high hardness, probably due to the ultra-fine microstructure as compared to conventional consolidated iron. Further investigations will be focussed on obtaining the exact stoichiometric ratio of the commercial M50 alloy, reducing the oxygen contamination in the compacts and examining the scale-up feasibility for production of these alloys.

Acknowledgments

Funding under grant ONR N00014-94-0579 is gratefully acknowledged and we thank CONACYT for the support of Dr. A. Garcia-Ruiz as a visiting scientist to the University of Connecticut. GMC would like to thank the support of the Office of Naval Research.

References

1. Seigel, R.W. *Mat. Sci. and Engg. A*, **1993**, *168*, 189.
2. Gleiter, H. *Nanostruct. Mater.* **1995**, *6*, 3.
3. Parker, J.C.; Seigel, R.W. *Nanostructr. Mater.* **1992**, *1*, 53.
4. Cahn, R. W. *Nature*, **1990**, *348*, 389.
5. Seigel, R. W. *Chem. Engg. News*, **1992**, *Nov. 23*, 18.
6. Hahn, H.; Averback, R. S. *J. Appl. Phys.* **1990**, *67*, 1113.
7. Davis, S. C.; Klabunde, K. J. *Chem Rev.* **1982**, *82*, 152.
8. Messing, G.; Zhang, S.; Jayanthi, G. *J. Am. Ceram. Soc.*, **1993**, *76*, 2707.
9. Xiao, T. D.; Gonsalves, K. E.; Strutt, P. R.; Klemens, P. *J. Mater. Sci.*, **1993**, *28*, 1334.
10. Cannon, W. R.; Danforth, S. C. *J . Am. Ceram. Soc.* **1982**, *65*, 324.
11. Xiao, T. D.; Gonsalves, K. E.; Strutt, P. R. *Nanostruct. Mater.* **1992**, *1*, 1.
12. Kameyama, T.; Sakanaka, K.; Motoe, A. *J. Mater. Sci.* **1990**, *25*, 1058.
13. Brinker, C. J.; Scherer, G. W. *Sol-Gel Science, The Physics and Chemistry of Sol-Gel Processing.* Academic press: San Diego, **1990**.
14. Phalippou, J. chapter in *Chemical Processing of Ceramics,* Eds. Lee, B. I.; Pope, E. J. A. Marcel-Dekker, New York, **1994**, p 265.
15. Chang, W.; Skandan, G.; Hahn, H.; Danforth, S. C.; Kear, B. H. *Nanostruct. Mater.* **1994**, *4*, 3.
16. Zheng, D.; Hampden-Smith. M. *J. Chem. Mater* **1993**, *5*, 681.
17. Bonneman, H.; Brijoux, W.; Brinkmann, R.; Fretzen, R.; Joussen, T.; Koppler, R.; Korall, B.; Neitler, P.; Richter, J. *J. Mol. Catal.* **1994**, *86*, 129.
18. Gonsalves, K. E.; Strutt, P. R.; Xiao, T. D.; Klemens, K. G. *J. Mater. Sci.* **1992**, *12*, 3231.
19. Dagani, R. *Chem. Eng. News*, **1992**, *72*, 18.
20. Gleiter, H. *Prog. Mater. Sci.* **1989**, *33*, 223.
21. Bridge, Jr., J. E.; Maniar, G. N.; Philip, T. V. *Metall. Trans.* **1971**, *2*, 2209.
22. Gonsalves, K. E.; Xiao, T. D.; Chow, G. M.; Law, C. C. *Nanostructr. Mater.* **1994**, *4*, 139.
23. Kayser, F.; Cohen, M. *Metal Progr.* **1952**, *61*, 79.
24. Pearson, W. B. *A Handbook of Lattice Spacing and Structures of Metals and Alloys,* vol. 1, Pergamon Press: New York, **1958**.
25. Kuo, K. J. *Iron Steel Inst.* **1953**, *173*, 363.
26. Suslick, K. S. *Science,* **1990**, *247,* 1439.
27. Suslick, K. S. *Ultrasonics,* **1992**, *30,* 171.
28. Suslick, K. S. in *Encyclopedia of Materials Science and Engineering,* 3rd suppl. Ed. Cahn, R. W. Pergamon Press, Oxford, **1993**, p. 2093.
29. Wonterghem, J. V.; Morup, S.; Charles, S. W.; Wells, S. *Phys. Rev. Lett.* **1985**, *55*, 410.
30. Gonsalves, K. E. U. S. Patent, 4,842,641, June 1989.
31. Suslick, K. S.; Fang, M.; Hyeon, T.; Chichowlas, A. A. *Mater. Res. Symp. Proc.* **1994**, *351,* 443.; Eds. Gonsalves, K. E.; Chow, G. M.; Xiao, T. D.; Cammarata, R. C.
32. Suslick, K. S. *Scientific American,* **1989**, *Feb.,* 80.
33. Schneider, J. *Acta Cryst.* **1987**, *A43,* 295.

RECEIVED December 29, 1995

Chapter 16

Chimie Douce Synthesis of Nanostructured Layered Materials

Doron Levin[1], Stuart L. Soled[2], and Jackie Y. Ying[1,3]

[1]Department of Chemical Engineering, Massachusetts Institute of Technology, Cambridge, MA 02139
[2]Exxon Research and Engineering Company, Route 22 East, Annandale, NJ 08801

A class of layered ammonium transition-metal molybdate materials were derived by a novel room-temperature *chimie douce* synthesis technique using calcined layered double hydroxides (LDHs) as precursors. The compounds obtained are highly crystalline, and retain the rhombohedral symmetry of the LDH precursors. The host structure consists of distorted divalent cation octahedra which share edges to form layers perpendicular to the c axis, analogous to the LDH precursor. The tetrahedral molybdate species, however, are not merely intercalated within the interlayer domain, but are bonded to the layers themselves through shared Mo-O-M bonds, where $M = Zn^{2+}$, Co^{2+}, Cu^{2+}, or Ni^{2+}. This arrangement results in the formation of a net negative charge on the host structure, leading to incorporation of ammonium ions between the layers for charge balancing. The applicability of this novel synthesis route is dependent on the composition of the LDH precursor, and it appears that metastability in the calcined LDH favors conversion to this phase.

Pillared layered structures (PLS) are nanocomposite materials prepared by linking molecules to a layered host. These structures are an excellent example of materials by design. Research in this area has been focussed on the design of new materials by intercalating nanostructures into layered precursors, the key issue being the engineering of the interlayer space between these two-dimensional precursors. There is considerable interest in the synthesis of new materials having interlayer dimensions on the nanometer scale (2-3 nm) which are thermally stable. In general, ionic lamellar solids are preferred hosts for the preparation of pillared intercalates. Synthesis of PLS can be accomplished by modification of the host-structure chemical composition, chemical or structural modification of the guest species domains, or both. There has been extensive research into the preparation of PLS, primarily based on structural modification of the guest species domains. Examples include the pillaring of smectites, such as montmorillonite,

[3]Corresponding author

0097–6156/96/0622–0237$15.00/0

by ion-exchange with polycationic species, e.g. Al_{13}-polyhydroxypolymer $(Al_{13}O_4(OH)_{24}(H_2O)_{12})^{7+}$ (*1-6*), Zr-hydroxypolymers $(Zr_4(OH)_8(H_2O)_{16})^{8+}$ (*7,8*), and other oligocations. One common feature of these materials, termed cross-linked smectite (CLS) molecular sieves, or pillared interlayered clays (PILC), is that they were prepared without chemical modification of the host composition. Other examples of PLS prepared by modification of only the guest species domains include the pillaring of layered double hydroxides (LDHs), also known as hydrotalcite-like materials, by various anionic species. Examples include the pillaring of LDHs with polyoxometalate (POM) anions of $[XM_{12}O_{40}]^{m-}$ or Keggin ion type by direct anion exchange (*9-14*), or utilizing an organic anion-pillared precursor that was subsequently exchanged with the appropriate isopolymetalate under mildly acidic conditions (*15-16*). To the best of our knowledge, however, no new layered structure has been prepared by the simultaneous modification of both the chemical composition of the host structure and the guest species domain. In this chapter, we report a novel room temperature *chimie douce* synthesis technique which produced a new class of layered transition metal molybdate (LTM) materials using calcined LDHs as precursors. These new materials, while being related structurally to the LDH precursor, have undergone a modification of the host chemical composition, as well as complete transformation of the interlayer domain.

Synthesis of LTM Materials

The synthesis of a LTM material occurs via the following three steps:
    i) Synthesis of the LDH precursor
    ii) Preparation of a metastable mixed oxide phase by calcination of LDH precursor
    iii) *Chimie douce* reaction of the mixed oxide phase to form the LTM
To illustrate these three steps, the synthesis of a Zn/Cu-LTM will be discussed.

Synthesis and Structure of the LDH Precursor. The precursor for the synthesis of Zn/Cu-LTM was a $Zn_{0.677}Cu_{0.098}Al_{0.225}(OH)_2(CO_3)_{0.1125}\cdot zH_2O$ LDH. The LDH was synthesized by coprecipitation at constant pH, under conditions of low supersaturation. The precursor was prepared by adding a solution of $Zn(NO_3)_2\cdot6H_2O$, $Cu(NO_3)_2\cdot6H_2O$ and $Al(NO_3)_3\cdot9H_2O$ to a solution of KOH and K_2CO_3, at relative rates such that a constant pH of 9.0 was maintained. The temperature of precipitation was 55±2 °C. Following precipitation, the material was aged in the mother liquor overnight at 70 °C. The material was then filtered, washed, and dried at 110 °C at atmospheric pressure. The X-ray diffraction pattern of this material is shown in Figure 1. The pattern was indexed using a hexagonal unit cell, with $a = 3.08(3)$ Å and $c = 22.83(3)$ Å, determined using a least squares method. The XRD pattern showed the presence of a secondary zincite (ZnO) phase. Peaks indicating the presence of this zincite phase are designated in Figure 1 with an asterisk. The presence of this secondary phase was expected due to the high zinc content (*17*), but proved to have no effect on the subsequent *chimie douce* synthesis.

The structure of the LDH precursor is very similar to that of brucite, $Mg(OH)_2$, where octahedra of Mg^{2+} (6-fold coordinated to OH⁻) share edges to form infinite sheets. These sheets are stacked on top of each other, and are held together by hydrogen bonding. Isomorphous substitution of a divalent cation in the lattice by a trivalent cation

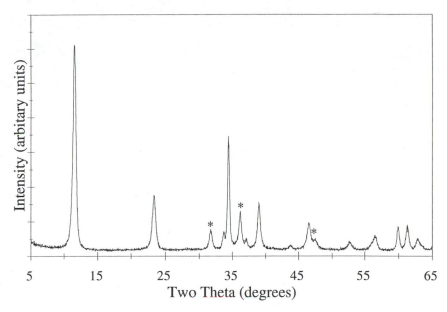

Figure 1.   Powder X-ray diffraction pattern of $Zn_{0.677}Cu_{0.098}Al_{0.225}(OH)_2$-$(CO_3)_{0.1125}·4H_2O$ (* indicates zincite phase).

having similar radius results in a positive charge generated in the hydroxy sheet. This net positive charge is compensated for by incorporation of hydrated anions in the interlayer region. Synthetic LDHs formed by coprecipitation from aqueous media crystallize in the $R\bar{3}m$ space group with three double-layers present per unit cell.

In general, the composition of the LDH precursors used for the synthesis of LTM materials can be represented by the formula:

$$[Zn_{(1-x-y)}M^{II}_yAl_x(OH)_2]^{x+}[(CO_3)_{x/2}·zH_2O],$$

where M^{II} is a divalent transition metal cation such as Ni^{2+}, Co^{2+}, or Cu^{2+}. Typically, the trivalent cation substitution parameter, x, given by $Al/(Zn + M^{II} + Al)$, is in the range of 0.2 to 0.33. The parameter y is constrained by the relationship $(1-x-y)/x > 1$, such that Zn is the dominant divalent transition metal cation.

Preparation of Metastable Mixed Oxide Phase. The key step in the synthesis of the LTM phase is the preparation of the metastable mixed oxide phase. This mixed oxide phase is prepared by calcination of the LDH precursor. At temperatures between initial LDH decomposition and spinel phase formation, a series of metastable phases, both crystalline and amorphous, can be formed (*18*). The properties of these phases depend on the cations constituting the original LDH, preparation and thermal decomposition

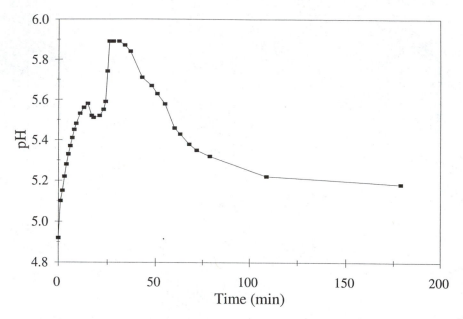

Figure 2. Change in pH during chimie douce reaction.

conditions, and the presence of impurities (*19*). The process of thermal decomposition of a carbonate LDH consists of dehydration, dehydroxylation and decarbonation, yielding a mixed oxide phase at temperatures less than 550 °C. This mixed oxide phase, when examined by X-ray diffraction, shows only broad peaks corresponding to a zincite (ZnO) phase. At this stage of the synthesis, the presence of any secondary zincite formed during precipitation of the LDH becomes indistinguishable. The importance of having zinc as the dominant transition metal results from the necessity of having the mixed oxide phase with a wurtzite structure. The Zn/Cu/Al-LDH precursor was calcined for 3.5 hours to prepared the metastable mixed oxide phase.

***Chimie Douce* Synthesis of Zn/Cu-LTM**. Following calcination, the mixed oxide phase was added to a room-temperature solution of ammonium heptamolybdate (0.05 M $Mo_7O_{24}^{6-}$ and 0.3 M NH_4^+). Progress of the reaction was followed by monitoring the pH. A typical pH profile over the course of a reaction is shown in Figure 2. The time to complete the reaction is dependent on the temperature of the mixed oxide as it enters the solution, and is typically between 40 minutes and two hours. Following completion of the reaction, the product was recovered by filtration, washed with deionized water, and dried overnight at 110 °C. The product, called Zn/Cu-LTM, was a fine, pale cyan powder.

Characterization of Zn/Cu-LTM

A powder X-ray diffraction pattern of this material, collected on a Rigaku diffractometer operated at 30 kV and 40 mA with CuKα radiation, is shown in Figure 3. This diffraction pattern is significantly different from ZnO, and bares no resemblance to that expected from a LDH pillared with the heptamolybdate anion (20,21). This diffraction pattern is almost identical to that of the Zn-LTM phase (22). A least squares refinement of the peak positions gave a hexagonal unit cell of $a = 6.104(9)$ Å, and $c = 21.69(9)$ Å. The crystal structure of the Zn/Cu-LTM is a derivative of the Zn-LTM, whose structure solution is reported elsewhere (22). The structure of the Zn-LTM was solved by Rietveld refinement of an isostructural ammonium nickel molybdate, $(NH_4)HNi_2(OH)_2(MoO_4)_2$, prepared by precipitation, whose own structure was determined, *ab initio*, from powder synchrotron data (23).

Crystal structure of Zn/Cu-LTM

The framework of the Zn/Cu-LTM consists of stacks of sheets built up from edge-sharing transition-metal octahedra to which tetrahedral molybdate groups are bonded. The zinc and copper atoms defining these layers are located at site 9(e), which can vary in occupancy from ½ to 1 (23). If the occupancy of this site is one, the arrangement of the transition metal atoms can be considered as a pattern of two alternating strings, one being M–M–M, as in LDH, the other being M–□–M, where M may be Zn or Cu, and □ represents an ordered cation vacancy. This ordered cation vacancy is independent of the occupancy of the 9(e) site. As the occupancy of this site deviates from unity, additional disordered vacancies appear in the sheet. The coordination of oxygens about the zinc and copper atoms is distorted octahedral, with a tetragonal contraction along the C_4 axis running through O–M–O, where M may be Zn or Cu, and O is a bridging oxygen. The zinc and copper atoms are linked to each other through double oxygen bridges. Each octahedron shares edges with four adjacent octahedra, thereby creating sheets of octahedra perpendicular to the c axis, analogous to the LDH precursor. The absence of an atom at the origin generates ordered vacancies in the sheet. These vacant octahedral sites are capped, both above and below, by tetrahedral molybdate groups. These tetrahedron above and below the octahedra sheet are related to each other by the center of inversion at the origin. The tetrahedra share the three oxygens forming the base with the zinc-oxygen octahedra, each oxygen in the base being shared with two different octahedra. Therefore, each molybdate tetrahedron shares corners with six different zinc octahedra, generating the hexagonal arrangement shown in Figure 4.

The three transition-metal molybdate layers defining the unit cell are stacked at a separation of $c/3$, analogous to the rhombohedral LDH precursor. In these layers, the position of the ordered vacancy capped by the molybdate groups follows the sequence A-B-C-A (where A, B, and C are the three threefold axes at x,y = 0,0; 2/3,1/3; and 1/3,2/3). This three layer arrangement is shown in Figure 5. Between each layer, in the space defined by the apical oxygens of six tetrahedral molybdates, lie ammonium ions. The positions of the nitrogen atoms follow the sequence C-A-B-C. The ammonium ions do not serve to connect the array of layers in the [00l] direction, the distance between the

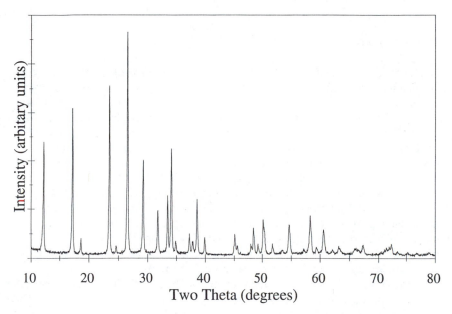

Figure 3. Powder X-ray diffraction pattern of Zn/Cu-LTM.

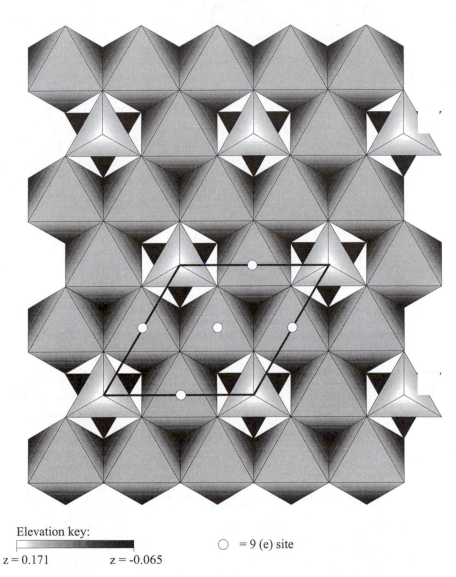

Elevation key:

z = 0.171   z = -0.065

○ = 9 (e) site

Figure 4. Basal plane of Zn/Cu-LTM, viewed along [001].

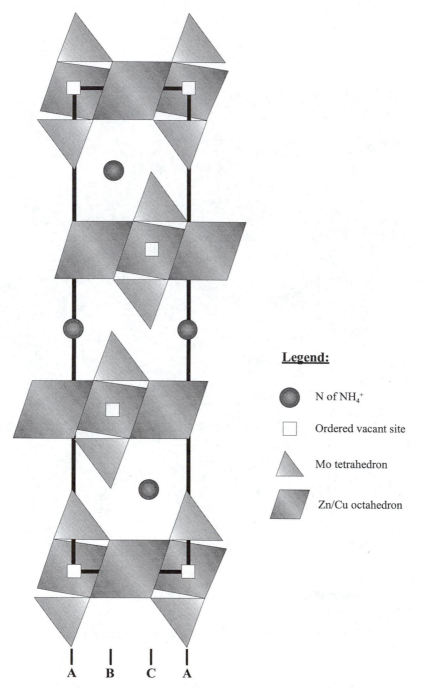

Legend:

⬤ N of NH_4^+

▢ Ordered vacant site

◣ Mo tetrahedron

▱ Zn/Cu octahedron

A   B   C   A

Figure 5. Crystal structure of Zn/Cu-LTM, viewed along [010], showing three layer arrangement in polyhedron representation.

nitrogen and the bridging oxygen in the adjacent layer being too long for hydrogen bonding. The vertical distance between the apical oxygen of the molybdate tetrahedra and the bridging oxygen of the adjacent zinc layer suggests that these two oxygens are involved in a hydrogen-bonding mechanism that serves to connect the layers (23). Figure 6 illustrates a three-dimensional polyhedron representation of the crystal structure of Zn/Cu-LTM.

Discussion

The crystal structure of Zn/Cu-LTM allows for an interesting comparison to its LDH precursor. Fundamentally, both structures have, as a basic framework, sheets of zinc, copper (and aluminum) octahedra running perpendicular to the c axis. The type and occupancy of the octahedral sites however, differ in the two structures. In the LDH framework, the metal cations are situated at site 3(a), and are fully occupied, such that a_0 equals the cation-cation distance within the sheet (Figure 7(a)). However, in the LTM structure, the metal cations are situated at site 9(e), and can have variable occupancy. Consequently, a for the LTM structure depends on the Mo-Mo distance, a distance equivalent to that between consecutive ordered vacancies at the origin (Figure 7(b)). Considering the case of a fully occupied 9(e) site in a LTM structure, there would be three metal cations per layer in the unit cell (Figure 7(b)). If we define a supercell of the LDH with $a' = 2 a_0$, $c' = c_0$, the area of the unit supercell would be almost identical to that of the LTM cell. In this equivalent area supercell however, there would be four metal cations per layer (Figure 7(c)). Therefore, the *chimie douce* synthesis reaction, while reconstructing the fundamental layered nature of the structure from the metastable zincite phase, resulted in a reduction of the number of cations in octahedral coordination in the layers.

A possible mechanism leading to this reduction in the number of cations has been determined on the basis of elemental analysis of the LTM phases, and ^{27}Al MAS-NMR data on the metastable mixed oxide phases (22). Elemental analysis had shown that the final product contained less than 1% Al. This suggested that, on reaction with the ammonium heptamolybdate solution, the Al in the zincite phase left the solid phase and entered into solution. The ^{27}Al MAS-NMR analysis of a calcined Zn_4Al_2-LDH showed a very high proportion (> 80%) of tetrahedrally coordinated Al (22). In comparison, ^{27}Al MAS-NMR analysis of a calcined Mg_4Al_2-LDH showed a relatively low proportion (<50%) of tetrahedrally coordinated Al (22). It is also known that no *chimie douce* reaction occurs for mixed oxide phases prepared from $[M^{II}_{(1-x)}Al_x(OH)_2]^{x+}[(CO_3)_{x/2} \cdot zH_2O]$ precursors, where $M^{II} = Mg^{2+}$ or Ni^{2+} (i.e. any M^{II} whose corresponding oxide, $M^{II}O$, is of the rock-salt structure.) This evidence suggested that a high proportion of tetrahedrally coordinated Al is a prerequisite for the *chimie douce* reaction. It also suggested that on reaction, the tetrahedrally coordinated Al enters into solution, thereby preventing the reconstruction of the LDH phase (which requires a positively charged framework resulting from trivalent cation substitution in the divalent lattice.) The loss of cations by this mechanism generates the vacancies in the octahedral sites of the oxide/hydroxide sheets which are required for the formation of the LTM structure.

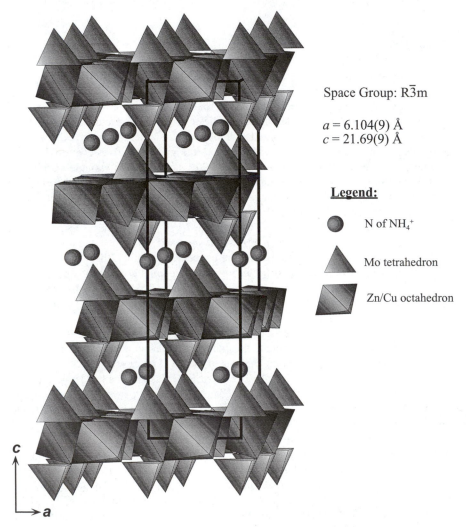

Space Group: R$\bar{3}$m

$a = 6.104(9)$ Å
$c = 21.69(9)$ Å

Legend:

⬤ N of NH_4^+

🔺 Mo tetrahedron

🔷 Zn/Cu octahedron

Figure 6. Crystal structure of Zn/Cu-LTM, viewed in polyhedron representation.

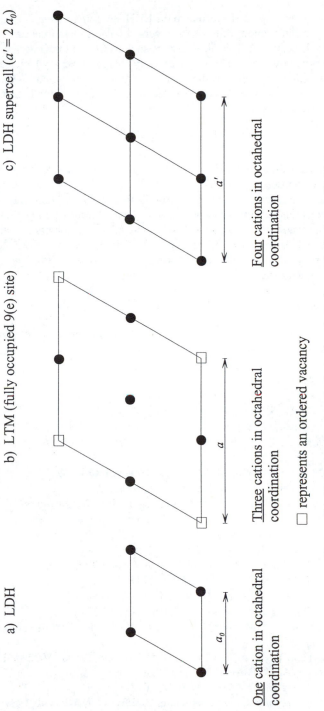

a) LDH

One cation in octahedral
coordination

b) LTM (fully occupied 9(e) site)

Three cations in octahedral
coordination

□ represents an ordered vacancy

c) LDH supercell ($a' = 2\,a_0$)

Four cations in octahedral
coordination

Figure 7. Variation in the number of cations in octahedral coordination resulting from chimie douce reaction.

When comparing the LTM structure to the LDH structure, it is interesting to note that the charge on the framework has been reversed. The LDH structure consists of positively charged sheets with intercalated anions such as CO_3^{2-} in the interlayer region. The LTM structure, however, consists of negatively charged layers, which leads to the incorporation of cations, e.g. NH_4^+, in the interlayer region. Therefore, this *chimie douce* reaction has changed not only the chemical and structural features of the LDH precursor, but its electronic nature as well.

Conclusions

We have illustrated the room-temperature synthesis of a crystalline structure prepared by the simultaneous modification of both the structure and chemical composition of the host and guest species domains of a layered precursor. This novel room-temperature *chimie douce* synthesis technique produced a new class of layered transition-metal molybdate (LTM) materials using calcined LDHs as precursors. These materials, being ionic lamellar solids themselves, may be suitable hosts for the synthesis of intercalated derivatives, thereby creating the potential for the preparation of a whole new class of molecularly designed materials.

Acknowledgments

This work was supported by the National Science Foundation (CTS-9257223) and the Exxon Summer Intern Program.

Literature Cited

(1)   Brindley, G. W.; Sempels, R. E. *Clay Minerals* **1977**, *12*, 229.
(2)   Lahav, N.; Shani, U.; Shabti, J. *Clays Clay Minerals* **1978**, *26*, 107.
(3)   Pinnavaia, T. J.; Tzou, M.-S.; Landau, S. D.; Raythatha, R. H. *J. Molec. Catal.* **1984**, *27*, 195
(4)   Occelli, M. L.; Tindwa, R. M. *Clays Clay Minerals* **1983**, *31*, 22.
(5)   Tokarz, M.; Shabtai, J. *Clays Clay Minerals* **1985**, *33*, 89.
(6)   Schutz, A.; Stone, W. E. E.; Poncelet, G.; Fripiat, J. J. *Clays Clay Minerals* **1987**, *35*, 251.
(7)   Jones, S.L. *Catalysis Today* **1988**, *2*, 209.
(8)   Sterte, J. *Catalysis Today* **1988**, *2*, 219.
(9)   Kwon, T.; Tsigdinos, G. A.; Pinnavaia, T. J. *J. Am. Chem. Soc.* **1988**, *110,* 3653.
(10)   Kwon, T.; Pinnavaia, T. J. *Chem. Mater.* **1989**, *1*, 381.
(11)   Doeuff, M.; Kwon, T.; Pinnavaia, T. J. *Synthetic Metals* **1989**, *34*, 609.
(12)   Kwon, T.; Pinnavaia, T. J. *J. Molec. Catal.* **74** (1992) 23.
(13)   Wang, J.; Tian, Y.; Wang, R.-C.; Colón, J. L.; Clearfield, A. *Mat. Res. Soc. Symp. Proc.* **1991**, *233*, 63.
(14)   Wang, J.; Tian, Y.; Wang R.-C; Clearfield, A. *Chem. Mater.* **1992**, *4*, 1276.
(15)   Drezdzon, M. A. *US Patent 4,774,212* **1988**.
(16)   Drezdzon, M. A. *Inorg. Chem.* **1988**, *27*, 4628.
(17)   Thevenot, F.; Szymanski, R.; Chaumette, P. *Clays Clay Minerals* **1989**, *37*, 396.

(18) Puxley, D. C..; Kitchener, I. J.; Komodromos, C.; Parkyns, N. D. *Preparation of Catalysts III*, Poncelet, G.; Grange, P.; Jacobs, P. A. (Eds), Elsevier Science, Amsterdam, **1983**, 237.
(19) Cavani, F.; Trifirò, F.; Vaccari, A. *Catalysis Today* **1991**, *11*, 173.
(20) Chibwe; K.; Jones, W. *Chem. Mater.* **1989**, *1*, 489.
(21) Misra, C.; Perotta, A. J. *US Patent 5,075,089* **1991.**
(22) Levin, D.; Soled, S. L.; Ying, J. Y. Submitted to *Chem. Mater.*
(23) Levin, D.; Soled, S. L.; Ying, J. Y. Submitted to *Inorg. Chem.*

RECEIVED December 14, 1995

Chapter 17

Clay-Reinforced Epoxy Nanocomposites: Synthesis, Properties, and Mechanism of Formation

Thomas J. Pinnavaia, Tie Lan, Zhen Wang, Hengzhen Shi, and P. D. Kaviratna

Department of Chemistry and Center for Fundamental Materials Research, Michigan State University, East Lansing, MI 48824

Epoxy-clay nanocomposites have been synthesized by exfoliation of organoclays during the epoxy thermosetting process. The extent of silicate layer separation is governed by the chain length of gallery cations, the clay layer charge density and the acidity of the gallery cations. By using different curing agents, we obtained both glassy and rubbery epoxy matrices. Reinforcement was substantially greater for the rubbery matrix. Both the tensile strength and the modulus of the rubbery epoxy-clay nanocomposite increased with increasing clay content. The reinforcement provided by the silicate layers at 15 wt% loading was manifested by a more than ten-fold improvement in tensile strength. The rubbery state of the matrix may allow alignment of the exfoliated silicate layers upon applying strain, thereby enhancing reinforcement.

The properties of hybrid organic-inorganic composite materials are greatly influenced by the length scale of component phases. Nanoscale dispersion of the inorganic component typically optimizes the mechanical properties of the composite. One successful approach to enhancing inorganic particle dispersion is the *in situ* polymerization of metal alkoxides in organic polymer matrices via the sol-gel process (1, 2). In these systems the inorganic components are formed by the hydrolysis and condensation of mononuclear precursors, such as tetraethoxysilane (TEOS).

Inorganic materials that can be broken down into their nanoscale building blocks can be superior alternatives to the sol-gel process for the preparation of nanostructured hybrid organic-inorganic composites (3). For instance, the exfoliation of the 10 Å-thick layers of smectite clays in polymer matrices is one such promising route to nanocomposite formation. Ion exchange of the Na^+ or Ca^{2+} gallery cations in the pristine mineral by organic cations allows modification

0097–6156/96/0622–0250$15.00/0

of the gallery surfaces for intercalation of organic polymer precursors. Also, the high aspect ratios of the exfoliated clay layers, typically in the range 200 - 2000, afford reinforcement properties comparable to fibers for certain polymer matrices.

In general, the dispersion of clay particles in a polymer matrix can result in the formation of three general types of composite materials (Figure 1). Conventional composites contain clay tactoids with the layers aggregated in unintercalated face - face form. The clay tactoids are simply dispersed as a segregated phase. Intercalated clay composites are intercalation compounds of definite structure formed by the insertion of one or more molecular layers of polymer into the clay host galleries and the properties usually resemble those of the ceramic host. In contrast, exfoliated polymer-clay nanocomposites have a low clay content, a monolithic structure, a separation between layers that depends on the polymer content of the composite, and properties that reflect those of the nano-confined polymer.

Intercalated polymer-clay nanocomposites have been synthesized by direct polymer intercalation (4-6), and *in situ* intercalative polymerization of monomers in the clay galleries (7, 8). Owing to the spatial confinement of the polymer between the dense clay layers, intercalated polymer-clay nanocomposites can exhibit impressive conductivity (6) and barrier properties (7). The exfoliation of smectite clays provides 10 Å-thick silicate layers with high in - plane bond strength and aspect ratios comparable to those found for fiber reinforced polymer composites.

Exfoliated clay nanocomposites formed between organocation exchanged montmorillonites and thermoplastic Nylon-6 have recently been described by Toyota researchers (9-11). Clay exfoliation in the Nylon-6 matrix gave rise to greatly improved mechanical, thermal and rheological properties, making possible new materials applications of this polymer. However, it is relatively difficult to achieve complete exfoliation of smectite clays into a continuous polymer matrix, because of the strong electrostatic attraction between the silicate layers and the intergallery cations.

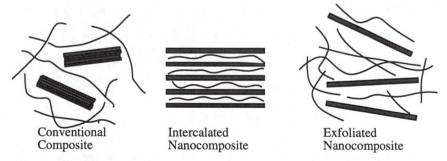

Conventional   Intercalated   Exfoliated
Composite   Nanocomposite   Nanocomposite

Figure 1. Schematic illustrations of the three possible types of polymer-clay composites.

Recently, exfoliated layered silicate-epoxy nanocomposites have been prepared from the diglycidyl ether of bisphenol-A and nadic methyl anhydride (12). The dynamic storage modulus of the nanocomposite containing 4 vol %

silicate was approximately 1.58 times higher in the glassy region and 4.5 times higher in the rubbery plateau region compared to the pristine polymer. Concurrently, we found that monolithic exfoliated clay nanocomposites can be formed by pre-swelling alkylammonium ion exchanged forms of the clays with epoxy resin prior to curing. Dramatic improvement in the tensile strength and modulus was realized, particularly when the matrix exhibited a sub-ambient glass transition temperature (13, 14). For instance, the reinforcement provided by the silicate layers at 16 wt% loading was manifested by a more than ten-fold improvement in both tensile strength and modulus.

In this paper, we describe the mechanism for formation of epoxy-clay nanocomposites and compare the tensile properties of the intercalated and exfoliated forms of these nanocomposites.

Interaction of organoclay by epoxy resin monomers

The exfoliation of the clay tactoids in a polymer matrix requires the driving force of polymerization to overcome the attractive electrostatic force between the negatively charged silicate layers and the gallery cations. Pre-intercalation of monomers into the clay galleries prior to the polymerization reaction should facilitate exfoliation. The organoclays used in this study are long chain alkylammonium exchanged montmorillonite. The organoclay can be easily prepared by ion exchange reaction using the desired onium salt in $H_2O/EtOH$ mixed solvent. Intercalation of epoxy resin monomers into smectite clay galleries is observed at 75 °C. The specific epoxy monomer of interest here, Epon-828, has the following structure:

$n = 0$ (88%); $n = 1$ (10%); $n = 2$ (2%).

The XRD basal spacings of organoclays under air dried and epoxy solvated conditions are listed in Table Ia. Note that the chain length of the alkylammonium ion greatly effects the extent of clay expansion upon epoxide solvation. The basal spacings for a series of epoxy-solvated $CH_3(CH_2)_{n-1}NH_3^+$-montmorillonites with $n = 4, 8, 10, 12, 16$ and 18 increase in proportion to the chain length of the onium ion. Assuming that the gallery cations upon epoxy solvation reorient from their initial monolayer, bilayer, or paraffin-like orientation to a vertical orientation relative in the clay basal surfaces, we expect the basal spacings (Å) to follow the relationship $d_{001} = (n - 1)*1.27 + d_A + r_M$, where $(n - 1)$ is the number of methylene groups in the onium ion chain, d_A is the basal spacing for NH_4^+-montmorillonite (12.8 Å), r_M is the van der Waals radius of the methyl end group (3.0 Å) and 1.27 Å is the contribution due to the $-CH_2-$ chain segments when the chain adopts an all trans configuration. As can be seen from the last entry in Table I, the observed values parallel the calculated spacings for a vertical orientation of the ions.

Alkylammonium ion with the same long chain but different head groups have been used to verify the change in the onium ion orientation upon the epoxy intercalation. Epoxy resin solvation in montmorillonites with primary, secondary, tertiary and quaternary alkylammonium species having the same alkyl chain length of 18 including: $CH_3(CH_2)_{17}NH_3^+$, $CH_3(CH_2)_{17}NH_2(CH_3)^+$, $CH_3(CH_2)_{17}NH(CH_3)_2^+$ and $CH_3(CH_2)_{17}N(CH_3)_3^+$ was studied to confirm the chain length control on the epoxy resin solvation degree. The basal spacings are listed in Table Ib. Although these montmorillonites have initial basal spacings, the epoxy solvated clays have essentially the same basal spacing which is controlled by the C_{18} long alkyl chain. The alkyl chain controlled clay intercalation has also been confirmed by using clays with different layer charge densities (13). Clays with different layer charge densities but with the same intergallery cations show the same swelling properties. Upon epoxide intercalation into the clay galleries, the alkylammonium ions change their original orientations, which is determined by the clay layer charge and alkyl chain length, to a more vertical orientation, as indicated by XRD studies.

Table Ia. Basal Spacings (Å) of Alkylammonium Exchanged Montmorillonites.

Gallery Cation	Initial Cation Orientation	Air-Dried	Epoxy Solvated	Calcd. Value
$CH_3(CH_2)_3NH_3^+$	monolayer	13.5	16.5	19.6
$CH_3(CH_2)_7NH_3^+$	monolayer	13.8	27.2	24.7
$CH_3(CH_2)_9NH_3^+$	monolayer	13.8	30.0	27.2
$CH_3(CH_2)_{11}NH_3^+$	bilayer	15.6	31.9	29.8
$CH_3(CH_2)_{15}NH_3^+$	bilayer	17.6	34.1	34.9
$CH_3(CH_2)_{17}NH_3^+$	bilayer	18.0	36.7	37.4

Table Ib.

Gallery Cations	Air-Dried	Epoxy Solvated
$CH_3(CH_2)_{17}NH_3^+$	18.0	36.7
$CH_3(CH_2)_{17}N(CH_3)H_2^+$	18.1	36.2
$CH_3(CH_2)_{17}N(CH_3)_2H^+$	18.7	36.7
$CH_3(CH_2)_{17}N(CH_3)_3^+$	22.1	36.9

The model illustrated in Figure 2 summarizes the overall mechanism for formation of epoxy polymer - clay nanocomposites. Upon solvation of the organoclay by the epoxide monomers, the gallery cations reorient from their initial monolayer, lateral bilayer, or inclined paraffin structure to a *perpendicular* orientation with epoxy molecules inserted between the onium ions. A related reorientation of alkylammonium ions has been observed previously for ε-caprolactam intercalated clay intermediates formed in the synthesis of Nylon-6 - exfoliated clay nanocomposites (9). Thus, the ability of the onium ion chains to reorient into a vertical position in order to optimize solvation interactions with the monomer may be a general prerequisite for pre-loading the clay galleries with sufficient monomer to achieve layer exfoliation upon intragallery polymerization.

Formation of Epoxy-Clay Nanocomposites

The addition of the curing agent m-phenylenediamine (mPDA) to epoxide-clay mixtures at room temperature resulted in little or no change in the clay basal spacings. However, upon initiating crosslinking at elevated temperatures, significant changes in clay basal spacings were observed.

A. Low charge density clay

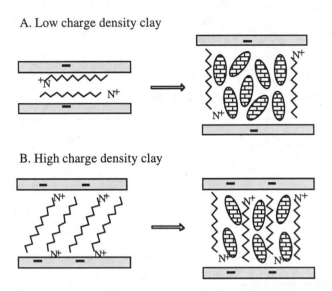

B. High charge density clay

Figure 2. Proposed model for the swelling of alkylammonium exchanged clay by Epon-828: A, low charge density clay with a lateral bilayer structure and B, high charge density clay with a paraffin structure. Cross hatched ellipses represent the intercalated resin molecules.

The optimized curing condition was found by using different curing cycles to carry out the resin thermosetting. XRD patterns for mPDA-cross linked epoxy-clay composites containing 5 wt% $CH_3(CH_2)_{15}NH_3^+$-montmorillonite cured under different thermal conditions indicate that the composites cured slowly at 75 °C for 4 h or rapidly at 140 °C for 4 h exhibit broad Bragg diffraction peaks at 32 Å and 34 Å, respectively. In contrast, the same compositions cured at intermediate rates by a two-stage curing process (first at 75 °C for 2 h and then at 100 °C for 4 h or at 125 °C for 2 h) do not exhibit diffraction peaks. The absence of Bragg scattering indicates that the clay tactoids have been completely exfoliated (delaminated) by the polymerization process into individual, 10 Å-thick layers to form a nanocomposite. Therefore, we used the curing circle of 2 h at 75 °C and additional 2 h at 125 °C to prepare epoxy-clay nanocomposites. The change in structure that occurs upon converting $CH_3(CH_2)_{15}NH_3^+$-montmorillonite into the epoxide solvated form and finally into a cured polymer-exfoliated clay nanocomposite is illustrated in Figure 3.

Because the amount of resin that can be loaded into the galleries of an organoclay is dependent on the chain length of the onium cation, the degree of organoclay exfoliation in a mPDA-epoxy matrix also depends on the gallery cations. XRD diffraction patterns for cured epoxy-clay composites containing 5 wt% of various alkylammonium exchanged forms of montmorillonites are shown in Figure 4. The results for NH_4^+- and Na^+-montmorillonite are included for comparison. Composites formed with $CH_3(CH_2)_7NH_3^+$-, $CH_3(CH_2)_{11}NH_3^+$- and $CH_3(CH_2)_{15}NH_3^+$-montmorillonite exhibit no Bragg scattering, indicating that

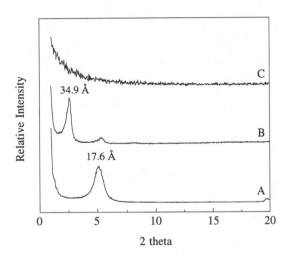

Figure 3.   X-ray diffraction patterns of $CH_3(CH_2)_{15}NH_3^+$-montmorillonite in different physical states: (A) pristine clay, (B) Epon-828 solvated clay, (C) exfoliated clay in a mPDA-cured epoxy nanocomposite.

the clay has been *exfoliated* into 10 Å-thick layers with no regular repeat distance between the layers. That is, the clay layer separation is large (>80Å) and highly disordered. It is noteworthy that $CH_3(CH_2)_3NH_3^+$-montmorillonite with an initial d_{001} of 13.5 Å characteristic of a monolayer of onium ion, gives rise to a diffraction peak at 16.6 Å under mPDA-cured conditions, indicating that a polymer monolayer is formed in the clay gallery. This latter material is an *intercalated* nanocomposite. No layer exfoliation or intercalation was observed for NH_4^+- and Na^+- exchanged montmorillonite. Thus, montmorillonites intercalated by alkylammonium cations with chain lengths larger than 8 carbon atoms yield *exfoliated* nanocomposites. Whereas the clays interlayered by shorter alkylammonium cations and simple inorganic cations tend to afford *intercalated* nanocomposite architecture and *conventional* phase segregated composites, respectively.

The dependence of nanocomposite formation on onium ion chain length can be explained by the following general reaction mechanism. Swelling of clay by the epoxy monomers controls the initial accessibility of the galleries for polymer formation. Long chain alkylammonium exchanged montmorillonites solvated by the epoxy provide a hydrophobic environment for mPDA to migrate into the clay interlayer region. Under appropriate curing conditions more epoxy and mPDA penetrate the gallery space and intragallery polymerization can occur at a rate that is comparable to extragallery polymerization. Consequently, the galleries continue to expand as the degree of polymerization increases and a monolithic exfoliated nanocomposite is formed. However, if the curing temperature is too low and the rates of epoxy and mPDA intercalation are too low, then extragallery polymerization will be faster than intragallery polymerization and intercalated nanocomposites will form. Intercalated nanocomposites also will

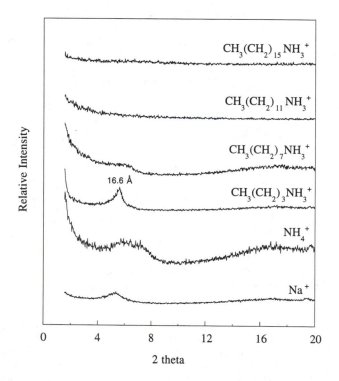

Figure 4. X-Ray patterns of epoxy-clay composite materials formed by the polymerization of a DGEBA resin, Epon-828, with a stoichiometric amount of mPDA as the curing agent in the presence of various cation exchanged forms of montmorillonite. The exchange ions is identified for each diffraction pattern.

form if the extragallery curing is favored by curing at high temperature. Thus, it is important to select curing conditions that balance the intra- and extragallery polymerization rates. Short alkylammonium exchange ions on the gallery surfaces make intragallery diffusion more restrictive, and it becomes more difficult to achieve the curing conditions needed to balance the rates of intra- and extragallery polymerization. Consequently, intercalated clay nanocomposite structures tend to be formed. Inorganic exchanged forms of montmorillonite are hydrophilic and not readily swelled by epoxy monomers, making intragallery polymerization impossible. Therefore, conventional phase-separated composites are formed for Na^+- and NH_4^+-montmorillonites.

Owing to the spatially restricted nature of the gallery environment, one might not expect the intragallery polymerization rate to be competitive with the extragallery rate under any curing conditions. However, the alkylammonium ions that favor clay layer exfoliation are acidic, and can catalyze the epoxy-amine polymerization reaction. The importance of intragallery catalysis is illustrated by the decrease in layer exfoliation with decreasing Bronsted acidity of the exchange ion in the following order: $CH_3(CH_2)_{17}NH_3^+ >$ C $H_3(CH_2)_{17}N(CH_3)H_2^+ >$ $CH_3(CH_2)_{17}N(CH_3)_2H^+ > CH_3(CH_2)_{17}N(CH_3)_3^+$. For these onium ions, the initial swelling of the clay by the epoxy monomer is essentially constant (d_{001} = 36.2 - 36.9 Å) simply because the epoxy intercalation is determined by the C_{18} chain length of the gallery cations. But the X-ray diffraction results for the amine-cured epoxy-clay composites, as shown in Figure 5, indicate the clays with primary and secondary onium ions form exfoliated nanocomposites, whereas the clays with tertiary and quaternary onium ions retain the clay tactoid structure typical of intercalated nanocomposites. Therefore, the formation mechanism of exfoliated epoxy-clay nanocomposite can be described as intragallery catalytic polymerization.

Mechanical Properties of Epoxy-Clay Nanocomposites

In order to assess the benefit of clay exfoliation in the epoxy matrix, the tensile strengths and moduli for the amine-cured epoxy-clay nanocomposites have been determined for loadings in the range 1 - 2 wt %. For the pristine amine-cured epoxy matrix, the tensile strength is 90 MPa and the tensile modulus is 1.1 GPa. Comparing the mechanical properties of the exfoliated and intercalated nanocomposite in Figure 6, we find that the exfoliated nanocomposites show improved performance, especially in the modulus, relative to the pristine polymer.

Due to the high glass transition temperature (~150 °C) of the epoxy matrix, the composites were in a glassy state at room temperature. Greatly enhanced properties might be expected for epoxy-exfoliated clay nanocomposites with sub-ambient glass transition temperatures. Thus, it was of interest to us to examine the properties of an epoxy-clay nanocomposite in a rubbery state. By using different curing agents, we can control the T_g of the epoxy matrix. The following polyetheramine (JEFFAMINE D2000, Huntsman) was used as the curing agent to achieve sub-ambient glass transition temperatures:

$$H_2NCHCH_2 \left[OCH_2CH \right]_x NH_2$$
$$\underset{CH_3}{|} \qquad \underset{CH_3}{|}$$

where x = 33.1 and the molecular weight is 2000.

The replacement of mPDA by a polyetheramine curing agent afforded rubbery epoxy clay nanocomposites by a mechanism analogous to that described earlier for brittle composites. Long chain acidic alkylammonium exchange forms of montmorillonite readily dispersed into the epoxy matrix during the thermosetting process, yielding exfoliated clay nanocomposites. However, short chain alkylammonium derivatives (n < 8) gave intercalated nanocomposites, as judged by XRD. For comparison purposes, we also prepared rubbery epoxy-clay nanocomposites using non-acidic alkylammonium exchanged montmorillonite.

XRD patterns of the pristine clays and epoxy-clay composites are shown in Figure 7. It is obvious that the epoxy-clay nanocomposite formed from non-acidic $CH_3(CH_2)_{17}N(CH_3)_3^+$- montmorillonite is an typical intercalated nanocomposite with a well expressed basal spacing of 41.2 Å. In contrast $CH_3(CH_2)_{17}NH_3^+$- montmorillonite affords a composite in which layers are almost completely exfoliated.

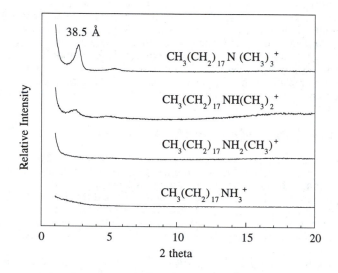

Figure 5.   X-ray powder diffraction patterns of amine-cured epoxy-clay nanocomposites formed from montmorillonite clays (5 wt%) containing primary, secondary, tertiary and quaternary onium ions with a n-C18 chain lengths.

The tensile strengths and moduli of our new epoxy-clay nanocomposites have been measured to evaluate the reinforcing effect of the exfoliated clay layers in the rubbery matrix. The presence of the organoclay substantially increases both the tensile strength and the modulus relative to the pristine polymer. The mechanical properties increase with increasing clay exfoliation in the order $CH_3(CH_2)_7NH_3^+$- < $CH_3(CH_2)_{11}NH_3^+$- < $CH_3(CH_2)_{17}NH_3^+$-montmorillonite. It is noteworthy that the strain at break for all of our epoxy-clay composites is essentially the same as the pristine matrix, suggesting that the exfoliated clay particles do not disrupt matrix continuity.

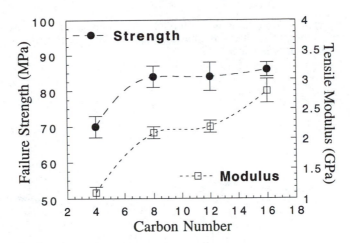

Figure 6. Dependence of tensile strength and modulus of amine-cured epoxy-clay nanocomposites on onium ion carbon number at clay loadings of 2 wt%.

Reinforcement of the rubbery epoxy-clay nanocomposites also was dependent on clay loading. The tensile strength and modulus for the $CH_3(CH_2)_{17}NH_3^+$-montmorillonite system increases nearly linearly with clay loading. More than a ten-fold increase in strength and modulus is realized by the addition of only 15 wt% (~ 7.5 vol%) of the exfoliated organoclay (14). The properties of the $CH_3(CH_2)_{17}NH_3^+$- and $CH_3(CH_2)_{17}N(CH_3)_3^+$- montmorillonite nanocomposites at different clay loadings are given in Figure 8. Both tensile strengths and moduli for these intercalated and exfoliated nanocomposites increase with increasing clay content in the composites. It is interesting that the tensile strengths and moduli of intercalated and exfoliated nanocomposites have very similar values at low clay loading (≤ 5 wt%). At high clay loadings (> 5 wt%), however, the exfoliated nanocomposites exhibit a much greater reinforcing effect.

Future studies of the mechanical properties of glassy and flexible epoxy-clay nanocomposites may elucidate the reinforcing mechanism. Owing to the increased elasticity of the matrix above T_g, the improvement in reinforcement may be due in large part to shear deformation and stress transfer to the platelet particles. In addition, platelet alignment under strain may also contribute to the improved performance of clays exfoliated in a rubbery matrix as compared to a glassy matrix. The most significant difference between glassy and rubbery polymers is the elongation upon stress. The rubbery epoxy matrix used in this work exhibits 40 ~ 60 % elongation at break, whereas for the glassy epoxy matrix we reported previously it is only 5 ~ 8 %. When strain is applied in the direction parallel to the surface, the clay layers will be aligned further. This strain - induced alignment of the layers will enhance the ability of the particles to function as the fibers in a fiber-reinforced plastics. Propagation of fracture across the polymer matrix containing aligned silicate layers is energy consuming, and the tensile strength and modulus is reinforced. In a glassy matrix, clay particle alignment upon applied stress is minimal and blocking of the fracture by the exfoliated clay is less efficient.

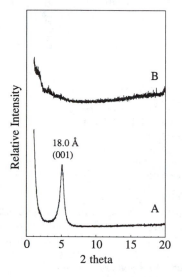
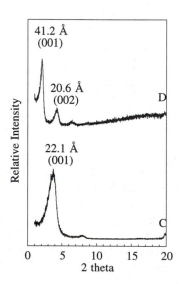

Figure 7. XRD patterns A: $CH_3(CH_2)_{17}NH_3^+$-montmorillonite, B: Epoxy nanocomposite containing 10 wt% of $CH_3(CH_2)_{17}NH_3^+$-montmorillonite, C. $CH_3(CH_2)_{17}N(CH_3)_3^+$-montmorillonite, D. Epoxy composite containing 10 wt% of $CH_3(CH_2)_{17}N(CH_3)_3^+$-montmorillonite.

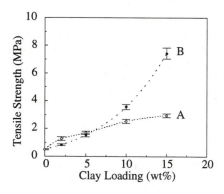
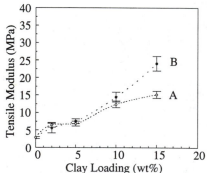

Figure 8. Comparison of tensile strengths and moduli of intercalated nanocomposites containing $CH_3(CH_2)_{17}N(CH_3)_3^+$-montmorillonite (curve A) and exfoliated nanocomposites containing $CH_3(CH_2)_{17}NH_3^+$-montmorillonite (curve B).

Summary

Smectite clays are promising materials for the preparation of nanostructured organic - inorganic composites by *in situ* intragallery polymerization processes. Epoxy-clay nanocomposites have been synthesized by pre-intercalating alkylammonium exchanged montmorillonites with epoxy monomer prior to the onset of polymerization. The extent of silicate layer separation achieved upon polymerization is governed by the chain length of gallery cations, the clay layer charge density and the acidity of the gallery cations. By using different curing agents, both glassy and rubbery epoxy composites may be formed. We have found that the mechanical reinforcement provided by the exfoliated clay layers is much more significant for a rubbery matrix than a glassy matrix. For a rubbery matrix, both the tensile strength and the modulus of the nanocomposite increased with increasing clay content. The reinforcement provided by the silicate layers at 15 wt% loading was manifested by a more than ten-fold improvement in tensile properties. The rubbery state of the matrix may allow alignment of the exfoliated silicate layers upon applying strain, thereby contributing to the reinforcement mechanism.

Acknowledgments

This research has been supported by the Michigan State University Center for Fundamental Materials Research and, in part, by NSF under grant CHE-92241023 and American Colloid Company. We thank the MSU Composite Materials and Structures Center for the use of the UTS system for mechanical testing.

References

1. Novak, B. *Adv. Mater.*, **1993**, 5, 422.
2. Philipp, G.; Schimdt, H. *J. Non-Crystalline Solids.*, **1984**, 283.
3. Pinnavaia, T.J. *Sci..* **1983**, 365.
4. Messersmith, P.B.; Giannelis, E.P. *Chem. Mater.* **1993**, 5, 1064.
5. Lan, T.; Kaviratna, P.D; Pinnavaia, T.J. *Chem. Mater.,* **1994**, 6, 573.
6. Ruiz-Hitzky, E.; Aranda, P.; Casal, B.; Galvan, C. *Adv. Mater.*, **1995**, 7, 180.
7. Kato, C.; Kuroda, K.; Misawa, M. *Clays and Clay Miner.* **1979**, 27, 129.
8. Vaia, R.A.; Ishii, H.; Giannelis, E.P. *Chem. Mater.* **1993**, 5, 1694.
9. Usuki, A.; Kawasumi, M.; Kojima, Y.; Okada, A.; Kurauchi, T.; Kamingato, O. *J. Mater. Res.* **1993**, 8, 1174.
10. Usuki, A.; Kojima, y.; Kawasumi, M.; Okada, A.; Fukushima, Y.; Kurauchi, T.; Kamigaito, O. *J. Mater. Res.* **1993**, 8, 1179.
11. Kojima, Y.; Usuki, A.; Kawasumi, M.; Okada, A.; Fukushima, Y.; Kurauchi, T.; Kamigaito, O. *J. Mater. Res.* **1993**, 8, 1185.
12. Messersmith P.B.; Giannelis, E.P. *Chem. Mater.*, **1994**, 6, 1719.
13. Lan, T.; Kaviratna, P.D.; Pinnavaia, T.J., *Preceedings of ACS, Div. of Polymeric Materials: Science and Engineering (PMSE)* **1994**, 71, 528.
14. Lan, T.; Pinnavaia, T.J., *Chem. Mater.*, **1994**, 6, 2216.

RECEIVED December 1, 1995

Chapter 18

New Polyphosphazene–Clay and Cryptand–Clay Intercalates

John C. Hutchison, Rabin Bissessur, and Duward F. Shriver

Department of Chemistry and Materials Research Center, Northwestern University, Evanston, IL 60208–3113

Poly(bis-methoxyethoxyethoxy phosphazene) (MEEP) and cryptand[2.2.2] (crypt) have been intercalated separately into sodium exchanged montmorillonite (Na-mont), and impedance spectroscopy indicates that ionic conductivities of the resulting nanocomposites (log $\sigma = -7.6$ $(\Omega \cdot cm)^{-1}$ for MEEP/Na-mont and log $\sigma = -8.1$ $(\Omega \cdot cm)^{-1}$ at 100°C) are enhanced by several orders of magnitude over the parent Na-mont. Intercalation is demonstrated by powder x-ray diffraction and FTIR spectroscopy. ^{23}Na and ^{29}Si MAS NMR spectra give evidence that the interlayer Na$^+$ are solvated by the intercalated ethers. FTIR spectroscopy, ^{13}C NMR spectroscopy and ^{1}H NMR spectroscopy indicate that cryptand intercalates preferentially to MEEP when a mixture of the two is stirred with Na-Mont.

Solid state electrolytes are the subject of increasing interest because of their potential in technical applications and for fundamental questions regarding conduction mechanism. Our group, as well as others, has been heavily involved in the exploration of polymer ion conducting systems such as poly(ethylene oxide) (PEO) / lithium triflate (LiSO$_3$CF$_3$) complexes (*1*), poly(bis-methoxyethoxyethoxy phosphazene) (MEEP) / LiSO$_3$CF$_3$ complexes (*2*), and poly(oligoether aluminosilicate) polyelectrolytes (*3*). The focus of study has shifted as understanding of structure of polymer electrolytes and mechanism of ion transport has developed. The topic of this paper, nanostructured clay intercalation compounds, represents a new direction in solid electrolytes (*4-6*).

Polymer-Salt Complexes and Polyelectrolytes

Research on solvent-free polymer-salt complexes accelerated after the pioneering studies of PEO-alkali metal salt complexes conducted by Armand (*7,8*) and Wright (*9-11*).

0097–6156/96/0622–0262$15.00/0

PEO-salt complexes are among the most thoroughly studied polymer electrolytes, and they provide a good general model for structure and conduction mechanism.

Poly(ethylene oxide) is a partially crystalline linear polyether (*12*).   When complexed with alkali metal salts, the oxygen atoms of a polyether solvate the metal cation. X-ray diffraction studies on oriented fibers of the partially crystalline PEO-salt complexes indicates coordination of the cation to the polyether oxygens and association of the cation with the anion (*13-15*). IR spectroscopy also confirms this structure in PEO and related oligoether comb polymers with evidence of cation vibrations within the ether cage (*16-18*). IR and Raman spectroscopies can also provide useful information as to degree of ion pairing and polymer conformation (*19-21*).   Berthier and co-workers and many subsequent investigators have demonstrated that conduction occurs preferentially in amorphous regions of the polymer (*22*). The temperature dependence of conductivity in polymer electrolytes provides important information about the nature of the conduction mechanism. The curvature in log σ vs. 1/T plots in variable temperature experiments indicates non-Arrhenius behavior (*12*). Above the glass transition temperature of the complex, conductivity data often fit very well to the Vogel, Tamman, Fulcher (VTF) equation (*23*):

$$\sigma = A exp\left(-\frac{B}{k(T-T_o)}\right)$$

in which A is a constant, T is temperature in degrees Kelvin, B is an empirical constant, k is Boltzmann's constant, and T_o is related to the glass transition temperature (T_g) of the polymer. Below the glass transition temperature, local high-amplitude motions of the polymer are lost and a huge drop in conductivity occurs. This corresponds to a change in macroscopic properties from a rubber above glass transition temperature to a glass upon cooling below glass transition temperature. VTF behavior, therefore, is consistent with a conduction mechanism in which ion motion is coupled with the thermal motion of the host polymer. Cheradame and co-workers discussed the VTF plots in terms of segmental motion of the polymer host (*24-27*), and the observed importance of glass transition temperature on conductivity led to theories which involved large amplitude segmental motion of the host polymer, such as free volume models and dynamic percolation models (*28-30*).

The importance of low glass transition temperature to achieve high conductivity led to the development of polymer hosts with high segmental mobility and lack of crystallinity such as the comb polymer MEEP $[(CH_3OC_2H_4OC_2H_4O)_2PN]_n$, $T_g = -80°C$ (*31*), and poly(oxymethylene oligo(ethylene oxide)), PEO with incorporated oxymethylene units to break up crystallinity (*32*).

Ion Pairing.   Another important influence on the conductivity in polymer-salt complexes is ion-ion interactions. Coulombic interactions between cations and anions pin the ions and reduce conductivity (*24,33*). Polymer complexes with weakly basic anions such as triflate ($SO_3CF_3^-$) or perchlorate (ClO_4^-) and larger cations display higher conductivity than polymer complexes with smaller cation-anion combinations where tight ion pairing occurs in the polymer. One strategy to reduce ion pairing is the introduction of cation complexing agent such as cryptand which surrounds the cation and thereby reduces the coulombic forces between ions (30,34,35).

Transport Numbers. The identity of the charge carriers in solid state electrolytes is an important issue for both fundamental and technological reasons. Polymer-salt complexes typically exhibit low cation transport numbers (t_+), below 0.4, (12) and current can be carried by low-charge, coloumbically-interacting multi-nuclear species (12,36). For unambiguous interpretation of conductivity data, and to minimize salt concentration gradients in the vicinity of electrodes in applications such as lithium batteries, a cation transport number of unity is desirable. Our research group and others have synthesized polymer polyelectrolytes in which the anions are covalently bonded to the polymer and therefore lack long-range mobility. Two such systems are sulfonate-modified poly(phosphazenes) (37,38), and aluminosilicate/poly(ethylene glycol) copolymers (3). Because the anions are immobile, t_+ is unity, and current is exclusively carried by cations. As with simple polymer-salt complexes, the polyelectrolytes exhibit VTF behavior, and the conductive mechanism involves large amplitude segmental motion of the polymer. Also, weakly basic anions and cryptands reduce ion-ion interactions and thereby increase conductivity.

Smectites as Electrolytes

We are attracted to the idea of using smectite clays as polyelectrolytes because the immobile anionic site can be buried in the aluminosilicate layer, thereby reducing ion pairing and enhancing cation transport. Smectites are low-charge hydrous layer silicates of the mineral family phyllosilicate with the general formula $C_{(x+y)}[A_{(a-x)}X_xSi_{(8-y)}Y_yO_{20}(OH)_4]$ with a = 6 if A is divalent and a = 4 if A is trivalent. Each layer consists of a central octahedral metal oxide sheet sandwiched between two sheets of SiO_4 tetrahedra bonded through oxygen. A slight negative charge on the silicate layer which arises from nonstoichiometric substitution in the oxide sheets with a lower valency species is compensated with interlayer cations. Smectites have been used as solid acid catalysts (39), and the poor basicity of the anions presumably leads to enhanced ion mobility. In addition, montmorillonite and hectorite have charge imbalance sites buried within the layers on the octahedral sheets which prevents close association with the interlayer cations. The fact that smectites exist in a variety of forms should aid in the identification of the factors affecting conductivity. The charge imbalance of smectites can vary from 0.6 to 1.2 equivalents per Si_8O_{20} unit. Since higher layer charges require more compensating interlayer cations, smectites such as hectorite have a higher interlayer charge carrier concentration than lower charged species such as montmorillonite. Table I shows the range of layer charges available in smectites. The ease with which interlayer cations can be exchanged allows different charge carriers to be studied in otherwise identical systems.

Smectites have been studied as ionic conductors and in fact can have quite high conductivities, often ranging up to 1×10^{-3} $(\Omega \cdot cm)^{-1}$ (40,41). However, these high

Table I. Smectites (55)

Subgroup	Species	Ideal Formula
Saponites	saponite	$Na_{0.6}[Mg_6(Al_{0.6}Si_{7.4})O_{20}(OH)_4]$
	hectorite	$Li_{1.6}[(Li_{1.6}Mg_{4.4})Si_8O_{20}(OH)_4]$
	fluorohectorite	$Li_{1.6}[(Li_{1.6}Mg_{4.4})Si_8O_{20}F_4]$
Montmorillonites	montmorillonite	$Na_{0.6}[(Mg_{0.6}Al_{3.4})Si_8O_{20}(OH)_4]$
	beidellite	$Na_{0.9}[Al_4(Al_{0.9}Si_{7.1})O_{20}(OH)_4]$

conductivities are dominated by protonic transport on the surface of the clay particles rather than contributions from interlayer cations. Ionic conductivity in undried smectites has a strong dependence on relative humidity, and the activation energy is *ca.* 0.2 to 0.3 eV regardless of interlayer cation (*41,42*). In order to study the mobility of interlayer cations, special precautions are required to exclude moisture.

Smectites display low ionic-conductivities due to interlayer cation mobility (*43*) because strong coulombic interactions between the cations and the anionic sheets tend to pin the cations. To minimize the coulombic forces between the cation and the aluminosilicate sheets, we have intercalated solvating species such as oligoether poly(phosphazene) and cryptand into the conduction plane. There is precedence in the literature both for the formation of intercalation compounds with smectites and for enhancing the mobility of interlayer cations with solvating compounds. Ruiz-Hitzky first reported intercalation of crown ethers, cryptands (*44*) and PEO (*45*) into smectites and reported that the mobility of interlayer cations was enhanced by several orders of magnitude (*4,43*). Giannelis demonstrated ionic transport in PEO/Li-montmorillonite composites (*5,46*). Lerner studied structure and ionic conductivity of poly(oxymethylene oligo(ethylene oxide)) montmorillonite intercalation compounds (*6*).

It is instructive to comment on the nature of smectite intercalation compounds. Because smectites have a variable crystallographic *c*-axis, the silicate sheets are able to expand and accommodate solvating species in the interlayer galleries. There are at least three general synthetic routes to smectite intercalation compounds. In the first of these, exfoliation of the smectite in a swelling solvent followed by addition of a neutral species results in an intercalation compound (*44*). Giannelis and coworkers recently demonstrated another route involving direct melt intercalation of poly(ethylene oxide) without a swelling solvent (*47*). A third strategy involves intercalation of a suitable monomer followed by polymerization inside the interlayer gallery of the smectite. The resulting composites are two-dimensional nanostructures of silicate layers alternating with cations and intercalated compound as verified by powder x-ray diffraction (XRD).

MEEP and Cryptand Montmorillonite Intercalates. Our studies have focused on the intercalation of cryptand[2.2.2], $N(C_2H_4OC_2H_4OC_2H_4)_3N$ (crypt) and poly(bis-methoxyethoxyethoxy phosphazene), $[(CH_3OC_2H_4OC_2H_4O)_2PN]_n$ (MEEP) (*31*) into sodium exchanged montmorillonite, $Na_{0.6}[(Mg_{0.6}Al_{(3.4)}Si_8O_{20}(OH)_4]$ (Na-mont). Crypt was chosen because it encapsulates Na^+ (*48,49*) which may lead to the enhancement of the conductivity in many polyelectrolytes (*34*). MEEP was chosen because its oligoether side chains solvate Na^+ and the high flexibility of the backbone (*2*). An additional motivation for using smectites as polyelectrolytes was that the MEEP intercalated Na-mont would have greater rigidity. As a consequence of its low glass transition temperature, MEEP has a low dimensional stability (*50*), and this is a problem in applications where polymer flow can lead to short circuits.

Preparation of Materials. Na-mont was stirred in either deionized H_2O or anhydrous CH_3CN, and measured amounts of MEEP (molecular weight= 10,300 by gel phase chromotography) or cryptand were added. The progress of the intercalation was monitored by powder XRD. After the mixture had a homogeneous appearance (*ca.* 3 days) and there was no evidence of unintercalated Na-mont by XRD, the reaction was considered complete. When H_2O was used as the solvent, the product was cast into a film and dried on a high-vacuum line (*ca.* 3×10^{-5} torr at 100°C.) When the synthesis was conducted in CH_3CN, the crypt, MEEP, and Na-mont were pre-dried on a high-vacuum line (*ca.* 3×10^{-5} torr at 100°C), and the CH_3CN was removed by evacuation or evaporation with a dry N_2 stream. Solvent removal was confirmed in all cases by FTIR spectroscopy of the dried products. Where appropriate in the synthesis and characterization, inert atmosphere techniques were employed to prevent adventitious H_2O from affecting the impedance measurements.

A MEEP to Na-mont ratio of one polymer repeat unit, $(CH_3OC_2H_4OC_2H_4O)_2PN$, to four clay Si_8O_{20} units, results in a material which exhibits an XRD powder pattern with two low-angle peaks. The smaller *d*-spacing of 13 Å is consistent with unintercalated Na-mont. The larger *d*-spacing of 18 Å is assigned to MEEP intercalated Na-mont (hereafter abbreviated MEEP/Na-mont). At MEEP to Si_8O_{20} units ratios of 1:2, 1:1, 2:1, 3:1, single-phases form as indicated by *d*-spacings of 19 Å, 20 Å, 21 Å, and 21 Å respectively. (Figure 1) Crypt intercalated Na-mont

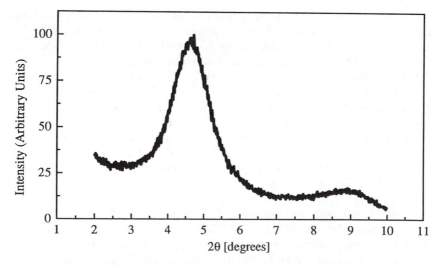

Figure 1. Powder x-ray diffraction pattern of (1:2) MEEP/Na-mont intercalate showing a *d*-spacing of 19 Å.

(abbreviated crypt/Na-mont) at loadings of 0.6 crypt to 1 Si_8O_{20} repeat un it and 1.2 crypt to 1 Si_8O_{20} repeat unit (*i.e.*: 1 crypt to 1 interlayer Na^+ and 2 crypt to 1 interlayer Na^+ respectively) have a *d*-spacing of 17.9 Å. FTIR spectra of the MEEP/Na-mont composites show absorption bands at frequencies corresponding to pure MEEP and pure Na-mont, but with changes in relative intensities and shape of the features associated with the polymer. (Figure 2) Sharpening of vibrational bands in the region of 1200-1400 cm^{-1} suggests interaction between the polymer and the clay. Similar spectroscopic phenomena have been observed for PEO/montmorillonite intercalation compounds by Ruiz-Hitzky (*43*), and Giannelis (*5*). FTIR spectra of crypt/Na-mont composites are a superposition of cryptand and montmorillonite bands.

The observed glass transition temperature for the (1:2) MEEP/Na-mont intercalate ($T_g = -70°C$) is 10°C higher than that of pristine MEEP. Melting transitions (T_m) and glass transitions associated with the intercalated polymer have not been observed in other polymer-clay intercalates (*51,52*). In particular, the related PEO-montmorillonite composites prepared independently by Giannelis (*5*) and by Lerner (*6*) do not display phase changes of the native polymer. Giannelis reported the absence of melting transition, and Lerner reported the absence of both glass and melting transitions. The shift in glass transition temperature observed in our system indicates that the local environment of the intercalated or absorbed MEEP is similar to that of the bulk polymer, but more constrained. This has important consequences for the conduction mechanism of the MEEP/Na-mont intercalates. If the mechanism is similar to that in bulk polymer electrolytes, high segmental motion is necessary for a high conductivity.

The ^{29}Si magic-angle spinning (MAS) nuclear magnetic resonance (NMR) spectrum of the (1:1) MEEP/Na-mont intercalate displays a feature that is shifted 1.5 ppm

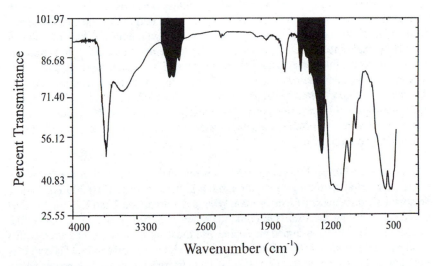

Figure 2. FTIR spectrum of (1:2) MEEP/Na-mont intercalate, thin film taped to holder. Shaded regions are due to MEEP.

upfield of pristine Na-mont. The Si resonance of the cryptand intercalate is intermediate between that of pristine Na-mont and the 1:2 MEEP/Na-mont intercalate, and the line width is narrower. The upfield shifts of the ^{29}Si resonance indicate that the Na^+ is separated from the silicate sheets by cation coordination to MEEP or cryptand. The ^{23}Na MAS NMR spectrum of pristine Na-mont consists of a diffuse feature centered at -17 ppm with respect to aqueous 1M NaCl. The (1:1) MEEP/Na-mont intercalate has a broad feature at -20 ppm, and the cryptand intercalate has a sharper resonance at -16 ppm. (Figure 3) This is similar to the solid state spectra of Na^+-cryptates and Na^+ salt-polyether complexes (53), which provides additional evidence for Na^+ solvation by the MEEP and crypt.

To elucidate the driving force for intercalation, mixtures of MEEP and crypt dissolved in anhydrous CH_3CN were added to a slurry of Na-mont in CH_3CN at a ratio of 1 $(CH_3OC_2H_4OC_2H_4O)_2PN$ repeat unit to 0.6 crypt to 1 Si_8O_{20} unit. The resulting composite has a powder XRD pattern which indicates a d-spacing of 17.9 Å, smaller than those seen in MEEP/Na-mont composites but consistent with (0.6:1) crypt/Na-mont intercalates. The reaction slurry of MEEP, crypt, and Na-mont was filtered, and FTIR, ^{1}H and ^{13}C NMR spectra of the filtrate show evidence of MEEP but no crypt. Control experiments with MEEP/Na-mont and crypt/Na-mont in the same relative ratios yielded no evidence of either MEEP or crypt in the filtrate. When crypt is present in amounts greater than 0.6 crypt : 1 Si_8O_{20} unit (*i.e.*: greater than a ratio of 1 crypt to 1 interlayer Na^+), FTIR, ^{1}H and ^{13}C NMR spectra indicate that crypt is present in the filtrate. This suggests a 1:1 crypt to interlayer Na^+ binding.

It is apparent that when the crypt is present in a one to one ratio with interlayer Na^+, it hinders the intercalation of MEEP, perhaps by preventing solvation of interlayer Na^+ by MEEP. This suggests that solvation of interlayer cation is necessary for intercalation of oligoether species into smectites. The driving force for solution intercalation of polymers into smectites has been attributed to the entropic gain from desorbtion of solvent molecules from the host out weighing the loss of polymer conformational entropy upon intercalation (54). Our results indicate that enthalpic gains from solvation by the species to be intercalated are also important, and in fact species can be intercalated preferentially according to cation solvation strength. Relevant to this conclusion are the reports from the Giannelis laboratory of direct melt intercalation of polymers, including poly(ethylene oxide) into Na-mont (5) (47). Direct intercalations of the molten polymer into anhydrous clay appear to be enthalpically driven (47).

Impedance Data. Impedance measurements were performed in the frequency range associated with long-range ion motion in these solids (1-10,000 Hz) with a Solartron 1250 Frequency Response Analyzer and a Solartron 1286 Electrochemical Interface. Varying the oscillation amplitude between 10mV and 1V produced no measurable deviations in impedance values, so most samples were measured with a oscillation amplitude of 500 mV to maximize the signal to noise ratio. The samples were prepared by pressing thin pellets (*ca.* 0.5 cm) from finely ground materials which had been dried on a high-vacuum line with heating (*ca.* 3×10^{-5} torr at 100°C) to

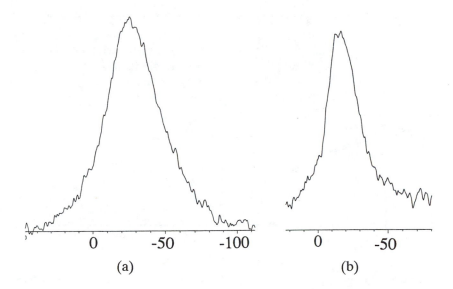

(a)   (b)

Figure 3. ^{23}Na MAS NMR spectra of pristine Na-mont (a) and (1.2:1) crypt/Na-mont (b). The broad feature in both spectra is assigned to interlayer Na$^+$. The full width at half maximum of (a) is 3300 Hz; that of (b) is 2000 Hz. Both samples were spun at 10,000Hz.

remove any residual solvent or water. Both faces of the pellet were then sputtered with gold, and the pellet was placed in a sealed symmetrical cell with stainless steel electrodes. Impedance spectra of Na-mont, (1.2:1) crypt/Na-mont, (2:1) MEEP/Na-mont, and (3:1) MEEP/Na-mont were fitted to a semi-circle, and DC-conductivities were assigned by low frequency touchdown according to a parallel resistor-capacitor equivalent circuit model. It is important to note that this treatment of the data ignores any anisotropy of the conductivity. The two dimensional structure of the clay likely gives rise to anisotropic conductivity, and conductivity parallel to the clay layers is expected to be greater than the conductivity perpendicular to the clay layers. There is probably some degree of preferential orientation of clay layers parallel to the electrodes in these experiments. Na-mont shows Arrhenius behavior with an apparent activation energy of 0.64 eV. (Figure 4) (1.2:1) crypt/Na-mont also shows Arrhenius behavior, but with a higher activation energy of 0.92 eV and enhanced conductivity in the experimental temperature range. Both of these apparent activation energies are higher than those for proton conduction in Na-mont where activation energies are typically around 0.2 eV (*42*). The behavior of the MEEP/Na-mont intercalate is more complex. The impedance spectra of MEEP/Na-mont intercalate thin films measured perpendicular to the layers show enhanced conductivity and deviations from Arrhenius behavior similar to those of polymer electrolytes are apparent at low temperatures.

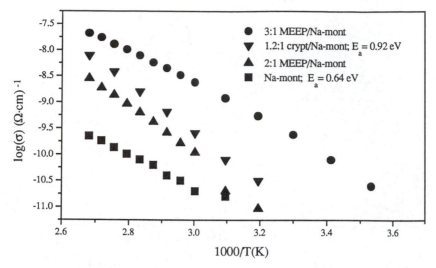

Figure 4. Log $\sigma(\Omega\cdot\text{cm})^{-1}$ vs. $1000/T(K)$ of (●) (3:1) MEEP/Na-mont; (▼) (1.2:1) crypt/Na-mont; (▲) (2:1) MEEP/Na-mont; (■) Na-mont.

Conclusion

Smectites can be readily intercalated with cation-solvating species such as cryptands and poly(oligoether phosphazenes). The driving force of the intercalation appears to be the solvation of interlayer cations, and the more strongly solvating cryptand intercalates preferentially to MEEP. The resulting composites are two-dimensional nanostructured materials as evidenced by powder XRD. ^{23}Na and ^{29}Si MAS NMR indicate that the intercalated MEEP or cryptand interacts with interlayer cations and shields them from the anionic aluminosilicate sheet, thereby greatly enhancing the ionic conductivity.

Acknowledgments

This research was supported by the Army Research Office under Grant No. DAAH04-94-6-0066, and made use of MRL Central Facilities supported by the National Science Foundation, at the Materials Research Center of Northwestern University under Award No. DMR-9120521. JCH gratefully acknowledges the NSF for a training grant, Award No. CHE-9256486, and Dr. Gary Beall and Dr. Semeon Tsipursky of American Colloid Company (1500 West Shure Drive, Arlington Heights, IL 60004-1434) for providing smectites and advice.

Literature Cited

1.   Armand, M. *Solid State Ionics* **1983**, *9-10*, 745.
2.   Blonsky, P. M.; Shriver, D. F. *Solid State Ionics* **1986**, *18 & 19*, 258.

3.   Rawsky, G. C.; Fujinami, T.; Shriver, D. F. *Chem. Mater.* **1994**, *6*, 2208.
4.   Ruiz-Hitzky, E.; Aranda, P.; Casal, B.; Galván, J. *Adv. Mater.* **1995**, *7*, 180.
5.   Vaia, R.; Vasudevan, S.; Wlodzimierz, K.; Scanlon, L.; Giannelis, E. *Adv. Mater.* **1995**, *7*, 154.
6.   Wu, J.; Lerner, M. *Chem. Mater.* **1993**, *5*, 835.
7.   Armand, M.; Chabango, J. M.; Duclot, M. *Extended Abstracts, Second International Conference on Solid Electrolytes*, St. Andrews, Scotland, 1978.
8.   Armand, M.; Chabango, J. M.; Duclot, M. *Fast Ion Transport in Solids*; Duclot, J. M.; Vashishta, P.; Mundy, J. N.; Shenoy, G. K., Eds.; North-Holland: Amsterdam, 1979.
9.   Wright, P. V. *Br. Polym. J.* **1975**, *7*, 319.
10.   Wright, P. V. *J. Polym. Sci., Polym. Phys. Ed.* **1976**, *14*, 915.
11.   Fenton, D. E.; Parker, J. M.; Wright, P. V. *Polymer* **1973**, *14*, 589.
12.   Ratner, M. A.; Shriver, D. F. *Chem. Rev.* **1988**, *88*, 109.
13.   Hibma, T. *Solid State Ionics* **1983**, *9/10*, 1101.
14.   Okamura, S.; Chatani, Y. *Polymer*, in press.
15.   Parker, J. M.; Wright, P. V.; Lee, C. C. *Polymer* **1981**, *22*, 1305.
16.   Papke, B. L.; Ratner, M. A.; Shriver, D. F. *J. Electrochem. Soc.* **1982**, *129*, 1694.
17.   Dupon, R.; Papke, B. L.; Ratner, M. A.; Whitmore, D. H.; Shriver, D. F. *J. Am. Chem. Soc.*, **1982**, *104*, 6247.
18.   Teeters, D.; Frech, R. *Solid State Ionics* **1986**, *18/19*, 271.
19.   Hardy, L. C.; Shriver, D. F. *J. Am. Chem. Soc.* **1985**, *107*, 3823.
20.   Spindler, R.; Shriver, D. F. *Macromolecules* **1986**, *19*, 347.
21.   Harris, C. H.; Ratner, M. A.; Shriver, D. F. *Macromolecules* **1987**, *20*, 1778.
22.   Berthier, C.; Gorecki, W.; Minier, M.; Armand, M.; Chabango, J. M.; Rigaud, P. *Solid State Ionics* **1983**, *11*, 91
23.   Vogel, H. *Phys. Z.* **1921**, *22*, 645. Tamman, G.; Hesse, W. *Z. Anorg. Allg. Chem.* **1926**, *165*, 254. Fulcher, G. S. *J. Am. Ceram. Soc.* **1925**, *8*, 339.
24.   Cheradame, H. In *IUPAC Macromolecules*; Benoit, H.; Rempp, P. Eds.; Pergamon Press: New York, 1982; p. 251.
25.   Killis, A.; LeNest, J. F.; Cheradame, H. J. *J. Polym. Sci., Makromol. Chem., Rapid Commun.* **1980**, *1*, 595.
26.   Killis, A.; LeNest, J. F.; Gandini, A.; Cheradame, H. J. *J. Polym. Sci. Polym. Phys. Ed.* **1981**, *19*, 1073.
27.   Killis, A.; LeNest, J. F.; Gandini, A.; Cheradame, H. J.; Cohen-Addad, J. P. *Solid State Ionics* **1984**, *14*, 231.
28.   Druger, S. D.; Ratner, M. A.; Nitzan, A. *Phys Rev. B.* **1985**, *31*, 3939.
29.   Tipton, A. L.; Lonergan, M. C.; Shriver, D. F. *J. Phys. Chem.* **1994**, *98*, 4148.
30.   Lonergan, M. C.; Ratner, M. A.; Shriver, D. F. *J Am. Chem. Soc.* **1995**, *117*, 2344.

31.   Allcock, H. R.; Austin, P. E.; Neenan, T. X.; Sisko, J. T.; Blonsky, P. M.;
        Shriver, D. F. *Macromolecules* **1986**, *19*, 1508.
32.   Lemmon, J.; Lerner, M. *Macromolecules* **1992**, *25*, 2907.
33.   Payne, V. A.; Forsyth, M.; Ratner, M. A.; Shriver, D. F.; deLeeuw, S. W.
        J. Chem. Phys. **1994**, *7*, 100.
34.   Doan, K.; Ratner, M. A.; Shriver, D. F. *Chem. Mater.* **1991**, *3*, 418.
35.   Chen, K.; Ganapathiappan, S.; Shriver, D. F. *Chem. Mater.* **1989**, *1*, 483.
36.   Lonergan, M. C.; Perram, J. W.; Ratner, M. A.; Shriver, M. A. *J. Chem.
        Phys.* **1993**, *98*, 4937.
37.   Ganapathiappan, S.; Chen, K.; Shriver, D. F. *Macromolecules* **1988**, *21*,
        2299.
38.   Chen, K.; Shriver, D. F. *Chem. Mater.* **1991**, *3*, 771.
39.   Vaughan, D. E. W. *Catalysis Today* **1988**, *2*, 187.
40.   Fan, Y. Q. *Solid State Ionics* **1988**, *28-30*, 1596.
41.   Wang, W. L.; Lin, F. L. *Solid State Ionics* **1990**, *40/41*, 125.
42.   Slade, R. C. T.; Barker, J.; Hirst, P. R. *Solid State Ionics* **1987**, *24*, 289.
43.   Aranda, P.; Galván, J.; Casal, B.; Ruiz-Hitzky, E. *Electrochem Acta* **1992**,
        37, 1573.
44.   Ruiz-Hitzky, E.; Casal, B. *Nature* **1978**, *276*, 596.
45.   Ruiz-Hitzky, E.; Aranda, P. *Adv. Mater.* **1990**, *2*, 545.
46.   Wong, S.; Vasudevan, S; Vaia, R.; Giannelis, E.; Zax, D. J. *J. Am. Chem.
        Soc.* **1995**, *117*, 7568.
47.   Vaia, R.; Hope, I.; Giannelis, E. *Chem. Mater.* **1993**, *5*, 1694.
48.   Lehn, J. *Acc. Chem. Res.* **1978**, *11(2)*, 1.
49.   Izatt, R. M.; Eatough, D. J.; Christensen, J. J. *Struct. Bonding* **1973**, *16*,
        161.
50.   Allcock, H. R. *Chem Mater.* **1994**, *6*, 1476.
51.   Messersmith, P.; Giannelis, E. *Chem. Mater.* **1993**, *5*, 1064.
52.   Mehrotra, V.; Giannelis, E. *Solid State Ionics* **1992**, *51*, 115.
53.   Rawsky, G.; Shriver, D. F. unpublished results.
54.   Theng, B. K. G. *Formation and Properties of Clay-Polymer Complexes*;
        Elsevier: New York, 1979.
55.   Mott, C. B. J. *Catalysis Today* **1988**, *2*, 199.

RECEIVED November 27, 1995

Chapter 19

Polyimide–Nylon 6 Copolymers: Single-Component Molecular Composites

Hong Ding[1] and Frank W. Harris[2]

Department of Polymer Science, Maurice Morton Institute of Polymer Science, University of Akron, Akron, OH 44325

Novel polyimide-g-nylon 6 and nylon 6-b-polyimide-b-nylon 6 copolymers were synthesized by polycondensation and subsequent anionic, ring-opening polymerization methods. The graft copolymers were prepared by initially incorporating an N-acylated caprolactam ring in a diamine monomer, i.e., 3,5-bis(4-aminophenoxy)benzoyl caprolactam The diamine was polymerized with dianhydrides to produce polyimides containing pendant acylated caprolactam moieties. The triblock copolymers were prepared by initially synthesizing N-acylated caprolactam end-capped polyimides. These multi- and difunctionalized polyimides were then used as polymeric activators for the subsequent anionic, ring-opening polymerization of ε-caprolactam. Thus, they were dissolved in molten ε-caprolactam at 140°C and treated with phenylmagnesium bromide. Nylon 6 chains grew from the activated sites on the polyimide chains to form well-defined graft and block copolymer structures. No nylon 6 homopolymer was produced. The effective reinforcement of the nylon 6 by only a small amount, i.e., 5%, of the polyimide components was demonstrated by the dramatic improvements in thermal and mechanical properties.

A molecular composite has been defined as a molecular dispersion of a reinforcing rigid-rod polymer in a random coil polymer matrix (1~3). Although such systems display enhanced thermal and mechanical properties over conventional fiber reinforced systems, phase separation is often encountered as the amount of the rigid-rod molecules increases and during film casting and melt processing (4, 5). A recent approach has been to chemically bond the rigid-rod molecules to the random coil matrix to form so called one-component molecular composites (6~9), as shown in Figure 1. This new concept involves novel synthetic approaches to preparing graft copolymers with rigid-rod polymer backbones and flexible side-chains and block copolymers containing rigid and flexible blocks. Previous syntheses of such copolymers have been carried out in harsh solvents, where competing homopolymerizations could not be avoided. Thus, the reactions provided low graft and block efficiencies.

[1]Current address: Calgon Corporation, P.O. Box 1346, Pittsburgh, PA 15230
[2]Corresponding author

0097–6156/96/0622–0273$15.00/0

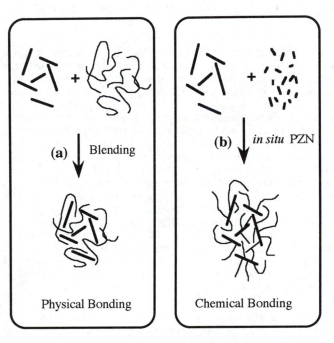

Figure 1.   Schematics of (a) two-component molecular composite.
(b) one-component molecular composite.

The objective of this research was to synthesize and characterize well-defined graft and block copolymers in which small amounts of extended chain polyimides were chemically-bound to flexible nylon 6 matrices (10, 11). Two approaches were to be used, as shown in Scheme 1. The first approach involved the synthesis of polyimide-g-nylon 6 copolymers, which was to begin with the preparation of a diamine monomer containing an N-acylated caprolactam moiety, i.e., 3,5-bis(4-aminophenoxy)benzoylcaprolactam (BAPBC). Polyimides were then to be prepared from the diamine and commercial dianhydrides. The polyfunctionalized polyimides were expected to act as both polymeric activators in subsequent anionic, ring-opening polymerizations of ε-caprolactam and as reinforcing agents in the copolymers formed. The second approach involved the synthesis of nylon 6-b-polyimide-b-nylon 6 copolymers. N-Acylated caprolactam-terminated polyimides were to be prepared from diamines, dianhydrides and the end-capping agent 4-aminobenzoylcaprolactam (12). The end-capped polyimides were to be used as difunctional activators for the anionic polymerization of ε-caprolactam to form nylon 6-b-polyimide-b-nylon 6 copolymers. It was speculated that only small amounts of polyimide would be needed to effectively reinforce and, thus, significantly enhance the thermal and mechanical properties of nylon 6.

Experimental

3,5-Bis(4-aminophenoxy)benzoylcaprolactam (BAPBC).   BAPBC (5) was prepared according to Scheme 2. Methyl 3,5-dihydroxybenzoate (84.00 g, 0.4999 mol) was treated with 1-flouro-4-nitrobenzene (141.12 g, 1.0000 mol) in 770 ml of DMAC containing K_2CO_3 (138.15 g, 1.0000 mol) at 80°C for 5 h to produce methyl 3,5-bis(4-nitrophenoxy)benzoate (1). After being recrystallized from ethyl acetate, compound 1 (41.00 g, 0.1000 mol) was hydrolyzed in a mixture of acetic acid, H_2SO_4, and H_2O (290:120:75 ml) at 120°C to produce 2. Compound 2 was collected by filtration, washed with water and with methanol. The white crystals 2 (39.62 g, 0.1000 mol) were then treated with thionyl chloride (365 ml) at reflux to produce 3,5-bis(4-nitrophenoxy)benzoyl chloride (3). After the excess thionyl chloride was removed by distillation, the residue (3) was dissolved in benzene (350 ml) and treated with ε-caprolactam (11.31 g, 0.1000 mol) at 90°C to form 3,5-bis(4-nitrophenoxy)benzoylcaprolactam (4). Compound 4 (9.83 g, 0.0200 mol) was hydrogenated under 1 atm of H_2 at 15°C in EtOH (100 ml) in the presence of Pd/C (5%, 0.40 g). When the theoretical quantity (0.120 mol) of hydrogen had been absorbed, the solution was filtered through Celite under nitrogen to remove the catalyst. After ethanol was removed from the filtrate under reduced pressure, the residue was dissolved in methylene chloride. The solution was filtered through Celite under nitrogen. The filtrate was evaporated to dryness under reduced pressure to yield 7.0 g (81%) of 3,5-bis(4-aminophenoxy)benzoylcaprolactam: mp 96-98°C, ^{1}H-NMR (CDCl3) δ 6.84 (m, 2H, ArH), 6.60 (m, 9H, ArH), 3.85 (d, 2H, RH), 3.49 (broad, 4H, NH2), 2.68 (δ, 2H, RH) and 1.70 ppm (s, 6H, RH); IR (KBr) 3454 and 3368 cm[-1] (NH2). Anal. Calcd. for $C_{25}H_{25}N_3O_4$: C, 69.59%; H, 5.84%. Found: C, 69.77%; H, 5.92%.

4-Aminobenzoylcaprolactam.   The end-capping agent was prepared from 4-nitrobenzoyl chloride by the known procedure (12).

Typical Procedure for the Preparation of Polyimides Containing Pendant N-Acylated Caprolactam Moieties.   After BAPBC (1.5000 g, 3.4772 mmol) was polymerized with 3,3',4,4'-biphenyltetracarboxylic dianhydride (BPDA) (1.0229 g, 3.4772 mmol) in N-methyl-2-pyrrolidinone (NMP) (14.5 ml) at 22°C for 12 h, pyridine (0.633 g, 8.00 mmol) and acetic anhydride (0.816 g, 8.00 mmol) were added. After stirring for another 24 h at 22°C under N_2, the solution was diluted with NMP and slowly added to vigorously stirred ethanol. The polymer that precipitated was

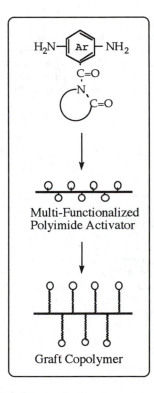
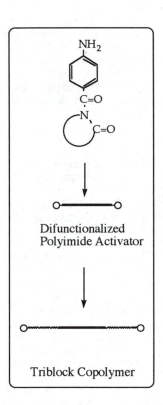

Scheme 1.   Approaches to novel polyimide-g-nylon 6 and
            nylon 6-b-polyimide-b-nylon 6 copolymers.

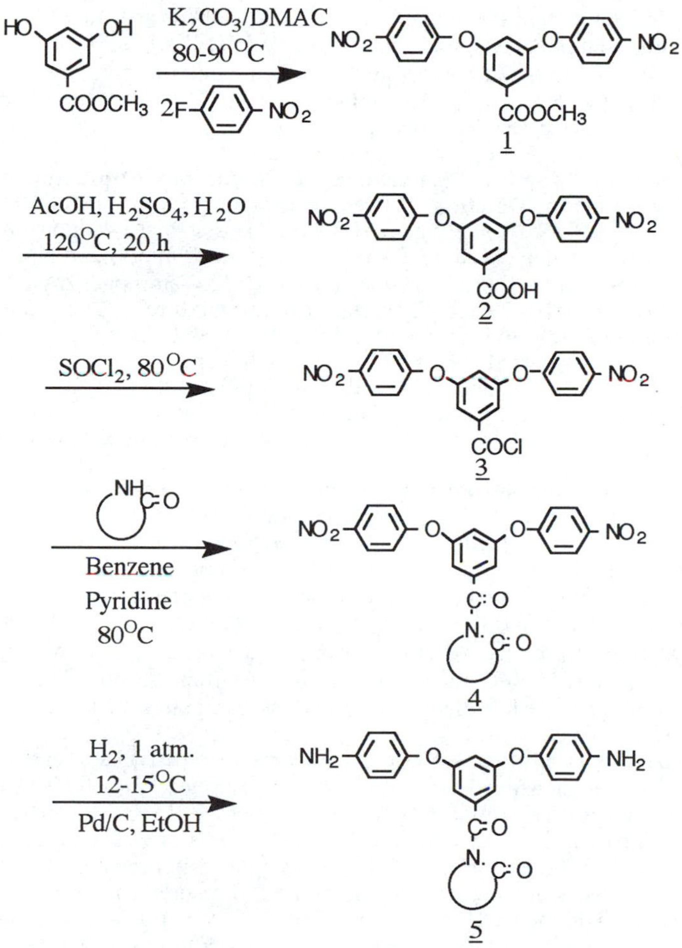

Scheme 2.   Synthesis of BAPBC

collected by filtration, washed with ethanol and ethyl ether before being dried at room temperature under reduced pressure.

Typical Procedure for the Preparation of the Polyimide-g-Nylon 6 Copolymers.
The polyimide (BAPBC/BPDA, 10.0 g) was gradually added to molten ε-caprolactam (150.0 g, 1.33 mol). The mixture was then stirred under N_2 at 140°C for 1 h to form a homogenous solution. An initiator solution was prepared separately by adding a 3.0 M solution of phenylmagnesium bromide (PhMgBr) in diethyl ether (4.8 ml, 14.5 mmol) to molten ε-caprolactam (40.0 g, 0.35 mol) at 140°C under N_2. After the polyimide activator solution was mixed with the initiator solution at 140°C, the mixture solidified within 5 min. The product was heated at 140°C for 1 h to afford a. tough, homogeneous, pale yellow solid mass.

Typical Procedure for the Preparation of Polyimides End-Capped with N-Acylated Caprolactam Moieties. After a mixture of o-tolidine (OTOL) (2.1231 g, 10.000 mmol), BPDA (0.9120 g, 3.100 mmol), 2,2'-bis(3,4-dicarboxyphenyl) hexafluoropropane dianhydryde (6FDA) (3.2870 g, 7.400 mmol) and NMP (35.83 g, 15 wt% solid) was stirred at 22°C under N_2 for 10 h, 4-aminobenzoylcaprolactam (0.2322 g, 1.000 mmol) was added, and the reaction mixture was stirred for another 10 h. Pyridine (1.7600 g) and acetic anhydride (2.2440 g) were added, and the reaction mixture was stirred for an additional 24 h. The solution was diluted with NMP and poured into ethanol. The end-capped polyimide (OTOL/BPDA/6FDA, 100/30/70 molar ratio, Mn=12,000) that precipitated was collected by filtration, washed with methanol and dried at room temperature under reduced pressure.

Typical Procedure for the Preparation of the Nylon 6-b-Polyimide-b-Nylon 6 Copolymers. The above end-capped polyimide (OTOL/BPDA/6FDA, 100/30/70 molar ratio, Mn=12,000, 10.0 g, 1.7 mmol of acylated caprolactam moieties) was slowly added to molten ε-caprolactam (150.0 g, 1.33 mol) under N_2 at 140°C to form a homogenous solution. An initiator solution was prepared separately by adding a PhMgBr ether solution (3.0M, 0.56 ml, 1.7 mmol) to molten ε-caprolactam (40.0 g, 0.35 mol) at 140°C under N_2. After the polyimide activator solution was mixed with the initiator solution at 140°C, the mixture solidified within 10 min. The product was heated at 140°C for 1 h to afford a tough, homogeneous, pale yellow solid mass.

Characterization Techniques. FTIR spectra were recorded on a Beckman FT-2100 infrared spectrometer using KBr pellets or film samples. ^{1}H-NMR spectra were obtained with a Varian Germini-200 Spectrometer at 200 MHz. Intrinsic viscosities were determined using a Cannon-Ubbedholde No. 150 or 200 viscometer at 25 $\pm$ 0.1°C. A 2.5 mg/ml solution of a polymer sample in m-cresol was prepared and filtered for Gel permeation chromatography (GPC) measurement. The solution (0.25 ml) was injected into the GPC instrument. The GPC was run with a flow rate of m-cresol of 1.0 ml/min at 100°C. Multi-angle light scattering, refractive index and viscosity detectors were used. Differential Scanning Calorimetry (DSC) data were recorded on a DuPont Model 910 thermal analyzer with a heating rate of 10 °C/min in N_2 or air. Thermogravimetric analysis (TGA) was performed on a DuPont 9900 Analyzer equipped with a 925 cell using a heating rate of 10 °C/min in N_2 or air. The dynamic mechanical properties of compression-molded or solution cast films were determined on a dynamic mechanical spectrometer (DMS) 200 (SII, Seiko Instruments, Inc.) in N_2 from -150 to 250°C with a heating rate of 1 °C/min. Nylon 6 homopolymer and the graft and block copolymer samples were compression-molded on a Mini-Max Injection Molder (Molder CS-183MM-99, Custom Scientific, Inc.) at 340-360°C into cylindrical dumbbell shapes with gage lengths of 7.0 mm and diameters of 1.5 mm. Nylon 6 homopolymer and nylon 6-b-polyimide-b-nylon 6 copolymer samples were melt spun into fibers at 250-265°C using an Instron

Capillary Rheometer, fitted with a capillary die with a diameter of 1.2 mm and a length/diameter ratio of 28.7. A plunger speed of 0.76 mm/min was used. The fibers were first drawn at room temperature, and then at 140°C to a total draw ratio of 4X. Tensile tests were conducted on an Instron Model 4204 at room temperature with a crosshead speed of 5 mm/min to break.

Results and Discussion

Monomer Synthesis. BAPBC was synthesized by the route shown in Scheme 2. The key step in the synthesis was the hydrogenation of the dinitro precursor. The control of both hydrogen pressure and reaction temperature was found to be important in producing high purity monomer. The monomer yield was monitored by ^{1}H-NMR of the product mixture. When the hydrogenation was carried out under 2-4 atm at 22°C, 25 mole percent of the N-acyl caprolactam rings were opend. By lowering the hydrogen pressure to 1 atm, the yield was improved from 75 to 90%. However, the monomer produced was still not pure enough to afford high molecular weight polyimides. Since the temperature of the hydrogenation solution increased from 22 to 50°C due to the exothermic reduction, an ice-water bath was used to control the temperature. The optimal hydrogenation conditions (1 atm H_2 at 15°C for 10 h) were finally identified and high monomer yield (≥97%) was achieved. The ^{1}H-NMR spectrum of the BAPBC monomer is shown in Figure 2. The monomer's purity and structure were also confirmed by HPLC, FTIR and elemental analysis.

Multi-functionalized Polyimides. As shown in Scheme 3, the polyimides containing pendant N-acylated caprolactam moieties were prepared by the two-step polycondensation of BAPBC with commercial dianhydrides. A polyamic acid (PAA) viscous solution was formed by stirring equimolar amounts of the derivatized diamine with the dianhydride in NMP at room temperature. The subsequent chemical imidization was carried out by adding pyridine and acetic anhydride to the PAA solution to produce the multi-functional polyimide. The amount of pendant groups incorporated and the rigidity of the polyimide chains were varied by copolymerizations with non-derivatized diamines such as o-tolidine (OTOL), m-tolidine (MTOL), or 2,2'-bis(trifluoro methyl)benzidine (PFMB). The incorporation of the N-acylated caprolactam moieties in the polyimide chains was confirmed by the FTIR absorptions at 2931 and 2864 cm^{-1} (v_{as} and v_s, CH_2 in pendant acylated caprolactam moieties) as well as the absorptions at 1778 and 1727 cm^{-1} (v_{as} and v_s, C=O in imide ring).

Good polymer solubility in NMP was an important prerequisite for producing high molecular weight derivatized polyimides. Good polyimide solubility in molten caprolactam was an equally important prerecquisite for forming homogeneous solutions for the subsequent graft and block copolymerizations. The initial copolyimides prepared from BAPBC, PFMB and BPDA had low inherent viscosities, presumably due to the low reactivity of PFMB at ambient temperature. Thus, the more reactive diamine MTOL was copolymerized with DAPBC and BPDA. The polymers became insoluble in NMP after imidization when the level of MTOL was greater than 20 mol percent. However, the copolyimides prepared from BAPBC, OTOL and BPDA showed very good solubility over a wide range of compositons in both NMP and molten ε-caprolactam. The inherent viscosities of the polymers ranged from 0.8 to 1.2 dl/g. The different solubility behavior of the OTOL' and MTOL' based polymers might be due to the twist between the imide ring and methyl-substituted phenyl rings produced by the o-methyl groups in OTOL (13). Polyimides prepared from 100/0/100, 70/30/100 and 50/50/100 mol ratios of BAPBC/OTOL/BPDA were chosen as the multi-functional activators for the anionic ring-opening polymerizations of caprolactam.

During this part of the work it was found that some polyfunctional polyimides would not dissolve in molten ε-caprolactam after being dried above 100°C under

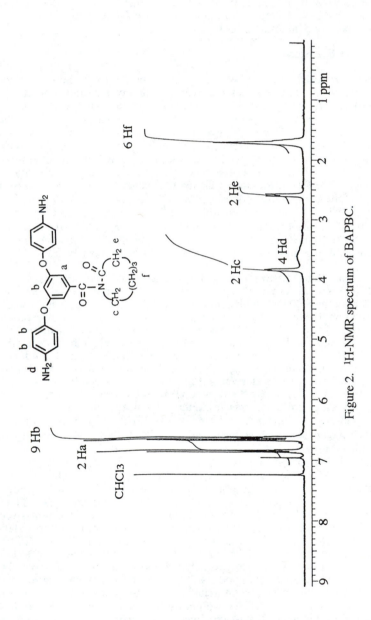

Figure 2. ¹H-NMR spectrum of BAPBC.

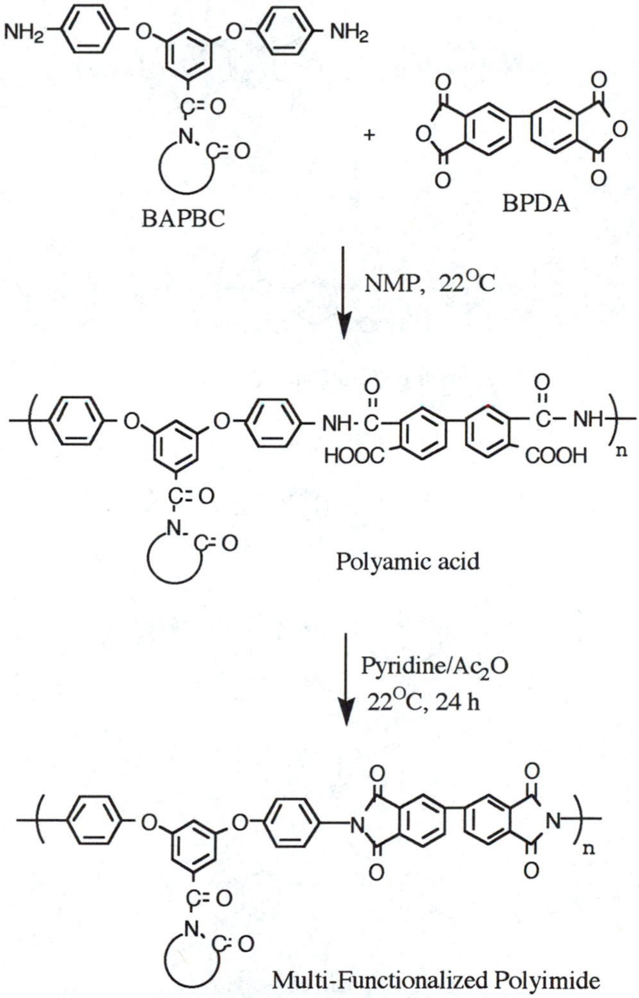

Scheme 3. Synthesis of a typical multi-functionalized polyimide.

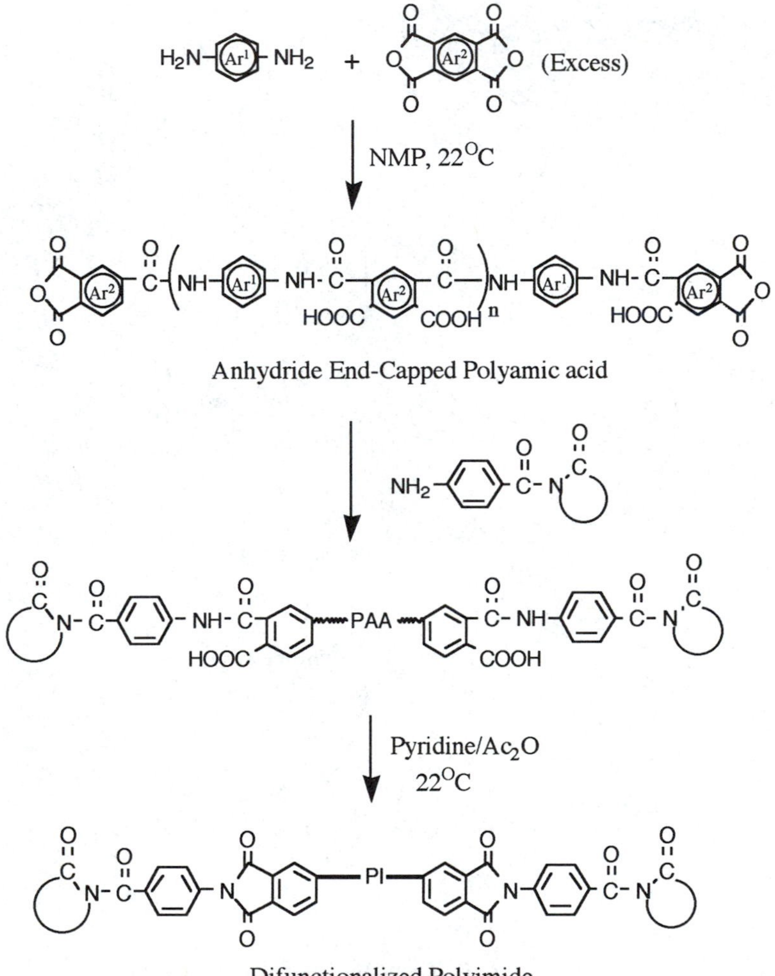

Scheme 4. Synthesis of difunctionalized polyimides.

reduced pressure. Since the polymers were soluble in ε-caprolactam prior to drying, the decreased solubility was attributed to further imidization. In fact, further imidization was a primary motive for the drying process. Chemical imidization is known to result in residual amic acid groups (13). Amic acid groups were to be removed to prevent them from terminating the subsequent anionic polymerization of caprolactam. In order to overcome this problem, the polyfunctional polyimides obtained from chemical imidization were first dried at room temperature and then further thermally imidized in the molten caprolactam at 140°C. Polyimides further imidized in this monomer remained in solution that were not soluble otherwise.

Difunctionalized Polyimides. As shown in Scheme 4, N-acylated caprolactam end-capped polyimides were prepared by the two-step polycondensation of a diamine, excess dianhydride and 4-aminobenzoylcaprolactam. The diamine was first treated with an excess amount of dianhydride at room temperature in NMP to produce an anhydride-terminated polyamic acid. This oligomer was allowed to react with the end-capping agent 4-aminobenzoylcaprolactam to form an N-acylated caprolactam end-capped polyamic acid, which was then converted to the corresponding polyimide with pyridine and acetic anhydride. The molecular weight of the end-capped polyimide was controlled by varying the molar ratio of the diamine to dianhydride.

A series of N-acylated caprolactam end-capped polyimides was prepared from OTOL, BPDA and 6FDA with various compositions and different calculated molecular weights. The balance of rigidity and solubility were adjusted by the relative ratio of BPDA to 6FDA. The solubility of the polyimides in molten caprolactam depended more on the polymer composition than on the molecular weight. The end-capped polyimides prepared from 100/100/0, 100/85/15, or 100/70/30 mol ratios of OTOL/BPDA/6FDA displayed poor solubility in both NMP and molten caprolactam. When the 6FDA content was increased, the solubility improved. The polyimide prepared from 100/50/50 mol ratio of OTOL/BPDA/6FDA was the most rigid polymer that dissolved in molten caprolactam. Polyimides with this composition and calculated molecular weights of 6,000 and 12,000 were chosen as the difunctional activators for the anionic ring-opening polymerizations of caprolactam.

Anionic Ring-Opening Polymerization of ε-Caprolactam in the Presence of Polymeric Activators. The anionic polymerization of ε-caprolactam in the absence of activators involves the slow initial nucleophilic addition of the caprolactam anion to the caprolactam monomer, followed by the fast nucleophilic addition of the caprolactam anion to the formed acylated caprolactam chain end (14). This polymerization usually requires a reaction temperature of 200-280°C due to the high activation energy of the initial step. However, the addition of an activator, i.e., an N-acylated caprolactam, allows rapid polymerization that by-passes the slow initial formation of the activated species and allows the polymerization to proceed at 140°C. At 140°C, nylon 6 chains grow only from these activator sites. This feature provides an effective means for attaching nylon 6 chains to the molecules containing the N-acylated caprolactam moieties.

As shown in Scheme 5, activators containing one functional group ensure the growth of the nylon 6 chain in one direction and lead to the formation of a linear polymer structure or a diblock copolymer. Similarly, nylon 6 chains can be made to grow in two or three directions by using di- or trifunctional activators to form triblock or star-shaped polymers. When the activating moieties are located along polymer backbones, graft copolymers with nylon 6 side chains can be prepared. In this research, multi- and difunctionalized polyimide activators were prepared and used to initiate the polymerization of caprolactam to produce polyimide-g-nylon 6 and nylon 6-b-polyimide-b-nylon 6 copolymers, respectively.

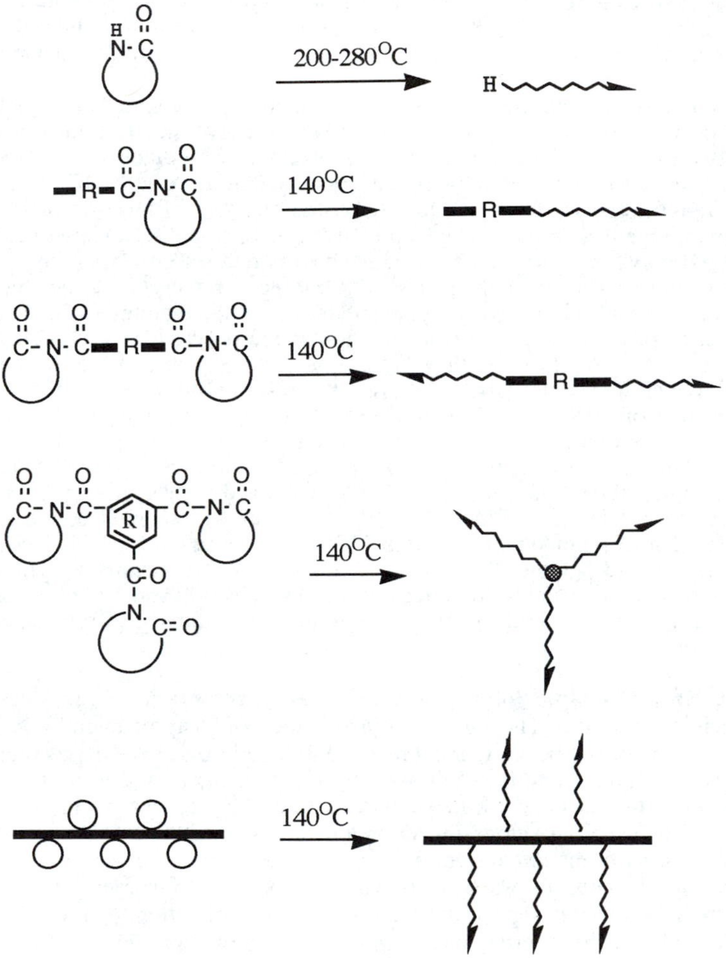

Scheme 5. Schematics of non-activated and activated anionic
ring-opening polymerization of caprolactam.

Polyimide-g-Nylon 6 and Nylon 6-b-Polyimide-b-Nylon 6 Copolymers. As shown in Scheme 6, the graft and block copolymers were prepared in a manner similar to that used in reaction injection molding (RIM) processes. Several weight percent of a multi- or difunctionalized polyimide was dissolved in molten caprolactam at 140°C under nitrogen to form a homogeneous polymeric activator solution. The anionic initiator solution was prepared separately by injecting a PhMgBr ether solution into molten caprolactam at 140°C under nitrogen. The molar ratio of the anionic initiator to the activating moieties in the polyimide was 1:1. No polymerization of caprolactam was observed in either the polymer activator solution or the initiator solution until the two were mixed. The reaction mixtures solidified within 5 min in the graft copolymerizations and within 10 min in the block copolymerizations. The graft and block copolymers containing 3 to 7 wt% polyimide were pale yellow, tough thermoplastic materials, which were compression-molded into dumbbell samples, cast from m-cresol solutions into films and melt spun into fibers for mechanical testing. No apparent phase separation occurred during any of the processing operations.

Characterization.   The different solubilities of the graft and block copolymers, compared to their homopolymer components, were the first indicaton of the formation of copolymers rather than homopolymer blends. Nylon 6 and the polyfunctional polyimides were soluble in formic acid and NMP, respectively. However, the graft and block copolymers were not soluble in formic acid or NMP. They were only soluble in m-cresol, indicating an enhanced chemical resistance compared to nylon 6.

The intrinsic viscosities of the graft copolymers were in the range of 1.3~1.9 dl/g, compared to 0.8~1.1 dl/g for their polyimide precursors. The block copolymers exhibited intrinsic viscosities of 2.2~4.6 dl/g, compared to 1.0~1.4 dl/g for the end-capped polyimides. The increase in the viscosity of the graft and block copolymers was the lowest when the most rigid polyimides were used.

The reaction of caprolactam with the polymeric activators was followed with FTIR spectroscopy. The incorporation of N-acylated caprolactam moieties in the polyimide chains was first confirmed by the appearance of the characteristic acylated caprolactam bands near 2931 and 2864 cm^{-1}. The polymer also displayed characteristic imide carbonyl absorptions at 1778 and 1727 cm^{-1}. As the graft or block copolymerizations proceeded, the intensities of the bands characteristic of nylon 6 segments at 3300, 3085, 1642 and 1545 cm^{-1} increased dramatically.

The graft and block copolymer samples were successively extracted with methanol, NMP and formic acid to determine the amounts of residual caprolactam, unreacted polyimides and nylon 6 homopolymer present. The residual caprolactam in the final products was less than 3 wt% and no nylon 6 homopolymer was detected.

The GPC elution peaks of the graft copolymers appeared at lower retention volumes than did those of the corresponding polyimide precursors. Both displayed similar polydispersities, ranging from 2.6 to 3.5 for the different polyimide structures. The end-capped polyimides and their corresponding block copolymers exhibited narrower polydispersities, ranging from 1.4 to 1.6. As the chain rigidity, molecular weight and amount of polyimide increased, the elution peaks of the corresponding block copolymers shifted to lower retention volumes, suggesting a more extended comformation. No nylon 6 homopolymer peak was detected.

Thermal Properties.   TGA showed that the graft copolymers, which contained 5 wt% polyimide, underwent 5% weight losses from 357 to 368°C when heated at 10°C/min in nitrogen. The block copolymers, which contained 3~7 wt% polyimide, underwent similar weight losses in the range of 373~380°C. For comparison, a sample of nylon 6, which was prepared by anionic polymerization, underwent a 5% wt% loss at 340~350°C when subjected to these conditions. DSC analysis showed that when the rigidity of the polyimide chain in the graft copolymers or the amount of polyimide incorporated in the block copolymers increased, the glass transition temperatures of the polymers increased. The increase was small, however, ranging from 2 to 4°C.

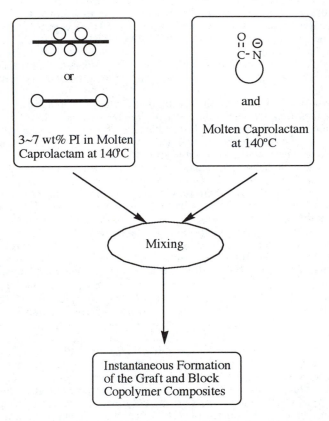

Scheme 6. Synthesis of polyimide-g-nylon 6 and nylon 6-b-polyimide-b-
         nylon 6 copolymers.

The crystallization temperatures also increased about 5~10°C, and their melting points decreased about 4°C compared to nylon 6. Typical DSC thermograms are shown in Figure 3.

Mechanical and Dynamic Mechanical Properties.   Tensile tests on compression-molded samples showed that the graft copolymers containing 5 wt% BAPBC/OTOL/BPDA (50/50/100 molar ratio) and the block copolymers containing 3~7 wt% OTOL/BPDA/6FDA (50/50/100 molar ratio, Mn=6,000 and 12,000) had tensile moduli and tensile strengths as high as 1700 MPa and 85 MPa, respectively, significantly higher than that of unmodified nylon 6. Figure 4 shows typical stress-strain curves of nylon 6, and the graft and block copolymers. The toughness (area under stress-strain curve) of the block copolymers was considerably higher than that of nylon 6. The tensile strengths of the block copolymer fibers were also as much as 40% higher than that of nylon 6 fibers.

Dynamic mechanical analysis (DMA) of films cast from m̲-cresol solutions revealed that the α-transition shifted to higher temperatures while the β-transition shifted to lower temperatures when the polyimides were incorporated into the nylon 6 matrix, as shown in Figure 5. The storage moduli (E') for the block copolymer films containing 3-5 wt% polyimide were about twice that of nylon 6 film, as shown in Figure 6. The block copolymers also showed good retention of their moduli at temperatures close to the nylon 6 melting point. Thus, the copolymers displayed improved high temperature mechanical stability relative to nylon 6.

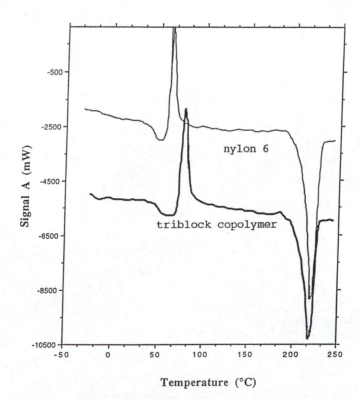

Temperature (°C)

Figure 3. Typical DSC thermograms of nylon 6 and the triblock copolymer prepared from 7 wt% OTOL/BPDA/6FDA (100/50/50 molar ratio; Mn=6,000).

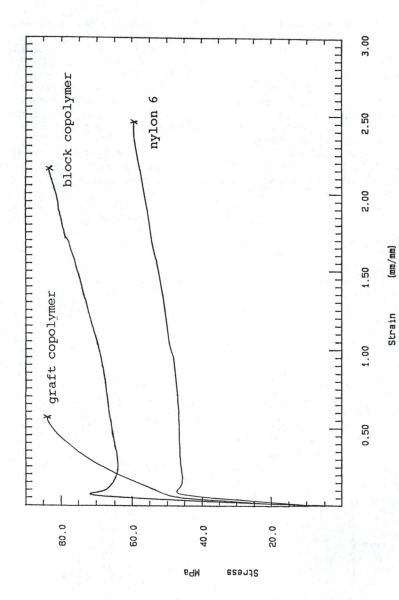

Figure 4. Typical stress-strain curves of nylon 6 and the graft and block copolymers.

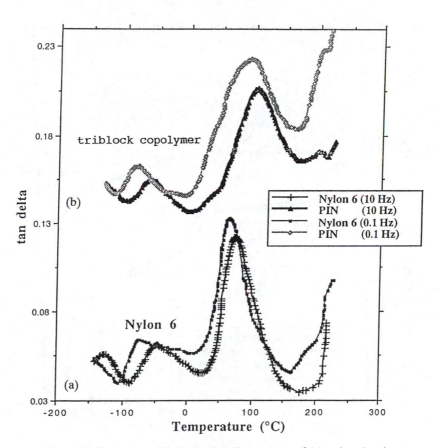

Figure 5. Typical mechanical relaxation spectra of (a) nylon 6 and
(b) the triblock copolymer prepared from 4 wt%
OTOL/BPDA/6FDA (100/50/50 molar ratio; Mn=12,000).

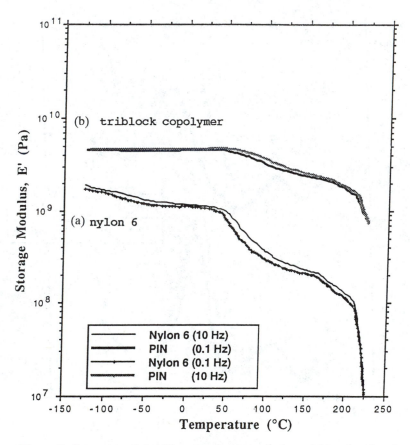

Figure 6. Storage modulus E' *versus* temperature plots of (a) nylon 6 and
(b) the triblock copolymer prepared from 4 wt%
OTOL/BPDA/6FDA (100/50/50 molar ratio; Mn=12,000).

Conclusions

Novel polyimide-g-nylon 6 and nylon 6-b-polyimide-b-nylon 6 copolymers were synthesized by polycondensation and subsequent anionic, ring-opening polymerization methods.   The graft and triblock copolymer structures were characterized with FTIR, GPC, selective extractions, DSC,TGA, tensile test and DMA. The effective reinforcement of nylon 6 with the incorporation of only a few wt% polyimide was demonstrated by the significant improvements in the copolymers' thermal and mechanical properties, and chemical resistance.

Acknowledgments

The financial support of this work by the State / Industry (EPIC) / University is also gratefully acknowledged.   We would like to thank Mr. B. P. Livengood for the thermal and DMA measurements.

Literature Cited

1. Takayanagi, M.; Ogata, T.; Morikawa, M.; Kai, T. *J. Macromol. Sci. Phys.*, **1980**, B17(4), 591.
2. Helminiak, T. E. *ACS Div. of Org. Coat. Plast. Chem., Preprints,* **1979**, *40*, 475.
3. Wang, C.S.; I. J. Goldfarb, I. J.; Helminiak, T. E. *Polymer*, **1988**, *29*, 825.
4. Chuah, H. H.; Kyu, T.; Helminiak, T. E. *Polymer*, **1987**, *28*, 2130.
5. Nishihara, T.; Mera, H.; Matsuda, K. *ACS Div. of Org.Coat. and Appl. Polym Sci., Proceedings,* **1986**, *55*, 821.
6. Takayanagi, M. *Polym. Composites*, **1986**, 4.
7. Dotrong, M.; Bai, S. J.; Evers, R. C. *37th International SAMPE Symposium*, **1992**, 1004.
8. Dotrong, M.; Dotrong, M. H.; Evers, R. C. *ACS Div. of Polym. Chem., Polym. Prepr.,* **1992**, *33*(1), 477.
9. Vakil, U. M.; Wang, C. S.; Lee, C. Y.-C.; Dotrong, M. H.; Dotrong, M.; Evers, R. C. *ACS Div. of Polym. Chem., Polym. Prepr.,* **1992**, *33*(1), 479.
10. Harris, F. W.; Ding, H. *US Patent*, Application, **1993**, 08/171,523.
11. Ding. H, Ph. D. Dissertation, *"Synthesis and Characterization of Novel Polyimid-g-Nylon 6 and Nylon 6-b-Polyimide-b-Nylon 6 Copolymers"*, The University of Akron, **1994.**
12. Donaruma, L. G.; Scelia, R. P.; Schonfeld, S. E. *J. Heterocyclic Chem.*, **1964**, *1*, 48.
13. Harris, F. W. in *"Polyimides"*, Wilsin, D.; Stenzenberger, H. D., Eds., Blackie & Son Ltd., New York, **1990**, p.1-37.
14. Salley, J. M.; Miwa, T.; Frank, C. W. in *"Materials Science of High Temperature Polymers for Microelectronics"*, Grubb, D. T.; Mita, I.; Yoon, D. Y., Eds., Materials Research Society, Pennsylvania, **1991**, Vol. 227, p.117-124.
15. Sebenda, J. *J. Macromol. Sci. Chem.*, **1972**, *A6*, 1145.

RECEIVED November 27, 1995

CERAMICS AND SOL–GELS

Chapter 20

Particulate Nanostructured Silicon Nitride and Titanium Nitride

R. A. Andrievski

Institute for New Chemical Problems, Russian Academy of Sciences, Chernogolovka, Moscow Region 142432, Russia

In recent years nanostructured (nanocrystalline, nanophase, or ultra-grained) materials (NM), which are commonly characterized by a grain size less than 100 nm, have attracted the most attention because of the hope of realizing the unique physical, mechanical, and chemical properties in the nanocrystalline state. In this connection the advanced ceramics as a whole, and silicon and titanium nitrides in particular because of their promising properties, are of great interest. Such properties of TiN and Si_3N_4 particulate NM as gas evolution at high-vacuum treatment and behavior during compaction, sintering, and high pressure sintering are described. The hardness values of these materials in the near dense state are discussed in detail. The necessity of further investigations is stressed.

Compared to the information for properties of some nanostructured metals (Cu, Ag, Ni, and Fe), intermetallics (TiAl and $NbAl_3$), and oxides (TiO_2 and ZrO_2), the data for high-melting compounds (nitrides, borides, and carbides) are very limited (*1-5*). It is common knowledge that silicon and titanium nitrides are the base of many tools and of structurally and functionally advanced materials. There is lot of information on properties of these nitrides in the conventional polycrystalline state and on single crystals (*6,7*). However, their investigation in the nanocrystalline state is only in beginning. Our selection of silicon and titanium nitrides was dictated not only by their practical applications but also by a desire to compare the behavior of compounds exhibiting different chemical bonding, i.e. covalent-ionic and metallic-ionic. We are focused only on particulate NM; thin film data will be shown for comparison.

Ultrafine Powders and Some of Their Characteristics

A large body of physical and chemical methods of preparation of high-melting compounds ultrafine powders (UFPs) has been examined in our papers(*4,5*). Thermal, plasma, and laser synthesis (decomposition), vaporization and condensation in

0097–6156/96/0622–0294$15.00/0

reactive   gases, and mechanical synthesis, as applied to Si_3N_4 and TiN, are the processes used very extensively for obtaining UFPs. It should be noted that the main part of works performed in recent years concerns the silicon nitride gas-phase synthesis (7). Making possible homogeneous nuclei formation, to the greatest degree gas-phase synthesis is accompanied by UFPs preparation with particle size less than 100 nm (for the most part $d<20$ nm). Reactions between silicon tetrachloride vapor (or silane) and their derivations with ammonia   conducted in plasma-chemical, laser-induced or conven-tional thermal interactions were the subject of many investigations (see, for example   Refs.(8-13)). The conditions of the thermal and laser pyrolysis of the crosslinked organosilicon polymers have also been analyzed in detail as applied to different gases (argon, oxygen, nitrogen, and helium) (14,15). Use of NMR and FTIR spectroscopy helped in the identification of different intermediate compounds formed. Plasma methods of UFPs preparation , fixed in to some extent as traditional, do not lose their importance (9,10,16,17) while using laser synthesis method is unique to the laboratory scale. Meanwhile, works for using gas-phase reactions in conditions of flowing thermal reactors are extending both for Si_3N_4 and TiN (13,18,19; Winter, G., H.C.Starck Co., personal communication, 1995).

Properties of Si_3N_4 and TiN UFP have been analyzed elsewhere (4,5,7,20). The most interesting features are the great enrichment of the particle surface in oxygen, carbon, and other admixtures and the poor compressibility of UFPs. According to data (21) the oxygen and carbon presence in surface layers of different Si_3N_4 UFPs is res-pectively 2-20 and 10^2-10^3 times higher than the bulk content. It is also interesting that there was no correlation between the overall chemical composition and the particle size of the powders in the experimental range of d values. The enrichment of the surface layers in oxygen and carbon was likewise weakly correlated with overall content of these impurities. It is also worthy of note that there are many other results which confirm the surface enrichment of TiN and Si_3N_4 UFPs in oxygen and carbon (for example, data obtained by XPS and AES methods (5,7)). Whereas the increased con-tent of oxygen on the surface can be naturally attributed to oxidation, in order to elucidate the causes of the enrichment in carbon it is useful to turn to the results of the study of gas evolution during the high-vacuum heat treatment of UFPs.

Figure 1 shows the influence of the high-vacuum treatment temperature on the rate of gas evolution for TiN UFP (d~70 nm) (5). One can see that there are two gas evolution peaks at ~250°C (for H_2O and N_2) and at ~650°C (for CO,CO_2 , and H_2). Carbon monoxide dominates appreciably in the evolving gases. XPS study has also revealed the enrichment of the surface layers in oxygen and carbon-containing components. A similar situation has also been observed in the case of Si_3N_4 UFP (7). Table I shows the content of evolved gases for amorphous plasma Si_3N_4 powders (d~20 nm) during high-vacuum heat treatment. Temperatures of maximum rate of gas evolution for CO, H_2 , and N_2 were 500-900°C and for H_2O and CO_2 ones were 300-400°C. Vacuum annealing results in significant decreasing desorbed gas quantity; however, the whole picture remains valid. It should be noted that it is not clear why the dominant   species in the desorbed gas below 1000°C is carbon monoxide, and not carbon dioxide, as could have been expected from the well known equilibrium of the Bell-Boudouard reaction. To gain a better understanding into this mechanism further investigations using different methods are badly needed.

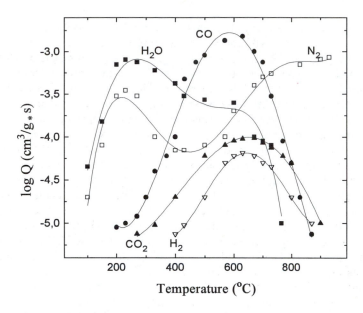

Figure 1. Temperature dependence of the rate of gas evolution Q for TiN UFP

Table I. Some Parameters of the Gas Evolution for Si_3N_4 UFP before (1) and after High-Vacuum Annealing for 1 h at 1200°C (2)

Powder type	Total gas volume (cm^3/g)	Predominant gas content (mass%)			
		CO	H_2	N_2	H_2O
initial (1)	625	59	11	21	9
annealed (2)	89	79	7	10	4

Another interesting feature of UFPs is their behaviour at compaction. It is well known that the compressibility of powders decreases markedly with decreasing particle size and so the UFPs compressibility is very poor. The last may be connected with severe interparticle friction of powders with high value of specific surface area. Beside that the poor ductility characteristics of TiN and especially Si_3N_4 are also responsible for the low compressibility of their powders. However, regarding UFPs the main factor determining compressibility is interparticle friction. As it was stressed in our paper (20) the compressibility of ductile nickel powder and brittle silicon nitride is nearly similar in the ultradispersed state. Table II demonstrates the comparison of the compressibility of silicon and titanium nitrides as well as nickel UFPs (4,5,20).

Table II. Influence of Particle size (d) and Compaction Pressure (P) on Relative Density of Si₃N₄, TiN, and Nickel UFPs (the Compact Diameter of ~10 mm)

Powder	$d(nm)$	Relative Density at P (GPa)			
		1	4	6	8
Si₃N₄	1000	0.64	0.74	0.79	0.82
	50	0.54	0.65	0.7	0.72
	17	0.47	0.55	0.58	0.62
TiN	80	0.64	0.8	0.84	0.85
	70	0.6	0.75	0.8	0.81
Ni	50	0.58	—	—	—
	15	0.49	—	—	—

From these data it is evident that the behavior of these powders during compaction seems to be roughly similar; however, the influence of pressure on relative density is more evident in the case of TiN. This can be explained by more covalent and respectively more brittle nature of silicon nitride. It has been pointed out that there is significant influence of chemical bond on compressibility for UFPs. It is also important to indicate that there is no possibility to obtain near dense compacts from UFPs in the conditions of cold compaction even at the compaction pressures up to 8-9 GPa.

NM Consolidation and Properties

One of the main problems in NM processing is consolidation with full densification and without sacrificing their nanocrystalline structure. Methods of UFPs consolidation, such as compaction and sintering, hot pressing, hot forging, shock compaction, electro-discharge compaction, high pressures and high temperatures technique, etc., have been discussed in review (*4*). Typical densification S-curves at pressureless sintering have been demonstrated in particular for metallic, metal-similar (like TiN), and covalent (like Si₃N₄) solids. The marked difference in the temperatures of active and full densification for metallic, metal-similar, and ionic solids, on the one hand, and for covalent ones, on the other was very striking. In first case, the sintering temperature of UFPs is substantially below that for conventional powders and ranges from 0.3 to $0.5T_m$ (T_m is melting point). However, for covalent solids (boron, silicon, Si₃N₄, etc.) the estimated active densification temperature during sintering is about $0.85\ T_m$ or higher. Si₃N₄ powders, even UFPs, are very difficult to densify without sintering aid, which is connected with the covalent origin of Si₃N₄ and its low diffusion and dislocation mobility.

For the most part intensive recrystallization accompanies high-temperature sintering obligatory and in these conditions the nanocrystalline structure disappearance is very likely. So only high-energy consolidation methods, such as shock compaction, high pressure and high temperature technique, etc., seem to be useful for effective NM processing. As applied to TiN and Si_3N_4 materials such experimental data are not comprehensive.

In our experiments with Si_3N_4 UFP (4,7,20), it was shown that increasing the compaction pressure from 1.5 GPa to 8.5 GPa results in a significant decrease of the densification temperature from ~1900°C to ~1100°C. The grain size of Si_3N_4 UFP compacted at 1200°C (P=8.5 GPa) was 0.5-1 μm (the initial particle size was ~50 nm). Shock-compacted Si_3N_4 powder was characterized by the grain size of 40-60 nm. In these experiments the compaction pressure and estimated temperature were respectively 40 GPa and 1200 K (22).

Figure 2 shows the change of relative density, hardness, and elastic modulus during pressure sintering of TiN UFP; the data for pressureless sintering are also demonstrating for comparison (23; Andrievski, R.A.; Urbanovich, V.S.; Kobelev, N.P.; Kuchinski, V.M. Proc. IV European Ceramic Society, Riccione, 2-5 October, 1995, in press). One can see that using the compaction pressure of 4 GPa results in a decrease of the active densification temperature on 600-800°C. It is also evident that the further pressure increase is not so effective. As is also clear from Figure 2, the elastic modulus and the hardness of specimens are practically changed monotonously as a density. As applied to the elastic modulus values obtained data seemed to be near those for the dense specimens (435-440 GPa) and some difference can be attributed to porosity influence. Some residual porosity of 1-3% was present in our near dense specimens. The fact that the hardness values were not changed in the interval from 1130°C to 1400°C may be explained in conjunction with the possible contribution between the porosity decrease and the recrystallization progress.

The XRD study has revealed only one phase of TiN in high-pressure sintered specimens. It was also detected that after heat and pressure treatment (T=1130-1200°C, P=4 GPa) the crystallite size and the magnitude of internal stresses derived from (422) line-broadening were 65-75 nm and $(2.2-2.7)10^{-3}$ respectively. The preliminary TEM observation of the specimen obtained at 1000°C (P=4 GPa) has revealed two kinds of crystallites with the sizes of 50-100 nm and of ~700 nm respectively. No additional phases were detected at the grain boundaries and triple junctions. At the same time, the optical microscopic and SEM study has revealed the inclusions presence. Their size varied from about micrometer parts to some micrometers but the main part had the size of 2-5 μm. The SEM investigations of fracture surfaces have also revealed the heterogeneous character of fracture. In some places of specimens the transgranular one has been observed. However, in many places the elucidation of structural peculiarities of fracture demands the more high resolution of SEM in comparison with our disposal. The inclusions hardness was lower than one for the main phase and was equal to 10-20 GPa. The largest inclusions had the least hardness value. The inclusions presence in the high pressure sintered specimens can be attributed both to their presence in initial UFP and to the result of beginning dynamic recrystallization. The detailed analysis of microstructure has been carried out elsewhere (Andrievski, R.A. Proc.6th Int.Symp. Fract.Mech.Ceram., Karlsruhe, 18-20 July, 1995).

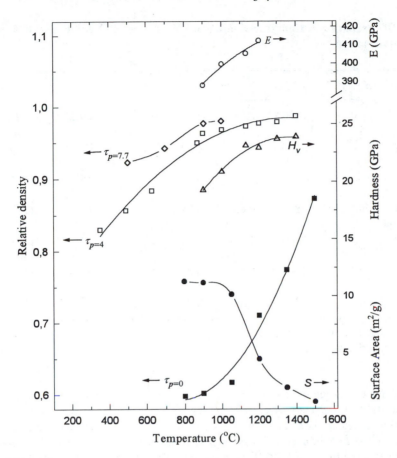

Figure 2. Effect of temperature on relative density (τ_p), microhardness (H_v), elastic modulus (E) at P=4 GPa and τ_p at P=7.7 GPa as well as on $\tau_{p=0}$ and specific surface area (S) during pressureless sintering.

It seems to be important to compare our densification and hardness data with literature information for TiN and Si₃N₄ materials. Table III summarizes this information. First of all it should be particularly emphasized that because of the different processing time, the distinct H_v and L measurements methods, etc., it is not easy to compare all these results. Some discrepancy is evident both for TiN and Si₃N₄. In the last case the enough modest results (*22*) may be connected with the low processing temperature. At the same time, however, it should be noted that mostly our H_v data for TiN are higher than those for the conventional sintered or hot-pressed materials which is likely connected with the presence of the nanocrystalline structure in our specimens.

Table III. Preparation Conditions and Some Properties of Near Dense TiN and Si₃N₄ Materials

Powder	d^a (nm)	T(°C)	P(GPa)	τ	L^b (nm)	H_v (GPa)	Source
TiN	~40	1500	0.04	0.96	~10^4	20-23	(24)
TiN	~10^3	2100	0.14	0.94	~10^4	10-11	(25)
TiN	~40	1300	–	0.99	~200	10	(26)
TiN	500-2 10^3	1850	0.03	0.94	2.5 10^3	12	(27)
TiN	~80	1100	4	0.98	65-75	24	(23)
Si₃N₄	~50	1200	8.5	0.99	500-10^3	32.4	(5)
Si₃N₄	~70	900	40	0.99	40-60	22	(22)

[a] d - the initial mean particles size.
[b] L - the specimens mean grain size.

It is also interesting to compare obtained results with ones for TiN films and Si₃N₄ single crystals. To our knowledge (4,5) the highest hardness values for TiN films are 45-75 GPa including data for nanocrystalline films (L~20 nm) and superlattice and multilayer ones (L~5-10 nm). The maximum hardness values of Si₃N₄ single crystal are also very high (up to 35-45 GPa) (6,7); at the same time, for monolithic polycrystalline Si₃N₄ microhardness is about 20-25 GPa (6). So our results are higher than those for the conventional dense polycrystalline TiN and Si₃N₄ specimens, however, they are smaller in comparison with TiN nanocrystalline films and Si₃N₄ single crystals. Therefore, the further investigations of the TiN and Si₃N₄ materials with the grain size of 10-20 nm seem to be very desirable.

Conclusions

To sum up this examination, it can been seen from the foregoing that only first steps have been taken in the study of silicon and titanium nitrides in the nanocrystalline state. The almost similar situation is in the case of another high-melting ceramic compounds. Problems, such as the selection of optimal UFPs preparation and conso-lidation methods in term of NM quality and cost effectiveness, the role of admixture, the origin of hardness and its relationship with grain size, nanophase equilibria in the systems Ti–N₂ , Si–N₂ , etc., and many others, remain unresolved and need further consideration and clarification.

Acknowledgments

The results described in this paper was made possible in part by the grants from the International Science Foundation and the Russian Government (NoMTF 300) and the Russian Basic Research Foundation (No 95-02-0318a).

I am sincerely grateful to Drs.G.V.Kalinnikov, M.A.Leontiev, R.A.Luitikov, and V.S.Urbanovich for their active assistance in this investigation.

Literature Cited

1. Gleiter, H. *Nanostr.Mater.* **1995,** *6,* 3.

2. Siegel, R. *Nanostr.Mater.* **1994,** *4,* 121.

3. Suryanarayana, C. *Int.Mater.Rev.* **1995,** *40,* 41.

4. Andrievski, R.A. *J.Mater.Sci.* **1994,** *29,* 614.

5. Andrievski, R.A. *Russ.Chem.Rev.* **1994,** *63,* 411.

6. Andrievski, R.A.; Spivak, I.I. *Strength of High-Melting Compounds and Materials on Their Base (in Russ.);* Metallurgia: Cheliabinsk, **1989**; pp.201-229.

7. Andrievski, R.A. *Russ.Chem.Rev.* **1995,** *64,* 291.

8. Lange, H.; Wotting, G.; Winter, G. *Angew.Chem.* **1991,** *30,* 1579.

9. Lee, H.J.; Eguchi, K.; Yoshida, T. *J.Am.Ceram.Soc.* **1990,** *73,* 3356.

10. Allaire, F.; Dallaire, S. *J.Mater.Sci.* **1991,** *26,* 6736.

11. Bauer, R.A.; Becht, J.G.M.; Kruis, F.F.; Scarlet, B., Schoonman, J. *J.Am.Ceram. Soc.* **1991,** *74,* 2759.

12. Danforth, S.C. *Nanostr.Mater.* **1992,** *1,* 197.

13. Chang, W.; Skadan, G.; Danforth, S.C.; Kear, B.H. *Nanostr.Mater.* **1994,** *4,* 507.

14. Gonsalves, K.E.; Strutt, P.R.; Xiao, T.D. *J.Mater.Sci.* **1992,** *27,* 3231.

15. Bahlout, D.; Pereira, M.; Goursat, P.; Choing Kwet Yive, N.S.; Corriu, R.J.B. *J.Am.Ceram.Soc.* **1993,** *76,* 1156.

16. Heidemane, G.M.; Grabis, Ja.P.; Miller, T.N. *Izv.Akad.Nauk SSSR. Neorg.Mater. (in Russ.)* **1979,** *15,* 596.

17. Batenin, V.M.; Klimovski, I.I.; Lysov, G.V.; Troitskii, V.N. *Superhigh Frequency Generators of Plasma;* CRS Press: Boca Raton, **1994;** pp.172-212.

18. Rabe, T.; Wasche, R. *Nanostr.Mater.* **1995,** *6,* 357.

19. Deccer, J.P.; van der Put, P.J.; Veringa, H.J.; Schoonman, J. *J.Mater.Chem.* **1994,** *4,* 689.

20. Andrievski, R.A. *Int.J.Powd.Metall.* **1994,** *30,* 59.

21. Szepvolgyi, J.; Bertoti, I.; Mohai-Toth, I.; Gilbart, E.; Riley, F.L.; Patel, M. *J.Mater.Chem.* **1993,** *3,* 279.

22. Hirai, H.; Kondo, K. *J.Am.Ceram.Soc.* **1994,** *77,* 487.

23. Andrievski, R.A.; Kalinnikov, G.V.; Potafeev, A.F.; Urbanovich, V.S. *Nanostr. Mater.* **1995,** *6,* 353.

24. Torbov, V.I.; Troitskii, V.N.; Rakhmatullina, A.Z. *Powd.Metal.(in Russ.)* **1979,** *(No12),* 28.

25. Moriyama, M; Komata, K.; Kobayashi, Y. *J.Cer.Soc.Jap.* **1991,** *99,* 286.

26. Ogino, Y.; Miki, M.; Yamasaki, T.; Inuma, T. *Mater.Sci.Forum* **1990,** *88-90,* 795.

27. Graziani, T.; Melaudri, C.; Bellosi, A. *J.Hard Mater.* **1993,** *4,* 29.

RECEIVED January 25, 1996

Chapter 21

Construction of Higher Order Silicate Structures from the $Si_8O_{20}^{8-}$ Silicate Species

Isao Hasegawa and Yasutaka Nakane

Department of Applied Chemistry, Faculty of Engineering, Gifu University, Yanagido 1–1, Gifu-City, Gifu 501–11, Japan

Organic-silica hybrids consisting of the double four-ring silicate ($Si_8O_{20}^{8-}$) structure as a building block have been produced through the reaction of the $Si_8O_{20}^{8-}$ silicate anion, which is dominantly present in a methanolic solution of tetramethylammonium silicate, with dimethyldichlorosilane in 2,2-dimethoxypropane. It has been found that the hybrids possess methoxyl groups which have been formed by esterification with methanol in the reaction mixture occurring simultaneously with the reaction of $Si_8O_{20}^{8-}$ with dimethyldichlorosilane. The hybrids have a BET surface area of 31 m^2 g^{-1}; the area increases to 339 m^2 g^{-1} upon calcining the hybrids at 350° C in air. The increase is ascribed to thermal decomposition of the methoxyl groups in the hybrids followed by condensation to develop the higher-order silicate structure of the hybrids. It is demonstrated that hydrolysis treatment of the hybrids also causes an increase in the BET surface area of the hybrids.

The building block approach is one of the procedures for rational construction of higher-order structures of inorganic materials at the nanometer level. In this procedure, compounds with well-defined structures such as clusters are usually used as starting materials and cross-linking reactions of such compounds are carried out to build up frameworks of the materials. Production of materials with structures as desired will be possible on the condition that the arrangement and cross-linking of the compounds will have been controllable.

Recently, we have reported on the synthesis of organic-silica hybrids by the building block approach from the cubic octameric silicate species ($Si_8O_{20}^{8-}$) and dimethyldichlorosilane [$(CH_3)_2SiCl_2$, DMDCS] (1,2). The higher-order silicate structure of the hybrids has been considered to consist of the $Si_8O_{20}^{8-}$ core as a building block, which is bridged by the $(CH_3)_2Si$ group, on the basis of their solid-state ^{29}Si NMR and IR spectra. In other words, both the building block and the degree of dispersion of the organic moiety in the hybrids would be regulated at the nanometer level.

0097–6156/96/0622–0302$15.00/0

Since the $Si_8O_{20}^{8-}$ structure corresponds to one of the secondary building units of zeolites (*3*), the hybrids are expected to show a high surface area. The as-synthesized hybrids exhibit a BET surface area of only 31 m^2 g^{-1}, while the value increases up to 339 m^2 g^{-1} upon heating the hybrids at 350° C in air for 1 h (*2*). The process responsible for this increase, however, has been obscure.

It is of importance for control over the framework of the hybrids to understand the process. Therefore, this study has been aimed at elucidating the process of increasing the surface area of the hybrids with the aid of the heat-treatment. Additionally, an alternative procedure is discussed for producing high surface area organic-silica hybrids.

Experimental

The $Si_8O_{20}^{8-}$ silicate species has been reported to be present dominantly in a methanolic solution of tetramethylammonium [$N^+(CH_3)_4$, TMA] silicate at a SiO_2 concentration of 1.0 mol dm^{-3}, a TMA ion concentration of 1.0 mol dm^{-3}, and a water concentration of 10.0 mol dm^{-3} (*4,5*). This methanolic solution was used as a source of $Si_8O_{20}^{8-}$.

The reaction of $Si_8O_{20}^{8-}$ with DMDCS was carried out by adding the methanolic solution dropwise to a 2,2-dimethoxypropane (DMOP) solution of DMDCS, yielding organic-silica hybrids comprised of the $Si_8O_{20}^{8-}$ structure as a building block. The detailed procedure has been described elsewhere (*2*).

The BET surface area of the hybrids was 31 m^2 g^{-1} (*1,2*). On the other hand, that of the hybrids subjected to heat-treatment at 350° C in air for 1 h was 339 m^2 g^{-1} (*2*) for which the temperature increasing rate was 10° C min^{-1}. Thus, the as-synthesized hybrids are referred to as "low-surface-area hybrids" and the hybrids after the heat-treatment as "high-surface-area hybrids" in this manuscript.

The low-surface-area hybrids were treated for hydrolysis by dispersing the hybrids (250 mg) in a mixture of acetone (*ca.* 5 cm^3), distilled water (2.5 cm^3), and 0.5 mol dm^{-3} hydrochloric acid (2 cm^3). The HCl concentration in the mixture is 0.1 mol dm^{-3}. The mixture was kept under stirring for 24 h at room temperature.

The BET surface area of the hybrids after the heat-treatment at 350° C was unchanged when the temperature increasing rate was 5° C min^{-1}. Then, the DTA-TG curves of the hybrids were measured in air over the range of room temperature to 800° C at a rate of 5° C min^{-1} with a MAC SCIENCE TG-DTA 2000S analyzer.

Solid-state ^{13}C and ^{29}Si NMR spectra were obtained on a Bruker MSL200 (CP-MAS $^{13}C\{^1H\}$: 50.3 MHz, a rotation frequency (r.f.): 3 kHz, a pulse width (p.w.): 3 μs, a repetition time (r.t.): 6 s, 6000-8000 scans, single-pulse (SP)-MAS $^{29}Si\{^1H\}$: 39.7 MHz, r.f.: 3 kHz, p.w.: 3.3 μs, r.t.: 60 s, 400-1000 scans) or a JEOL GSX-400 spectrometer (CP-MAS $^{13}C\{^1H\}$: 100.5 MHz, r.f.: 4.66 kHz, p.w.: 6 μs, r.t.: 5 s, 711 scans, SP-MAS $^{29}Si\{^1H\}$: 79.4 MHz, r.f.: 4.7 kHz, p.w.: 4.2 μs, r.t.: 60 s, 1000 scans). Chemical shifts are given with reference to an external standard of tetramethylsilane.

The silicate structure of the hybrids is represented using the notations of D, T, and Q, which correspond to $(CH_3)_2\underline{Si}(O^-)_2$, $CH_3\underline{Si}(O^-)_3$, and $\underline{Si}(O^-)_4$ siloxane units, respectively. Superscripts of the notations indicate the number of Si atoms adjacent to a siloxane structural unit.

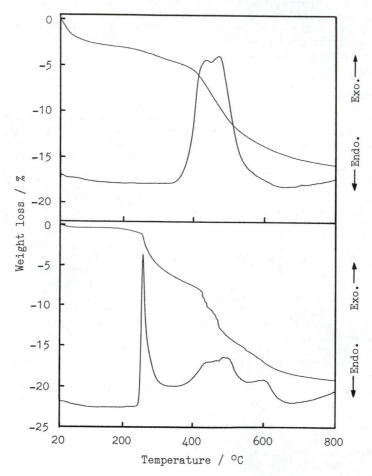

Figure 1. The DTA-TG curves of (a) the low-surface-area hybrids and (b) organic-silica hybrids formed with the hydrolysis treatment of the low-surface-area hybrids.

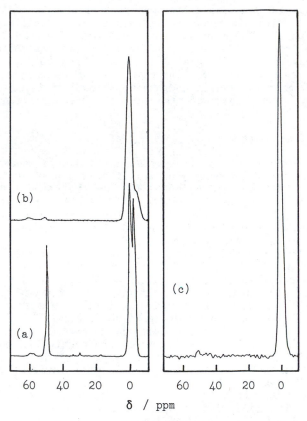

Figure 2. The solid-state CP-MAS ^{13}C NMR spectra of (a) the low-
and (b) high-surface-area hybrids (obtained at a field of 50.3MHz),
and (c) organic-silica hybrids formed with the hydrolysis treatment
of the low-surface-area hybrids (obtained at a field of 100.5 MHz).

FT-IR spectra of the samples in the form of KBr pellets were recorded on a Perkin-Elmer FT-IR 1640 spectrometer. BET surface area measurements were carried out at 77 K using nitrogen with a BEL JAPAN BELSORP 36 apparatus. Pretreatment of samples for the measurements was performed at 150° C for 3 h *in vacuo*. Chemical compositions of the hybrids were also measured. The samples were dried by heating at 150° C for 30 min before the chemical analysis.

Gas chromatography was performed with a Shimadzu GC-8A gas chromatograph equipped with a hydrogen flame-ionization detector and a CLH-702 split sample injector for connecting to a capillary column (a stationary phase: ULBON HR-1, i.d.: 0.25 mm, length: 25 m). The injection temperature was maintained at 70° C and the column temperature was programmed to rise at 10° C min^{-1} from 30 to 150° C. The carrier gas was nitrogen. For the detection of trimethylsilyl derivatives of low-molecular-weight compounds composed of D and/or Q units, the analytical conditions were the same as those described elsewhere (*6*).

Results and Discussion

In the DTA-TG curves of the low-surface-area hybrids shown in Fig. 1 (a), exothermic peaks accompanying weight loss appear in two temperature regions, 245-295 and 345-646° C. On the other hand, no peak appears in the range of 245-295° C in the DTA curve of the high-surface-area hybrids (data not shown). This implies that the exothermic reaction at 245-295° C is related to the increase of the BET surface area of the organic-silica hybrids.

The C-to-Si atomic ratio of the low-surface-area hybrids (Anal. Found: C, 16.61; H, 4.36; Si, 39.92 %) is 0.97, whereas that of the high-surface-area hybrids (Anal. Found: C, 12.16; H, 3.33; Si, 40.81 %) is 0.70. This suggests that an organic component in the low-surface-area hybrids is removed upon the heat-treatment at 350° C, in other words, during the exothermic reaction at 245-295° C.

Solid-state ^{13}C NMR spectra of the low- and high-surface-area hybrids are shown in Fig. 2 (a) and (b), respectively. An intensive signal at -2.2 ppm is assigned to the methyl group in D^1 units and that at 0.1 ppm to the methyl group in D^2 units in the hybrids (Hasegawa, I. *J. Sol-Gel Sci. Technol.*, in press.).

The most significant difference in the spectra lies in the intensity of the signal at 49.4 ppm, which is assigned to the methoxyl group in the CH$_3$-O-Si≡ (*7*). The signal is intensive in the spectrum of the low-surface-area hybrids, while the signal is hardly seen in that of the high-surface-area hybrids.

The CH$_3$-O-Si≡ bond has been reported to give rise to a band at 2840 cm^{-1} in IR spectra (*8*). The band is visible in the FT-IR spectrum of the low-surface-area hybrids [Fig. 3 (a)], whereas the band hardly appears in that of the high-surface-area hybrids [Fig. 3 (b)].

As additional evidence for the presence of the methoxyl group in the low-surface-area hybrids, formation of methanol is expected upon hydrolysis. To test this, the hybrids were treated for hydrolysis in a mixture of acetone and water containing HCl at a concentration of 0.1 mol dm^{-3}. A peak due to methanol is revealed together with a peak due to acetone in the gas chromatogram of the mixture after the treatment. These facts suggest that the CH$_3$-O-Si≡ bond is present in the low-surface-area hybrids.

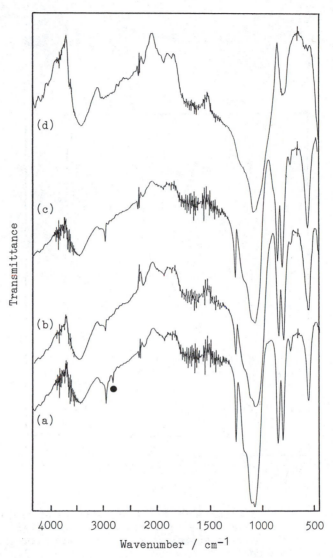

Figure 3. FT-IR spectra of (a) the low-, (b) high-surface-area hybrids, (c) hybrids after the hydrolysis treatment, and (d) products
formed on heating the low-surface-area hybrids at 800° C in air.
The closed circle indicates a band corresponding to the CH_3-O-Si≡
bond (2840 cm^{-1}). Assignments of intensive bands are as follows:
3448 cm^{-1}, O-H str.; 2965 cm^{-1}, C-H str.; 1263 cm^{-1}, Si-CH$_3$ def.;
1082 cm^{-1}, Si-O-Si str.; 850 and 804 cm^{-1}, Si(-CH$_3$)$_2$ str.

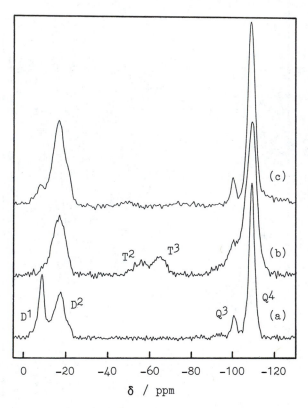

Figure 4. The solid-state SP-MAS ^{29}Si NMR spectra of (a) the low-
and (b) high-surface-area hybrids (obtained at a field of 39.7
MHz), and (c) the hybrids after the hydrolysis treatment (obtained
at a field of 79.4 MHz).

In addition, no low-molecular-weight compounds consisting of D and/or Q units are found in the solution by means of the trimethylsilylation technique (*9,10*) combined with gas chromatography. This would indicate that hydrolysis takes place at the $CH_3-O-Si\equiv$ group but not at siloxane bonds in the hybrids.

Figure 2 (c) shows the solid-state ^{13}C NMR spectrum of the hybrids after the hydrolysis treatment. From comparison with the spectrum shown in Fig. 2 (a), it is clear that the signal at 49.4 ppm drastically decreases in intensity with the hydrolysis treatment. In addition, the intensity of the band at 2840 cm^{-1} also decreases in the FT-IR spectrum of the hybrids after the treatment [Fig. 3 (c)]. These findings mean that the $CH_3-O-Si\equiv$ bond in the low-surface-area hybrids can be removed with the hydrolysis treatment.

Formation of the methoxyl group in the low-surface-area hybrids would be due to the esterification reaction with methanol, which occurs simultaneously with the reaction of $Si_8O_{20}^{8-}$ with DMDCS in DMOP. The methanolic solution of TMA silicate was used as a source of the $Si_8O_{20}^{8-}$ silicate species in this study. In addition, DMOP used as a solvent reacts with water in the methanolic solution to give methanol and acetone. Therefore, a considerable amount of methanol is present in the reaction mixture. The reaction of $Si_8O_{20}^{8-}$ with DMDCS proceeds under acidic conditions where the esterification reaction is also catalyzed, which would result in the formation of methoxyl groups in the hybrids.

Comparing the DTA-TG curves of the hybrids after the hydrolysis treatment [Fig. 1 (b)] with the curves shown in Fig. 1 (a), it is evident that exothermic peaks at 245-295 and 536-646° C are not revealed in the DTA curve of the hybrids after the hydrolysis treatment. This implies that these exothermic peaks are due to thermal decomposition of $CH_3-O-Si\equiv$ bonds in the low-surface-area hybrids and there are two kinds of $CH_3-O-Si\equiv$ bonds which decompose at different temperatures. Thus, the heat-treatment at 350° C would cause thermal decomposition of lower-temperature-decomposable $CH_3-O-Si\equiv$ groups in the hybrids.

The broad exothermic peaks at 345-536° C, seen in Fig. 1 (b), correspond to thermal decomposition of $CH_3-Si\equiv$ bonds in the hybrids. This is because bands due to C-H (2965 cm^{-1}) and CH_3-Si (1263 cm^{-1}) vibrations are not seen in the FT-IR spectrum of products formed upon heating the hybrids at 800° C in air [Fig. 3 (d)].

The solid-state ^{29}Si NMR spectra of the low- and high-surface-area hybrids are shown in Fig. 4 (a) and (b), respectively. From comparison of these spectra, it can be stated that a signal at -8.5 ppm due to D^1 units in the hybrids decreases in intensity and signals due to T^2 and T^3 units newly appear at -52.7 and -64.6 ppm, respectively, upon heat-treatment. The appearance of signals due to T^2 and T^3 units implies oxidation of a portion of the D units in the hybrids occurs to form T units upon the heat-treatment.

The decrease in intensity of the signal due to D^1 units indicates that condensation of D^1 units in the low-surface-area hybrids takes place by the treatment. This fact is also confirmed from the solid-state ^{13}C NMR spectra shown in Fig. 2 (a) and (b), in which the signal at -2.2 ppm due to D^1 units decreases in intensity upon calcining at 350° C.

On the basis of these interpretations, the heat-treatment of the hybrids causes the following reactions: (i) Thermal decomposition of

$CH_3-O-Si\equiv$ bonds, (ii) Condensation of D^1 units, and (iii) Thermal decomposition of a portion of the D units to form T units. These reactions would lead to development of the framework of the hybrids and, as a result, an increase in the surface area.

Among these reactions, reaction (ii) would take place as a result of reaction (i). Then, the hybrids subjected to hydrolysis are expected to show a higher BET surface area than the low-surface-area hybrids, since it is assumed that silanol groups formed by the hydrolysis are reactive to cause their condensation, that is, reaction (ii). The BET surface area of the hybrids after hydrolysis is measured to be 61 m^2 g^{-1}. The area actually increases with the aid of the hydrolysis treatment, although the change is small.

The solid-state ^{29}Si NMR spectrum of the hybrids subjected to hydrolysis is shown in Fig. 4 (c). In consequence of the hydrolysis treatment, the intensity of a signal due to D^1 units decreases. Recalling that low-molecular-weight compounds comprised of D and/or Q units are not formed in the solution after the hydrolysis treatment, it is considered that condensation of the D^1 units takes place together with the hydrolysis of $CH_3-O-Si\equiv$. However, a small signal due to D^1 units is revealed in Fig. 4 (c). This would indicate that the hydrolysis and condensation are not complete, which may result in a slight increase in the surface area of the hybrids upon hydrolysis.

The most significant difference in the spectra shown in Fig. 4 (b) and (c) is that no signal due to T units is seen in Fig. 4 (c). This means that the surface area of the hybrids can be increased without degradation of the D units by the aid of the hydrolysis treatment. In addition, signals due to D^2, Q^3, and Q^4 units do not shift with the hydrolysis treatment, which would mean that the D-Q linkage and the $Si_8O_{20}^{8-}$ structure as a building block are preserved in the hybrids subjected to hydrolysis.

It is still obscure at present what extent reaction (iii) contributes to an increase of the surface area of the hybrids. On the basis of the fact that the increase is achieved only with reactions (i) and (ii) caused by the hydrolysis treatment, however, it is considered that this treatment would be effective for synthesizing high surface area organic-silica hybrids without thermal degradation of an organic component.

Conclusions

The organic-silica hybrids produced by the building block approach from $Si_8O_{20}^{8-}$ and DMDCS were found to possess methoxyl groups. The heat-treatment of the hybrids at 350° C in air caused thermal decomposition of the methoxyl group followed by condensation. These reactions would develop higher-order silicate structures of the hybrids, resulting in the increase of their surface area. Hydrolysis and condensation of the groups in the hybrids were found to have potential as alternative procedures for increasing the surface area of the hybrids.

Acknowledgement

The authors acknowledge support of this work by IZUMI SCIENCE AND TECHNOLOGY FOUNDATION.

Literature Cited

1. Hasegawa, I.; Ishida, M.; Motojima, S. In *Proceedings of the First European Workshop on Hybrid Organic-Inorganic Materials*; Sanchez, C.; Ribot, F., Eds.; Bierville, France, 1993; pp. 329-332.
2. Hasegawa, I.; Ishida, M.; Motojima, S.; Satokawa, S. In *Better Ceramics Through Chemistry VI*; Cheetham, A.K.; Brinker, C.J.; Mecartney, M.L.; Sanchez, C., Eds.; Materials Research Society Symposium Proceedings 346; Materials Research Society: Pittsburgh, PA, 1994; pp. 163-168.
3. Meier, W.M. *Molecular Sieves*; Society for Chemical Industry: London, 1968, p. 10.
4. Hasegawa, I.; Sakka, S.; Kuroda, K.; Kato, C.; *J. Mol. Liq.* **1987**, *34*, 307.
5. Hasegawa, I.; Sakka, S.; Sugahara, Y.; Kuroda, K.; Kato, C. *J. Chem. Soc., Chem. Commun.* **1989**, 208.
6. Hasegawa, I.; Ishida, M.; Motojima, S. *Synth. React. Inorg. Met.-Org. Chem.* **1994**, *24*, 1099.
7. Taylor, R.B.; Parbhoo, B.; Fillmore, D.M. In *The Analytical Chemistry of Silicones*; Smith, A.L., Ed.; Chemical Analysis 112; John Wiley and Sons, New York, NY, 1991; p. 364.
8. Lipp, E.D.; Smith, A.L. In *The Analytical Chemistry of Silicones*, Smith, A.L., Ed.; Chemical Analysis 112; John Wiley and Sons, New York, NY, 1991; p. 325.
9. Lentz, C.W. *Inorg. Chem.* **1964**, *3*, 574.
10. Hasegawa, I.; Sakka, S.; Kuroda, K.; Kato, C. *J. Chromatogr.* **1987**, *410*, 137.

RECEIVED December 1, 1995

Chapter 22

Application of Fourier Transform Infrared Spectroscopy to Nanostructured Materials Surface Characterization

Study of an Aluminum Nitride Powder Prepared via Chemical Synthesis

Marie-Isabelle Baraton[1], Xiaohe Chen[2], and Kenneth E. Gonsalves[2]

[1]Laboratory of Ceramic Materials and Surface Treatments, Unité de Recherche Associée 320, Centre National de la Recherche Scientifique, University of Limoges, 123 Avenue Albert Thomas, F–87060 Limoges, France
[2]Polymer Science Program at the Institute of Materials Science and Department of Chemistry, University of Connecticut, Storrs, CT 06269

Due to unsaturation or strain, equilibrium of the interatomic forces on a surface is reached by adsorption of surrounding molecules. Therefore, the chemical composition of the first atomic layer may be different from that of the bulk. The importance of the surface with respect to the bulk in nanostructured powders makes the exact knowledge of the surface composition critical. Fourier transform infrared (FT-IR) spectrometry is a powerful tool to determine the nature of the chemical surface species as well as the reactive sites. As an example of an FT-IR surface study, a nanostructured aluminum nitride powder was analyzed and its surface was compared with the γ-alumina surface.

The surfaces of oxide nanopowders have been extensively studied either for characterization purposes or for potential catalytic properties investigation. As for non-oxide ceramic powders, references on their surface studies are scarce. Due to hydrolysis in ambient atmosphere, the unavoidable presence of oxygen in the first atomic layer of these nanostructured powders may drastically modify their expected properties. After briefly introducing our characterization technique of nanosized powder surfaces, we will present as an example, the Fourier transform infrared (FT-IR) surface analysis of a nanostructured aluminum nitride powder obtained via sol-gel type chemical synthesis.

General Scope

Importance of the surface characterization for nanostructured powders. In the bulk of materials, the interatomic forces are balanced whereas on the surface dangling bonds and defects lead to unsaturation or strain. As a consequence, equilibrium may be reached either by rearrangement of the surface atoms or by adsorption of surrounding molecules. Thus the chemical composition of the first atomic layer of any material is different from that of the bulk.

The surface atoms can be present as free bonds, free bonding orbitals with affinity to electrons or occupied bonding orbitals with a low ionization potential (*1*). Thus a wide range of different types of adsorption centres may result on a given surface. Moreover these centres determine the nature of the adsorption and, in return, the adsorbed molecules may change the properties of the surface.

On the other hand, the nature of these adsorption centres and their relative concentration depend on the synthesis and collection conditions of the materials. All these remarks bring emphasis on the complexity of a surface structure and the close interdependence of the influencing parameters. In the case when the specific surface area of the powder is increased, the surface plays a non negligible part in the overall powder properties, and the need for a specific surface characterization then becomes critical. Many technological processes depend on the chemical composition of these surfaces and the increasing production of nanostructured powders raises the demand for well controlled surfaces.

FT-IR Surface Spectrometry. Fourier transform infrared spectrometry is one of the most convenient techniques for the surface characterization of ceramic nanopowders. Indeed, the vibrational spectra bring information on the nature of the bond formed between the surface and the adsorbed molecules and consequently on the nature of the adsorption centres. Moreover, chemical species irreversibly grafted on the surface have specific absorption bands in the powder spectrum, and are considered as intrinsic probes since they may be perturbed by molecular adsorption.

To fully characterize a surface, the experimental process summarized in Flow Chart 1 must be followed. During the first step, the surface is activated. The activation consists in heating the sample under dynamic vacuum for one or two hours. This thermal treatment cleans the surface from physisorbed and weakly chemisorbed species according to the temperature.

Since hydrolysis is the most probable reaction to produce saturated bonds and balanced forces on any surface, many surface bonds involve hydrogen atoms. Consequently, an isotopic exchange H/D by deuterium addition will discriminate between hydrogen-containing groups located on the surface and the ones in the bulk. Moreover the vibrational frequencies of the exchanged groups shift toward lower values due to the higher molecular weight of deuterium. In other words, deuterium acts as a marker of the hydrogen vibrations.

Methanol is a useful probe molecule to check the lability of OH surface groups. The reaction of methanol on these OH groups leads to the formation of methoxy groups. But other adsorption processes can also occur when acidic and basic sites are present on the surface.

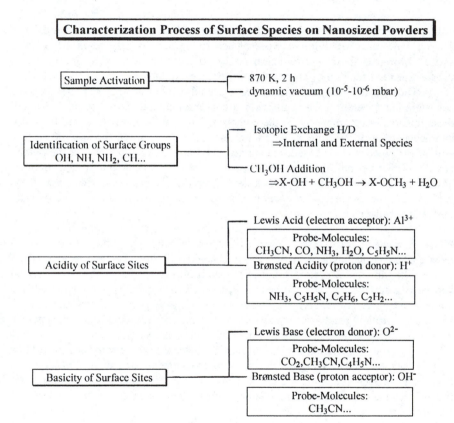

Characterization Process of Surface Species on Nanosized Powders

Sample Activation ——————————— 870 K, 2 h
                                                            dynamic vacuum (10^{-5}-10^{-6} mbar)

Identification of Surface Groups —————— Isotopic Exchange H/D
OH, NH, NH_2, CH...   ⇒Internal and External Species

                                                            CH_3OH Addition
                                                            ⇒X-OH + CH_3OH → X-OCH_3 + H_2O

Acidity of Surface Sites ——————— Lewis Acid (electron acceptor): Al^{3+}

| Probe-Molecules: |
| CH_3CN, CO, NH_3, H_2O, C_5H_5N... |

Brønsted Acidity (proton donor): H^+

| Probe-Molecules: |
| NH_3, C_5H_5N, C_6H_6, C_2H_2... |

Basicity of Surface Sites ——————— Lewis Base (electron donor): O^{2-}

| Probe-Molecules: |
| CO_2, CH_3CN, C_4H_5N... |

Brønsted Base (proton acceptor): OH^-

| Probe-Molecules: |
| CH_3CN... |

Flow Chart 1. FT-IR surface characterization process.

Lewis acidic and basic sites, respectively, correspond to electron acceptor and electron donor sites whereas Brønsted acidic and basic sites correspond to proton donor and proton acceptor sites. For a complete characterization, different molecules should be used to probe these sites, but it must be kept in mind that a molecule can probe different types of sites at the same time. A striking example is the CH_3CN probe-molecule which can form at least four different species by adsorbing on different sites (hydrogen bonding, coordination on metallic ion, coordination on OH^- ion, reaction with O^{2-} ion). Moreover, acidic and basic characteristics are interdependent, since their formation depends on the stoichiometry, the crystalline phase, the synthesis conditions, and the impurities or the contaminants. Therefore, several experiments have to be successively run with different probe molecules to get a good knowledge of the surface reactivity.

In our experimental procedure the transmission spectra are recorded *in situ* using a specially designed cell. This cell enables heating the sample from room temperature to 873 K under atmosphere, vacuum or controlled pressures of various gases (*2*). This cell is placed in the sample chamber of the spectrometer so that it is possible to exactly follow the spectrum modification at any step of the experiment. Another advantage of this *in situ* analysis is the possibility of performing difference spectra. The difference between two spectra recorded at different steps can make the evolution of the surface species clearer. However, to avoid temperature effects, the two spectra must be recorded at the same temperature. It must also be noted that, even for nanosized powders, the contribution of the surface species to the spectrum is minor. Therefore the large amount of powder needed to get a good surface spectrum obscures the wavenumber region of the bulk vibrations and thus prevents simultaneous surface and bulk analysis.

Experimental

Synthesis conditions. The synthesis method for the nanostructured AlN powders was similar to that reported previously (*3*). Aluminum chloride hexahydrate and urea in equimolar ratio were thoroughly dissolved in deoxygenated water. Under vigorous stirring, anhydrous ammonia was bubbled through the aqueous solution. The reaction temperature was then increased gradually to 363 K over a period of 24 hours. A white gel formed and the solution viscosity increased. The reaction was completed by heating the mixture to 393 K for another 24 hours. The precursor gel was obtained by removing the solvent under vacuum. Pyrolysis of the precursor gel was accomplished by high temperature processing at up to 1373 K under a continuous flow of anhydrous NH_3 gas for 10 hours. Prior to pyrolysis, the chamber was evacuated to about 10^{-5} mbar, then repeatedly flushed and back filled with ultra pure nitrogen. The as-synthesized white powders were stored and handled under argon.

Bulk characterization tools. The formation of nanostructured AlN powders have been confirmed by FT-IR spectroscopy (Nicolet 60SX), Raman spectrometry (Dilor microprobe), X-ray diffraction (XRD) (Norelco X-ray diffraction unit with wide range goniometer), and transmission electron microscopy (TEM), along with the corresponding electron diffraction in a JEOL 200CX microscope.

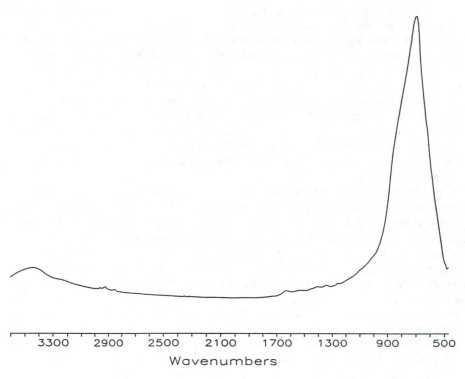

Figure 1. FT-IR spectrum of the as-synthesized nanostructured AlN.

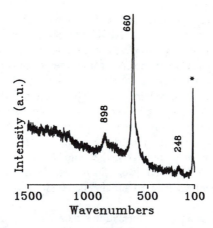

Figure 2. Raman spectrum of the as-synthesized nanostructured AlN (the * mark refers to a plasma line).

For IR bulk characterization, the AlN powder was dispersed in KBr and pressed according to the conventional sampling technique. The powder was analyzed as received under the Raman microscope. The XRD test sample is made by adhering the AlN powder onto an adhesive tape and for the TEM analysis the grid is dipped in a suspension containing the AlN powder.

XPS and surface FT-IR measurements. X-ray photoelectron spectroscopy (XPS, Perkin-Elmer Physical Electronics PHI 5300) was also used to extract information about the surface species. The XPS instrument is equipped with a monochromatic Al K_α X-ray source (1486.6 eV) and hemispherical analyzer. The sample was prepared for XPS analysis by sprinkling the powders on an adhesive tape so as to obtain uniform and complete coverage.

The IR surface spectra were recorded using a FT-IR Nicolet 5DX spectrometer from 4000 to 400 cm^{-1} with a 4 cm^{-1} resolution. The AlN powder was slightly pressed into a grid supported pellet (*2*) and placed inside the cell described above for transmission analysis. All the spectra were recorded at room temperature unless otherwise stated. Deuterium (99% pure) was from Air Liquide and underwent no further purification. Methanol and pyridine from Merck-Uvasol were dried over molecular sieves.

Bulk Characterization.

FT-IR characterization. The strong absorption band (Figure 1) at 690 cm^{-1} is characteristic of the aluminum nitride transverse optical (TO) mode (*4*). Another broad band was observed in the range of 3200 to 3400 cm^{-1}. This corresponds mainly to the stretching vibrations of hydroxyl and amine groups present on the powder surface. The band was carefully investigated and will be discussed below.

Raman analysis. Raman peaks (Figure 2) were observed at 898, 660, and 248 cm^{-1}. According to the literature (*5*), the peak at 660 cm^{-1} (strong) corresponds to amorphous aluminum nitride, while the peaks at 898 and 248 cm^{-1} (weak) correspond to those observed for polycrystalline aluminum nitride. This strongly suggests that the nanostructured AlN sample contains a mixture of amorphous and polycrystalline phases.

XRD measurements. Only aluminum nitride peaks were observed in the XRD patterns of the as-synthesized powders (Figure 3). The spectra also indicated that hexagonal crystals were obtained. Average crystallite diameter was estimated as 60 nm through a line broadening calculation of the XRD data (*6*).

TEM results. The bright field TEM micrograph and the selected area electron diffraction of the powder (Figure 4) identified the crystalline phase of the nanostructured AlN powder to be hexagonal. The results matched what was observed from the XRD data (*6*).

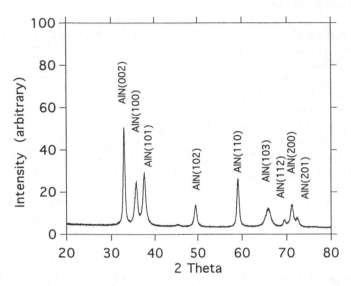

Figure 3. Powder X-ray diffraction pattern of the nanostructured AlN.

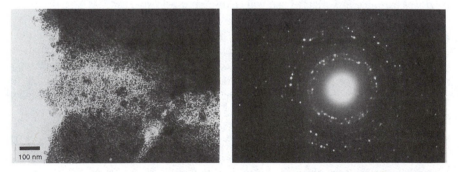

Figure 4. Bright-field TEM micrograph and the corresponding electron diffraction pattern of the nanostructured AlN sample.

Surface characterization: XPS and FT-IR results.

XPS results. The XPS survey spectrum (Figure 5) of the AlN powder surface revealed a combination of species containing aluminum, nitrogen, and oxygen. All XPS spectra were referenced to the adventitious carbon 1s photoelectron peak at 284.6 eV. The N(1s) peak (Figure 6) at 396.0 eV could be assigned to nitride because nitride peaks in XPS spectra occur in the range of 396.4±0.3 eV (*7,8*). Although the Al(2s) or Al(2p) may be used for analyzing the oxidized powders (*8*), it has been indicated in the previous literature (*5,8*) that the separation of the aluminum photoelectron peaks for aluminum nitride and aluminum oxide was insufficient to allow for unambiguous chemical state determination. When, however, surface FT-IR technique is applied for the same system, XPS becomes a good way of comparing to the observed information. For instance, there was no N(1s) peak observed in the range of 400.8±0.3 eV, suggesting that no dinitrogen species exist (*7*). This observation proves to be very instructive for peak identification in later surface FT-IR studies.

Activation. The IR spectrum of a pure as-synthesized AlN pellet is presented in Figure 7a. As mentioned above, the vibrational modes of the bulk obscure the wavenumber region below 1000 cm^{-1}. The other spectra (Figure 7b-d) were recorded during a step by step activation. Figure 8 depicted the difference spectra for each activation step. The negative bands correspond to disappearing species whereas positive bands correspond to appearing species. The difference spectrum (Figure 8d) showed the changes occurring during the complete activation at 873 K. The broad negative band centered around 3300 cm^{-1} and the negative shoulder at 1630 cm^{-1} indicated the desorption of water from the ambient atmosphere (*9-11*). In this case, ammonia originating from the synthesis process can also desorb, contributing to the above-mentioned bands. Other disappearing species were identified as monodentate carbonate groups whose ν(CO) absorption frequencies are located at 1534 and 1390 cm^{-1} (*12*). They proved a surface contamination by atmospheric carbon dioxide. A positive band at 2157 cm^{-1} corresponded to the formation of a new species on the surface. Since there was no evidence of N(1s) peak in the XPS spectrum, the possibility of formation of dinitrogen species observed in aluminum nitride films (*7,13-15*) can be discarded. This 2157 cm^{-1} frequency is very close to the ones of the ν(Si-H) and ν(B-H) stretching vibrations in surface groups (*16-18*). By comparison, we propose the assignment of this 2157 cm^{-1} band to the ν(Al-H) stretching vibration (*19*). However, this frequency is high, compared to the 1800 cm^{-1} absorption range usually given in the literature for the ν(Al-H) modes (*20*). This frequency assignment will be discussed further in this paper. But, an explanation for the high value of this frequency could be the different electronic distributions in the bulk and on the surface.

After activation (Figure 7d) the spectrum of the AlN powder presented a complex band in the highest wavenumber region (3780-3730 cm^{-1}) assigned, by comparison with γ-alumina (*11,21,22*), to the ν(OH) stretching vibrations. The broad band at 3205 cm^{-1} corresponded to the ν(NH) stretching vibrations of NH$_2$ groups as in nitrided alumina (*23,24*).

The step by step activation revealed the complex phenomena in the 2300-2000

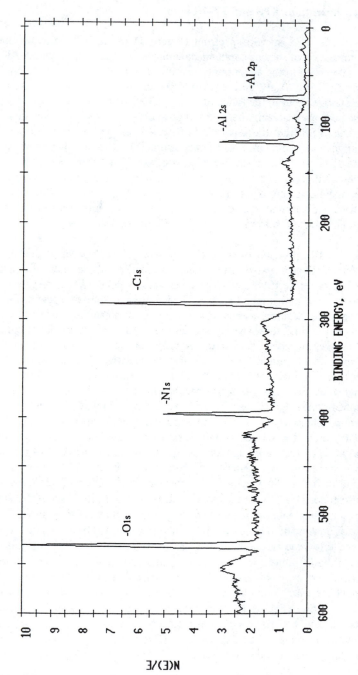

Figure 5. XPS survey spectrum of the as-synthesized nanostructured AlN.

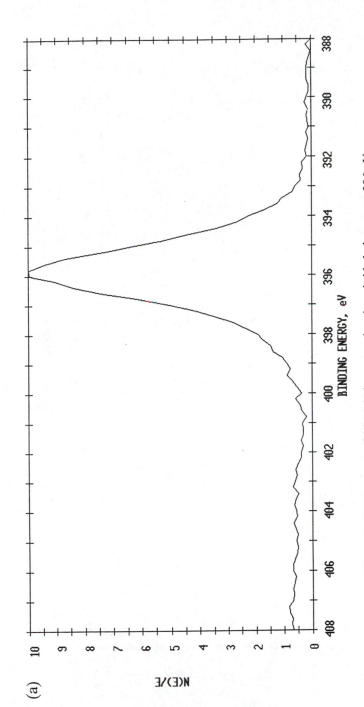

Figure 6. XPS spectra of (a) N (1s) spectra showing nitrided nitrogen at 396 eV, whereas no peak existed in the range of 400.8±0.3 eV; and (b) Al (2p) peak at 74 eV.

Continued on next page

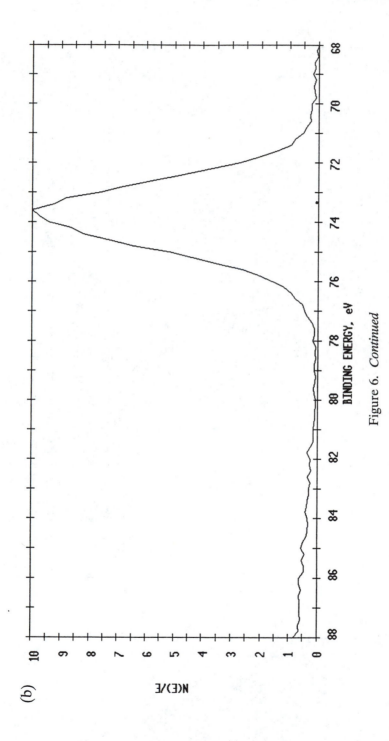

(b)

BINDING ENERGY, eV

N(E)/E

Figure 6. *Continued*

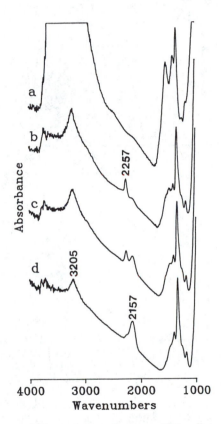

Figure 7. FT-IR transmission spectra of a pure AlN pellet at room temperature (a) and activated at different temperatures: 573 K (b); 673 K (c); 873 K (d).

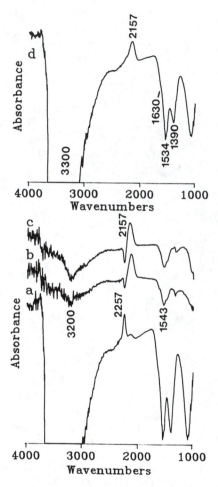

Figure 8. Difference spectra between two activation steps (to be compared with Figure 7): from room temperature to 573 K, 7b-7a (a); from 573K to 673K, 7c-7b (b); from 673 K to 873 K, 7d-7c (c); from room temperature to 873 K, 7d-7a (d).

cm^{-1} region (Figure 8a-c). By heating under vacuum from room temperature to 573 K, a band appeared at 2257 cm^{-1}. Subsequent heating made this band decrease whereas a band at 2157 cm^{-1} concurrently appeared. At the end of the activation process, the former 2257 cm^{-1} band no longer existed and only the 2157 cm^{-1} band was visible. We propose the assignment of the 2257 cm^{-1} band to the $\nu(Al\text{-}H)$ stretching vibration in dihydride $Al\text{-}H_2$ groups. These groups may originate from ammonia used for the synthesis. Indeed, at the high temperature of the synthesis, a part of ammonia decomposes into H_2 and N_2 and then H_2 possibly adsorbs on the surface (*18*) leading to AlH_2 groups. On the other hand, from 673 to 873 K, a negative band appeared in the difference spectrum (Figure 8c) at 1543 cm^{-1}. This band has already been assigned in nitrided aluminum oxide to the $\delta(NH_2)$ bending vibration of the $Al\text{-}NH_2$ groups (*23,24*). The two corresponding $\nu(NH)$ stretching bands are joined in the broad feature at ~3200 cm^{-1}. These $\nu(NH)$ and $\delta(NH_2)$ negative bands showed that a part of the $Al\text{-}NH_2$ groups was released from the surface. An explanation of this evolution could be the release of ammonia formed from $Al\text{-}NH_2$ and $Al\text{-}H_2$ surface species leaving $Al\text{-}H$ monohydride groups behind.

Deuteration. As previously explained, the surface species containing hydrogen atoms were probed by deuterium addition. To the sample activated at 873 K, 130 mbar of D_2 were added (Figure 9a,b). The H/D exchange took place immediately (within 5 minutes) on the OH and NH_2 groups. The $\nu(OH)$ bands shifted from 3780-3730 cm^{-1} to 2780-2750 cm^{-1} while the $\nu(NH)$ band shifted from 3205 to 2373 cm^{-1}. On the difference spectrum (Figure 9c), the changes caused by the isotopic exchange were clearly visible including the $\delta(NH_2)$ bending vibration at 1508 cm^{-1} shifted toward the obscure region of the spectrum.

Like the Si-H surface groups on Si-containing ceramics (*17*), the Al-H groups were not exchanged, even after addition of several subsequent doses. A plausible explanation can be that these groups are not accessible to deuterium because they are located inside the bulk. However, an alternate explanation is that the strength of the Al-H bonds makes the exchange difficult. Thus we must be aware of the limitations of the isotopic exchange.

The $\nu(OD)$ absorption revealed several bands (Figure 10). As on $\gamma\text{-}Al_2O_3$ surface (*11,21,22*), several hydroxyl groups existed leading to several OD groups. Given that the $\nu(OH)/\nu(OD)$ ratio is 1.356, it is possible to calculate the frequencies of the different $\nu(OH)$ absorption bands. This method is more accurate than a direct measurement of the $\nu(OH)$ frequencies. Indeed, in the highest wavenumber region, the infrared spectra are usually noisy due to the absorption by atmospheric water. In the present AlN sample, the $\nu(OH)$ stretching absorption bands are located at 3782 cm^{-1} ($\nu(OD)=2789$ cm^{-1}), 3738 cm^{-1} ($\nu(OD)=2757$ cm^{-1}) and 3706 cm^{-1} ($\nu(OD)=2733$ cm^{-1}). Thus, three types of hydroxyl groups are clearly identified whereas five types are well-known on $\gamma\text{-}Al_2O_3$ surface. By comparison with the assignment of the $\nu(OH)$ bands on $\gamma\text{-}Al_2O_3$ given in literature (*21*), the highest frequency 3782 cm^{-1} is assigned to the OH groups bonded to a tetrahedral Al^{3+} ion while the band at 3738 cm^{-1} is attributed to the OH groups bonded to an octahedral Al^{3+} ion. As for the lowest frequency, 3706 cm^{-1}, it corresponds to hydroxyl groups bonded to two Al atoms.

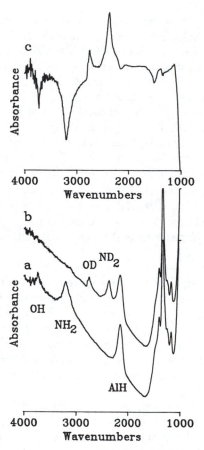

Figure 9. FT-IR transmission spectra of the AlN pellet activated at 873 K (a); after deuterium addition (b); difference spectrum b-a (c).

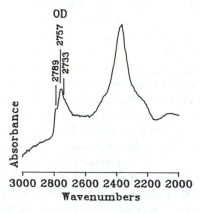

Figure 10. ν(OD) absorption range (to be compared with Figure 9c).

The $v(OH)$ frequencies in AlN are slightly lower than the ones in γ-Al_2O_3. This is possibly a consequence of the inductive effect modification when an oxygen atom is replaced by a less electronegative nitrogen atom (*17,25*). For the same reason, the $v(NH)$ absorption is observed at lower frequencies than on a nitrided Al_2O_3 surface (*9,26*).

Methanol addition.   It is well-known that methanol can adsorb on an activated alumina surface and form at least three types of surface complexes (*27*), namely hydrogen bonded alcohol, alkoxide species and carboxylate species. Since the AlN surface presents analogies with γ-Al_2O_3 surface, the comparison between both surfaces was carried out with the adsorption of methanol on the AlN activated surface.

The spectrum recorded after adding 8 mbar of methanol (Figure 11b), as well as its difference to the bare activated surface spectrum (Figure 12a) brought evidence of hydrogen bond formation, as shown by the broad positive band centered around 3200 cm^{-1} and the negative bands around 3740 cm^{-1}. The first band was assigned to hydrogen-bonded OH groups while the latter ones corresponded to decreasing free OH groups. The group of bands in the 2900 cm^{-1} region were assigned to the $v(CH)$ stretching vibrations in the methanol molecule. An evacuation at room temperature (Figures 11c and 12b) removes a part of the hydrogen bonded methanol. But the negative 3740 cm^{-1} band and the positive 3200 cm^{-1} band are still visible. In this new species, the hydrogen bond is indeed strengthened by a secondary interaction with Al^{3+} acidic surface sites as on alumina (*28*). At the same time, in the 1700-1000 cm^{-1} region, complex modifications of the spectra showed both increasing and decreasing species. Successive evacuations at increasing temperatures were performed to characterize the thermal stability of these species.

First of all, the band appearing at 1603 cm^{-1} after evacuation at 423 K (Figure 12c) which persists up to 723 K (Figure 12d) is characteristic of the $v_{as}(CO)$ stretching vibration in the formate ion. Its formation due to a dissociation reaction of methanol involves both acidic and basic surface sites, as already explained on an alumina surface (*12,29*). Other bands around 2905 ($v(CH)$), 1405 ($\delta(CH)$) and 1380 cm^{-1} ($v_s(CO)$) also correspond to this formate ion. The second species easily identified and also observed on alumina surface (*28,29*) are the methoxy Al-OCH_3 groups resulting from the reaction of CH_3OH and Al-OH groups. These latter species are quite stable up to 723 K (Figure 12d) and their vibrational frequencies can be located at 2952 ($v_s(CH_3)$), 2855 ($v_{as}(CH_3)$), 1467 ($\delta(CH_3)$), 1197 ($r(CH_3)$) and 1095 cm^{-1} ($v(CO)$). However, another species can be identified by the sharp band at 2828 cm^{-1} (Figure 12d) which does not appear in the spectrum of methanol adsorbed on alumina. Moreover, a broad band at 3218 cm^{-1} proves that new N-H groups are formed. As already noticed on the silicon nitride surface (*9*), basic sites corresponding to surface Al_3N groups could possibly exist here and react with acidic methanol leading to new imido bridges. Since several species are simultaneously present, it is hard to describe the exact mechanism of this latter adsorption. But the absence of $\delta(OH)$ vibration in the 1420 cm^{-1} region may suggest a dissociative adsorption opening Al_3-N bridges and yielding new N-H and Al-OCH_3 groups whose $v(CH)$ stretching frequencies would be at 2828 cm^{-1} and 2950 cm^{-1}. As for the other vibrational modes, they would be mixed with the ones of the previously described Al-OCH_3 species.

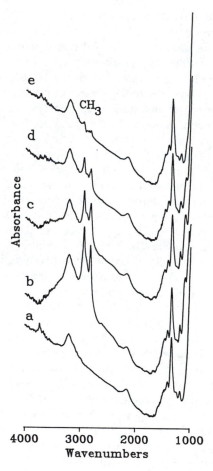

Figure 11. FT-IR transmission spectra of the AlN pellet activated at 873 K (a); after addition of 8 mbar methanol at room temperature (b); after evacuation at room temperature (c); after evacuation at 423 K (d); after evacuation at 723 K (e).

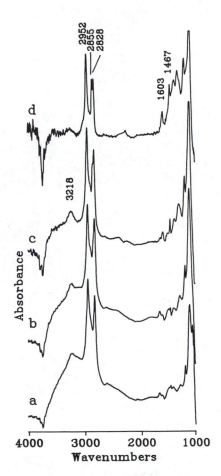

Figure 12. Difference spectra between two steps of the methanol adsorption (to be compared with Figure 11): methanol addition, 11b-11a (a); after evacuation at room temperature, 11c-11a (b); after evacuation at 423 K, 11d-11a (c); after evacuation at 723 K, 11e-11a (d). The Y scale has been increased for the last two spectra (c and d) for clarity's sake.

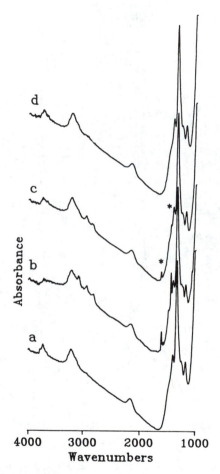

Figure 13. FT-IR transmission spectra of the AlN pellet activated at 873 K (a); after addition of 5 mbar pyridine and subsequent evacuation at room temperature (b); after evacuation at 423 K (c); after evacuation at 873 K (d).

This example of methanol adsorption showed the complexity of the phenomena occurring on a surface when several species simultaneously exist or when some species possibly transform into new ones under particular conditions.

Pyridine adsorption. From Flow Chart 1 we can see that pyridine can either coordinate on a Lewis acidic site or react with a Brønsted acidic site. In this latter case a pyridinium ion is formed with characteristic absorption frequencies (*30*). Two fundamental modes of the aromatic ring v_{19b} and v_{8a} are sensitive to the coordination state of the pyridine. Thus they are good sensors of the Lewis acid strength since the frequency shifts are directly related to it.

On the activated AlN surface, 5 mbar of pyridine were added at room temperature and then evacuated at room temperature in order to remove both the gaseous phase and the hydrogen-bonded molecules (Figure 13a,b). On the difference spectrum (Figure 14a), in addition to a group of bands in the 3000 cm^{-1} region corresponding to $v(CH)$ stretching vibrations, the two intense sharp bands at 1614 and 1448 cm^{-1} are clearly visible (cf. also * marks on Figure 13b). After evacuation at 423 K these two bands shift to higher frequencies at 1621 and 1452 cm^{-1} (Figure 14b and * marks on Figure 13c). Then they were eliminated by subsequent evacuation at higher temperatures (Figures 13d and 14c). These bands were respectively assigned to the v_{8a} and v_{19b} modes shifted with respect to the free molecule vibrations. By comparison with the pyridine adsorption on an alumina surface, we assign these bands to pyridine coordinated on Al^{3+} Lewis acidic sites. The shift toward higher wavenumbers at 423 K showed the formation of stronger Lewis acid sites by heating (*31,32*). As on alumina, no pyridinium ion was formed, thus these results indicated a weak protonic acidity.

In literature the two types of acidic sites on alumina surface have been discussed (*33,34*) as follows:
i- the site of weaker acidity, corresponding to the 1614 and 1448 cm^{-1} bands, is formed by a pair of coordinatively unsaturated (cus) cations, namely Al^{3+} in octahedral and tetrahedral coordination.
ii- the site of higher acidity is a cus tetrahedral Al^{3+} corresponding to the 1621 and 1452 cm^{-1} bands.

Consequently, the AlN surface presents Al^{3+} sites in tetrahedral and octahedral coordination.

Conclusion

The surface FT-IR study of a nanosized aluminum nitride powder definitely brought evidence of the specific chemical composition of its first atomic layer. The unavoidable contamination, mainly by atmospheric water, implies the presence of oxygen and hydrogen in this first layer. As a result, a comparison with the alumina surface appeared quite reasonable. Indeed, methanol and pyridine used as probe molecules showed the same behavior on both material surfaces. The acidity of the AlN surface was proven by the presence of two types of Al^{3+} Lewis sites. However for AlN, a specific dissociative adsorption mechanism of acidic methanol could be possibly explained by the presence of weakly basic Al$_3$N sites. Besides, the isotopic exchange

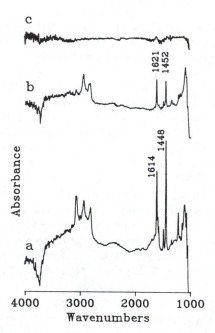

Figure 14. Difference spectra between two steps of the pyridine adsorption (to be compared with Figure 13): pyridine addition and subsequent evacuation 13b-13a (a); after evacuation at 423 K, 13c-13a (b); after evacuation at 873 K, 13d-13a (c).

H/D by deuterium addition showed that in addition to the hydroxyl group also present on alumina, Al-NH$_2$ groups exist on the AlN surface. Al-H$_2$ groups are also identified when the activation temperature was kept below 573 K. These groups transform into Al-H species when the temperature increases. Moreover, heating the AlN powder under vacuum caused the release of H$_2$O, NH$_3$ and carbonate groups.

As a conclusion to this study, it must be kept in mind that all the properties involving the surface of nanostructured powders require a specific characterization and a thorough understanding of the first atomic layer.

Acknowledgments.

We thank Dr. Gan-Moog Chow of the U.S. Naval Research Lab, D.C., for the TEM studies. KEG also acknowledges partial support from the Office of Naval Research.

References

1.   Davydov, A.A. In *Infrared Spectroscopy of Adsorbed Species on the Surface of Transition Metal Oxides*; Rochester, C.H., Ed.; John Wiley & Sons: Chichester, England, 1990.
2.   Baraton, M.-I. *High Temp. Chem. Process.* **1994**, *3*, 545.

3.   Xiao, T.D.; Gonsalves, K.E.; Strutt, P.R. *J. Am. Ceram. Soc.* **1993**, *76*, 987; Xiao, T.D.; Gonsalves, K.E.; Strutt, P.R.; Chow, G.M.; Chen, X. *Ceram. Eng. Sci. Proc.* **1993**, *14*, 1107.
4.   Collins, A.T.; Lightowlers, E.C; Dean, P.J. *Phys. Rev.* **1967**, *158*, 833.
5.   Dupuie, J.L.; Gulari, E. *J. Vac. Sci. Technol.* **1992**, *A10* (1), 18.
6.   Chow, G.M.; Xiao, T.D.; Gonsalves, K.E.; Chen, X. *J. Mater. Res.* **1994**, *9*, 168.
7.   Liu, H.; Bertolet, D.C.; Rogers, J.W. Jr. *Surface Science* **1994**, *320*, 145.
8.   Wang, P.S.; Malghan, S.G.; Hsu, S.M. *J. Mater. Res.* **1995**, *10*, 302.
9.   Busca, G.; Lorenzelli, V.; Porcile, G.; Baraton, M.-I.; Quintard, P.; Marchand, R. *Mater. Chem. Physics* **1986**, *14*, 123.
10. Busca, G.; Lorenzelli, V.; Baraton, M.-I.; Quintard, P.; Marchand, R. *J. Mol. Struct.* **1986**, *143*, 525.
11. Hair, M.L. *Infrared Spectroscopy in Surface Chemistry*; Marcel Dekker: New-York, US, 1967.
12. Busca, G.; Lorenzelli, V. *Mater. Chem.* **1982**, *7*, 89.
13. Mazur, U. *Langmuir* **1990**, *6*, 1331.
14. Wang, X.D.; Hipps, K.W.; Mazur, U. *Langmuir* **1992**, *8*, 1347.
15. Wang, X.D.; Hipps, K.W.; Dickinson, J.T.; Mazur, U. *J. Mater. Res.* **1994**, *9*, 1449.
16. Baraton, M.-I.; Merle, T.; Quintard, P.; Lorenzelli, V. *Langmuir* **1993**, *9*(6), 1486.
17. Baraton, M.-I. *Nanostruct. Mater.*, **1995**, *5*(2), 179.
18. Chesters, M.A.; Sheppard, N. In *Spectroscopy of Surfaces, Advances in Spectroscopy*; Clark, R.J.H.and Hester, R.E., Eds.; J. Wiley & Sons: New-York, US, 1988, Vol. 7; pp. 377.
19. Merle, T.; Baraton, M.-I.; Quintard, P.; Laurent, Y.; Lorenzelli, V. *J. Chem. Soc. Faraday Trans.* **1993**, *89*(16), 3111.
20. Bertolet, D.C.; Liu, H.; Rogers, J.W. *J. Appl. Phys.* **1994**, *75* (10), 5385.
21. Busca, G.; Lorenzelli, V.; Sanchez Escribano, V.; Guidetti, R. *J. Catalysis* **1991**, *131*,167.
22. Ballinger, T.H.; Yates, J.T. *Langmuir* **1991**, *7*(12), 3041.
23. Knözinger, H. *Adv. Catal.* **1976**, *25*, 184.
24. Bartram, M.E.; Michalske, T.A.; Rogers, J.W.; Mayers, T.M. *Chem. Mater.* **1991**, *3* (5), 953.
25. Sanderson, R.T. *Chemical Bond and Bond Energy*; Academic Press: London, England, 1971.
26. Merle, T.; Baraton, M.-I.; Goeuriot, D.; Quintard, P.; Lorenzelli, V. *J. Mol. Struct.* **1992**, *267*, 341.
27. Deo, A.V.; Chuang, T.T.; Dalla Lana, I.G. *J. Phys. Chem.* **1971**, *75*(2), 234.
28. Busca, G.; Rossi, P.F.; Lorenzelli, V.; Benaissa, M.; Travert, J.; Lavalley, J.C. *J. Phys. Chem.* **1985**, *89*, 5433.
29. Greenler, R.G. *J. Chem. Phys.* **1962**, *37*, 2094.
30. Parry, E.P. *J. Catalysis* **1963**, *2*, 371.
31. Abbattista, F.; Delmastro, S.; Gozzelino, G.; Mazza, D.; Vallino, M.; Busca, G.; Lorenzelli, V.; Ramis, G. *J. Catalysis* **1989**, *117*(1), 42.
32. Riseman, S.M.; Massoth, F.E.; Dhar, G.M., Eyring, E.M. *J. Phys. Chem.* **1982**, *86*(10), 1760.
33. Morterra, C.; Coluccia, S.; Garrone, E.; Ghiotti, G. *J. Chem. Soc. Faraday Trans. I* **1979**, *75*, 289.
34. Morterra, C.; Chiorino, A.; Ghiotto, G.; Garrone, E. *J. Chem. Soc. Faraday Trans. I* **1979**, *75*, 271.

RECEIVED January 16, 1996

Chapter 23

Biomolecular Delivery Using Coated Nanocrystalline Ceramics (Aquasomes)

Nir Kossovsky

Biomaterials Bioreactivity Characterization Laboratory, Department
of Pathology and Laboratory Medicine, University of California,
Los Angeles, CA 90095–1732

We have designed a self-assembling composition comprised of a
nanocrystalline solid ceramic core and a glassy polyhydroxyloligomeric
surface coating. Biochemically active molecules of different classes of
activity have been non-covalently bound to this coated nanocrystalline
ceramics bio/molecular delivery system. Data accumulated suggest that
bound enzymes and drugs exhibit prolonged biological activity. Data
also suggest that bound antigens elicit strong immune responses. The
application potentials for this platform bio/molecular delivery
technology, aquasomes, are explored.

Most biochemically important molecules, specifically polypeptides, tend to be
relatively labile.[1,2,3,4,5,6] Their stabilization in active conformations and the prevention
of degradation and denaturation between the time and place of synthesis and the time
and place of use has been the singular focus many chemists working with such diverse
products as pharmaceuticals, agrochemicals, pigments, dyes, specialty chemicals, and
even explosives. Vehicles to protect labile molecules, known by various names such
as drug delivery systems, molecular stabilizers, targeted therapeutics and activity
enhancers, constitute the family of bio/molecular delivery technologies. Many
different nanoscaled delivery vehicles have been proposed over the years including
aggregated natural organic macromolecules such as albumin, polysaccharides such as
dextran, polymeric particulates such as those fabricated from polystyrene or
methylmethacrylate, and even novel lipid composites such as liposomes.[7]

The effort to find an optimal system has been driven by the widely held belief
that such a system would offer the biochemical community a number of benefits.
Such an enablement is exemplified by the variety of applications that appear to be
possible with a three layered self assembling nanocrystalline composition we have
developed. First, effective delivery of viral antigens could enable vast numbers of new

vaccinating compounds for therapeutic, industrial and diagnostic applications. Using a ceramic carbon nanocrystalline particlate core coated with glassy cellobiose, we have delivered antigens for Epstein Barr virus[8] and Human Immunodeficiency virus[9] to elicit immunity, muscle adhesive protein to elicit polyclonal antibodies for affinity chromatography,[10] and rgp120 (HIV) to elicit polyclonal antibodies for ELISA-based immunassays for HIV infection. Second, effective delivery of drugs would enable both targeted drug delivery as well as facilitate the drive to reduce drug toxicity by improving drug performance.[11] It could also enable the patentability of effective but off-patent agents. Using a degradable calcium phospate nanocrystalline particulate core coated with glassy pyrdoxal-5-pyrophosphate, we have delivered insulin into both non-diabetic and diabetic animals for the purpose of demonstrating the improved efficacy of the immobilized drug.[12] We have used the same technology to deliver *in vitro* the therapeutic enzyme, DNase.[13] Last, effective bio/molecular delivery could protect the labile polymorphic oxygen binding molecule, hemoglobin for the purposes of fabricating synthetic blood. Again using a degradable calcium phospate nanocrystalline particulate core coated with glassy pyridoxal-5-pyrophosphate, we have shown the successful immobilization of hemoglobin while preserving the allosteric oxygen binding activity activity of the complex molecule.[14]

A detailed review of a recently completed study of the activity of immobilized DNase is exemplary.

In Vitro Delivery of DNase Using Coated Nanocrystalline Ceramics

DNase, a well characterized and biologically specific enzyme with therapeutic value in the treatment of cystic fibrosis,[15] was chosen as a model agent for studying the effects on an enzymatic bio/molecule of surface immobilization on nanocrystalline ceramic particulates. As the data show, and consistent with previous observations, DNase exhibits a marked retention of biological activity when surface immobilized on the solid phase of a colloid comprised of polyhydroxyl oligomeric films investing degradable calcium-phosphate nanoparticles.

Experimental Methods

The research objective was to determine if enzyme activity is affected favorably or adversely by the state of being immobilized on a polyhydroxyloligomeric film investing a nanoparticle. DNase I (Sigma) was the enzyme, and its activity was measured by measuring the nicks produced in the substrate, linear double stranded DNA. The nanoparticles, aquasomes, were calcium phosphate ceramics produced through colloidal processing, and the polyhydroxyloligomeric films were comprised of adulterated cellobiose.

DNase I was passively adsorbed over 24 hours, 4°C, in situ to calcium phosphate nanoparticles that had been modified with a lyophilized film of adulterated

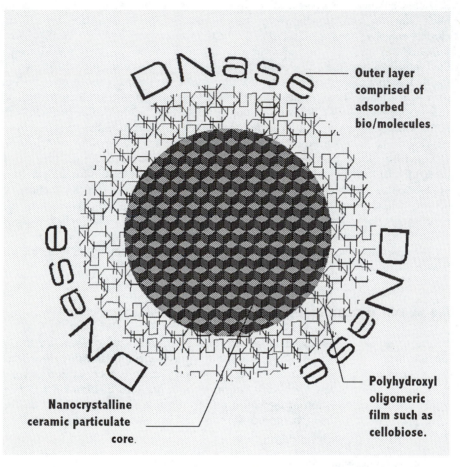

Figure 1. The synthetic product of self-assembling chemistry using non-covalent forces exclusively is a three layered composition comprised of the ceramic nanocrystalline core, the polyhyroxyloligomeric film coating, and the non-covalently bound layer of the drug/enzyme DNase.

cellobiose.* The quantity of adsorbed enzyme was determined by pelleting the mass (from a parallel reaction) and assaying residual DNase activity in the supernatant. Free enzyme and immobilized enzyme activity on the substrate [present in huge excess] were then assayed. Enzyme activity was measured by the change of absorption at 260nm over a five minute period with measurements taken at one second intervals.

The following experimental conditions were tested:

1. Free fresh enzyme (day 0) activity, 133U, on substrate, 20°C, 5 minutes. This control established the baseline, and presumably, optimal level of activity for 133U of DNase.
2. 1, 3, 5, 7 day free solution aged (4°C), 133U, on substrate, 20°C, 5 minutes. This control established the expected activity decay for the standard form of the enzyme.
3. Enzyme immobilized on adulterated cellobiose film coated nanoparticles, corrected to 133U, 1, 3, 5, 7 day aged, on substrate, 20°C, 5 minutes. This is the experimental lot. A time 0 sample could not be tested because the immobilization process required at least several hours.
4. Enzyme first inactivated covalently with iodoacetate and then immobilized on adulterated cellobiose film coated nanoparticles, corrected to 133U, 1, 3, 5, 7 day aged, on substrate, 20°C, 5 minutes. Iodoacetate fully inactivates the enzyme as we first confirmed with the free form. This control helped establish that the signals being observed spectrophotometrically were not artifactual.
5. Freshly made nanoparticles coated with adulterated cellobiose film, only. A not-necessarily applicable control since coated ceramic particles will adsorb substrate from the assay and produce a negative signal. As a point of interest, adsorption artifacts may also be generated by materials #3 and #4 but there is no practical way to isolate the signal.

The spectrophotometric data for the immobilized enzymes colloidal systems showed marked light scattering effects. The data were therefore subjected to a smoothing function by curve fitting a (best fit) fourth order polynomial to the data set from 18 to 80 seconds (the time period investing Vmax as determined from the data set of fresh (day 0) free enzyme activity). After curve fitting, the derivatives were calculated and enzyme velocity (V18-80) was determined for each of the four systems (free enzyme, surface immobilized, inactivated surface immobilized, and immobilized supernatant) over the four time intervals (day 1, 3, 5 and 7). The data were then normalized to the activity of the free enzyme and plotted.

Results

The synthetic product, a three layered composition comprised of the ceramic nanocrystalline core, the polyhyroxyloligomeric film coating, and the non-covalently bound layer of the drug/enzyme DNase, is shown in Figure 1. Typical spectrophotometric data for the first 200 seconds are shown in Figure 2. The "free" enzyme solution shows a typical time/absorption profile while the "immobilized"

---

* A proprietary solution of cellobiose and other solutes. Patent pending to the Regents of the University of California.

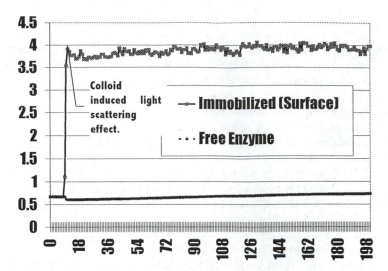

Figure 2. Typical spectrophotometric data for the first 200 seconds. The nanocrystalline particulates induced marked light scattering which necessitated mathematical smoothing of the data before a slope (enzyme velocity) could be calculated.

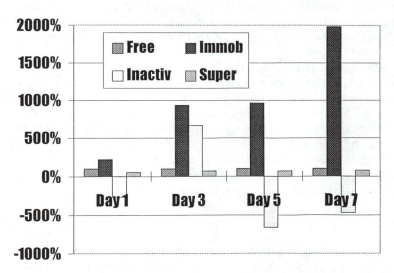

Figure 3. As shown in Figure 3, the relative activity of the immobilized enzyme compared to the free enzyme was almost 2000% after 7 days. In absolute terms, these data reflect the combined effects of progressively increasing enzyme activity of the surface immobilized agent and progressively decreasing enzyme activity of the free solution-based material.

enzyme shows marked light scattering effects and a significant elevation in the baseline optical density. The y-axis is absorption units (AU) at 260nm. Examination of the curves showed that the "noise" created by the mixing of the colloid into the substrate solution was absent after 18 seconds, and so the velocity calculations were temporally truncated. After curve fitting, V18-80 was calculated from the respective curve derivative functions and the 62 values for each enzyme group-day data set were averaged.

As shown in Figure 3, the relative activity of the immobilized enzyme compared to the free enzyme was almost twice as great on day one, and almost 20 times as great on day 7. In absolute terms, these data reflect both the combined effects of progressively increasing enzyme activity of the surface immobilized agent and the simultaneous progressively decreasing enzyme activity of the free solution-based material. The relative activity of the inactivated enzyme showed a negative value, and we believe this represents a net reduction in optical density owing to concomitant adsorption of substrate onto the particle surfaces. Last, the curious behavior of the inactive enzyme at day 3 has not been explained to our satisfaction and these experiments will be repeated with a different enzyme/substrate pair.

Overall, these data, in accordance with previous studies, show that surface adsorption to aquasomes may enhance, protect or otherwise augment the biological activity of a bound agent. To appreciate fully the significance of the implications of this statement, the rationale for bio/molecular delivery is reviewed.

Discussion of the Rationale for and Synthesis of the Bio/Molecular Delivery Using Coated Nanocrystalline Ceramics

Bio/molecular delivery systems consist of two key attributes: the physico/chemical nature of the transporting vehicle and the method by which the biochemically active molecule is coupled to the vehicle. The former attribute is represented by an extraordinarily wide range of physico/chemical units ranging from integrated parent molecules to submicron particulates. The latter attribute is represented primarily by the process of covalent bonding and more recently non-covalent interactions. [16,17,18,19]

Physico/Chemical Nature of Vehicles In General

There are three basic approaches to delivery systems. The simplest approach is chemical conversion of an biochemically active agent to an inactive or pro-agent form. Reversion to an active agent is controlled by intrinsic physiological or environemental processes, and thus inappropriate biochemical activity is theoretically reduced. The second approach employs simple soluble macromolecules to which an agent is immobilized. The macromolecular carriers exhibit intrinsic targeting properties. Delivery of an active agent bound to the macromolecular carriers is controlled by processes intrinsic to the carrier and endogenous ligands. The third approach employs more complex particulate multicomponent carriers within which the agent is shielded from degradative environmental processes during transit. Delivery of an active agent is controlled by enhanced agent survival and may be targeted by the addition of specific ligands to the carrier.

The three competing approaches are not mutually exclusive, and many specific systems based on one approach employ some principles of the other two. Overall, they differ in their carriage capacities, structure, stability *in vivo* and in storage, modes of administration and possible applications. Each of them has favorable attributes and limitations. Major technical challenges include agent loading, size, cost and stability. Major biological challenges include immunity, agent denaturation, targeting and non-specific toxicity (Table I).

Table I: Technical Challenges of Biomolecular Delivery

Class	Issue	Examples
Chemical	Degradation	Deamidation, oxidation, isomerization, proteolysis, aberrant disulfide bond formation, and beta elimination
Physical	Degradation	Denaturation, aggregation, precipitation, adsorption
Biological	Drug Load	Soluble macromolecules are limited in their delivery capacity
Biological	Targeting	Particulate systems especially are subject to early filtration in the lung, liver, spleen and kidney.
Biological	Immunity	Delivery vehicles may adjuvantize the agent
Commercial	Cost	Liposomes and recombinant engineering products use expensive raw materials. Aquasomes, in contrast, are relatively inexpensive

The vehicle format we have used to produce aquasomes is the complex particulate multicomponent system. In general, complex particulate delivery systems are assemblies of simple polymers, complex lipid mixtures or ceramic materials that tend to measure individually between 30 and 500 nm in diameter. Being solid or glassy particles dispersed in an aqueous environment, they exhibit the physical properties of colloids; their mechanism of action is controlled by their surface chemistry. They may deliver agents through a combination of specific targeting, molecular shielding, and slow release processes.

Their large size and their inherently active surfaces allow them to be loaded with substantial amounts of agent through non-covalent processes. [20] Also owing to their size and relative structural stability, they may be "cloaked" to avoid being quickly filtered by the reticuloendothelial system or otherwise degraded by environmental processes.

Aquasomes Deliver Antigens. An antigen's efficacy at eliciting a conformationally correct antibody mediated immunological reaction is, in part, a function of the ability of the antigen's carrier to present the antigen "properly." Although B-cells usually recognize protein antigens in their native conformational state[21], conventional methods for raising methods for raising antibodies tend to shield potentially critical determinants.[22]   This is particularly true for the lipoidal micelle systems such as Freund's adjuvant[23] or the newer saponins. Cognizant of the ability of surfaces to direct synthetic process,[24] and congnizant of the very high intrinsic surface energy of ceramic carbon,[25] a hydrophilic polyhydroxyloligomeric film of the dissacharide, cellobiose, was allowed to adsorb to carbon ceramic surface to yield a coated solid antigen delivery vehicle.

In detail, 1 gram quantities of diamond powder [General Electric, Worthington OH.] was cleaned by 400 Watt sonication at 4°C [Branson, Danbury CT.] in 12 N HCl for 16 hours and washed with ultrafiltered water until the pH was near 7. The resultant opaque dispersions were layered over glass plates and dessicated in a vacuum oven for 2 days at 185°C. Dried diamond was then rehydrated once more and acid washed as previously described. Activated diamond dispersions in ultrafiltered deionized water were adjusted to 1.0 mg/ml and then introduced to 250mM cellobiose [Sigma, St. Louis Mo.] and lyophilized for 24 hours. Unadsorbed cellobiose was removed by ultrafiltration dialyses against sterile water in a 100 kd nominal molecular weight cutoff stir cell [Filtron, Northborough Ma.] at room temperature.

Aquasomes Deliver Insulin or Hemoglobin. Recognizing that at the application level, a substantial mass of "vehicle" would be administered in conjunction with biochemically active hemoglobin, we explored synthesizing a degradable system.[26] We therefore opted to assess the ability of degradable calcium-phosphate ceramic nanoparticles to safely deliver hemoglobin. Also, recognizing the need to incorporate an allosteric modifier into the hemoglobin to enable its characteristic oxygen induced molecular cooperativity, a film of pyridoxal-5-pyrophosphate was used as the polyhydroxyloligomer.

In detail, nanocrystalline calcium phosphate [brushite] was prepared by streaming equal volumes of 0.75 M $CaCl_2$ and 0.25 M Na_2PO_4 against each other. The immediately formed precipitate was sonicated [Branson 450 watt sonicator] for 60 minutes at 4.0°C and washed with 500 ml of distilled water. By transmission electron microscopy [TEM], such preparations yielded stable dispersions that range from 5 to 250 nm. To passivate the highly reactive ceramic surface, arrest further particulate growth, and provide an allosteric modifying nidus for hemoglobin binding, pyridoxal-5-phosphate [PLP] was then added: Typically, a 250 mg pellet of freshly prepared nanocrystalline brushite was reconstituted in 1.0g of PLP in water. The mixture was then lyophilized overnight, washed with distilled water, and sonicated at 450 Watts, 4°C, for 30 min.

Immobilization of a Bio/Molecule on a Delivery Vehicle

The traditional approach to coupling a biochemically active molecule to a delivery vehicle is through covalent chemistry. This is very much the case with the various macromolecular vehicles, but has also been used to couple agents to the larger particulate carriers. The alternative approach, the employment of non-covalent forces to promote "self-assembly" of the biochemically active molecule to the carrier, has seen use in liposome and some complex polymeric carrier systems.

Antigen carriers and adjuvants, selected ostensibly to enhance the immunogenicity of proteins antigens, are an intrinsic component of a functional antigen system by virtue of their close physical association with the antigen.[22] Since molecular flexibility is an intrinsic component of viral activity and may be critical for viral infectivity as well as immune recognition,[27] a non-constraining coupling process based exclusively on non-covalent interactions was employed. Proteins from Epstein-Barr virus and Human Immunodeficiency virus, purified mussel adhesive protein, and recombinant gp120 (HIV) were all immobilized on the coated ceramic carbon surfaces. The method for isolating HIV proteins and immobilizing them on the ceramic carbon is examplary of the general method.

Briefly, HTLV IIImn (HIV-1) (10 5.75 transforming units per ml based on TCID50 titer assay) (1.0 ml) [Advanced Biotechnologies Incorporated, Columbia, Maryland] was added to envelope extraction buffer (1.0% triton X 100/0.25 mM DTT/10 mM Tris pH 7.4/1.0 mM MgCl) (5.0 ml) and allowed to incubate for 1.0 hour at room temp. The viral nucleocapsid was cleared from the supernatant by ultracentrifugation (100 K g accomplished with the Beckman SW50.1 rotor revolving 35K rpm), 4° C.

The volume of protein extract solution was adjusted to 5.0 ml with a phosphate reaction buffer consisting of 20 mM phosphates at pH 7.4 and then incubated with 5.0 mg ultra clean carbon/cellobiose core (0.5 ml of stock solution) at 4o C for 24 hours. The final colloid consisted of carbon ceramic particulates coated with a layer of cellobiose and a more superficial layer of adsorbed HIV proteins. These colloidal ceramic viral decoys were prepared for use by clearing unadsorbed material by ultrafiltration dialysis with a stir cell [Filtron, Northborough, MA] mounted with a 100 kd filter and flushed with 100 ml of sterile phosphate buffered saline (injection grade) at 4°C under a N_2 pressure head of 10 psi.

Non-covalent interactions create less significant three dimensional molecular constraints. Because insulin must be free to bind its cellular receptor while hemoglobin must rearrange itself under the stress of oxygen tension to exhibit its characteristic sigmoidal affinity, non-covalent interactions were also used to immobilize these bio/molecules to the aquasomes.

Briefly, to immobilize hemoglobin onto the coated brushite cores, four ml [50.0 g/ml] of brushite was dripped over 5.0 ml of human hemoglobin[30.0 g/dl] as calculated by a standard calibration curve] in a jacketed stir cell at 4.0°C for 36 hours. Phospholipids were then added [phosphotidyl choline, phosphotidyl serine, and cholesterol] in a molar ratio of 1:1: 0.1 respectively, and extruded over cores by propelling the preparation through an 22 gauge needle.

Characterization and Properties of Various Proteins Adsorbed to Surface Modified Nanocrystalline Ceramics

Physical and biological characterization of aquasomes coupled with a range of bio/molecules has been carried out over the past five years. In general, the data provide support for the hypothesis that conformationally specific biochemical activity is preserved when a biochemically active molecule is immobilized on the polyhydroxyloligomeric coated ceramic coares.

Properties of Epstein-Barr Viral Proteins Immobilized on Aquasomes. Doppler electrophoretic mobility studies were conducted between the pH range of 4.5 to 9.0. Both the native EBV and aquasome decoy, constructed from surface modified ceramics and EBV proteins, retained virtually identical mobilities of approximately - 1.4 µm-cm / V-s throughout the pH range of this experiment. Untreated tin oxide demonstrated a mobility of approximately -1.0 µm-cm/ V-s at pH of 4.5 which then rose rapidly to -3.0 µm-cm / V-s at pH values of 5.0 and higher. Finally, surface-modified tin oxide treated with cellobiose only retained a mobility of approximately - 1.5 µm-cm/ V-s until it increased rapidly to -2.5 µm-cm/ V-s at a pH of 7.5.

Properties of HIV Viral Proteins Immobilized on Aquasomes. Physical characterization showed that aquasomes with immobilized antigen ("HIV decoys") measured 50 nm in diameter (HIV=50-100 nm) and exhibited the same ζ potential as whole (live) HIV. *In vitro* testing showed that the HIV decoys were recognized by both conformationally non-specific and specific monoclonal antibodies, were recognized by human IgG from HIV antibody postive patients, and could promote surface agglomeration among malignant T-cells similar to live HIV. Last, *in vivo* testing in three vaccinated animal species showed that the HIV aquasomes decoys elicited humoral and cellular immune responses similar to that evoked by whole (live) HIV.

Properties of Mussel Adhesive Proteins Immobilized on Aquasomes. The presentation of antigen on aquasomes yielded a strong and specific antibody response. The binding affinity of antibodies elicited in the rabbit against MAP was conformationally specific; antibodies bonded 37% more readily to MAP immobilized on the standard ELISA polystyrene substrate than to MAP immobilized on a more hydrophobic surface.

Properties of rgp120 Immobilized on Aquasomes. Polyclonal antibodies were raised against rgp 120 for use in ELISA testing to quantify the amount of gp120 or HIV virus produced in culture, and to identify and recognize gp120 produced in infected monkey serum. Our studies show that the polyclonal antibodies raised in the rabbit by aquasomes bind avidly to cultured HIV and extracted HIV proteins as confirmed by Western Blot analysis.

Properties of Insulin Immobilized on Aquasomes. When insulin was adsorbed to aquasomes and administered intravenously in rabbits and in diabetic dogs, the blood glucose concentrations were reduced (149.31 - 156.99 mg/dL). Aquasome-delivered insulin showed a time-activity profile, (AUC) that was between 110%-160% of the time-activity profile of insulin alone. Furthermore, insulin aquasomes that had been aged at 4°C for up to seven months still showed full activity months after the expiration date of the insulin lot.

Properties of Hemoglobin Immobilized on Aquasomes. Doppler electrophoretic light scatter analysis of the zeta potential (surface charge) of the bright red colloid demonstrated stable attachment of the hemoglobin to the brushite carrier over a broad pH range. Transmission electron microscopy demonstrated particles in the 50-100 nm size range with approximately 2% unbound hemoglobin. Spectrophotometric analysis indicated a hemoglobin concentration of approximately 10 g/dl. Compared to whole blood which has a hemoglobin concentration of approximately 13-16 g/dL, the synthetic blood formulation fabricated from the coated nanoparticles is strikingly similar. Oxygen affinity curves of these preparations showed a p50 ranging from 26-32 torr. *In vivo* characterization of the current formulation is pending.

Molecular Stabilization by Polyhydroxyloligomeric Films - The "Secret" of Aquasomes

As the name implies, aquasomes are "water bodies." While not actually water, they are water-like, and it is the water-like interactive properties of aquasomes that appear to enable these delivery vehicles to protect and preserve fragile biological molecules such as polypeptides and proteins.

Water-protein interactions are important in maintaining proteins in conformations that are vital for biological activities.[28] The presence of water molecules affects the overall conformations of proteins by affecting the electrostatic interactions between charged and polar groups. At the active site crevices, water molecules are likely to affect the interactions between substrates and proteins, thereby affecting the detailed mechanism of substrate binding and enzymatic activity. Water acts as a molecular plasticizer (softener) that "lubricates" the local dynamics of proteins and renders them sufficiently flexible to assume the varying conformations that are a feature of substrate, ligand or antigen binding. More generally, on a molecular level, water facilitates the spatial recognition of one molecule by another as the means by which energy and information are transmitted, products are generated, responses are initiated and complex biological structures are built.

Ironically, while water is critical in maintaining molecular shape, the aqueous state is not one in which proteins are long resistant to denaturation. A variety of environmental changes such as temperature, pH, salts and solvents can cause protein inactivation in the aqueous state, and the mechanisms of irreversible protein inactivation often follow common pathways. These include cystein destruction, thiol-catalyzed disulfide interchange, oxidation of cystein residues, deamidation of asparagine and glutamine residues and hydrolysis of peptide bonds at aspartic acid residues.

In general therefore, proteins are more stable in the solid state. For example, lyophilized (freeze dried, powdered) ribonuclease in the dry state or suspended in anhydrous solvents exhibits appreciable stability even in temperatures well above 100°C whereas in an aqueous solution undergoes rapid inactivation.   Even mild humidity will lead to the aggregation and inactivation of lyophilized tissue-type plasminogen activator (tPA) and growth hormone.

As nature would have it, however, the solid state within which proteins are stable does not include the solid state created when a drug is coupled to a delivery vehicle. Indeed, in general, various solid-state protein delivery vehicles and even the surfaces of storage containers tend to inactivate bound or otherwise adsorbed drugs. This is because surface adsorption from an aqueous phase to the surfaces of the delivery vehicles or storage containers is generally driven by the entropic gain associated with dehydration. Dehydration, the loss of water molecules critical in maintaining molecular shape, tends to produce significant conformational changes (Figure 4). An analogy may be made with regard to the significant conformational changes induced by an automobile windscreen upon contact with an organized biomolecular structure such as an insect - the so called "brick wall" effect (Figure 5).

There is an interesting exception to the above generalizations. Experiments of nature coupled with basic microbiology research have disclosed a family of molecules which, while dry and in the solid state, exhibit water-like properties that enable this family of molecules to stabilize the "aqueous" conformation of biochemically active molecules.   Commonly written as $[C(H_2O)]_n$, oligomeric carbohydrates and related polyhydroxyloligomers have been shown to exhibit dehydroprotection properties and preserve molecular conformation in the dry solid state.

In 1945, Lewis noted that the ergot producing *Claviceps purpurea* grew naturally on host rye in the presence of heavily glycosylated sap droplets.[29,30,31] He then   showed that a 40% sucrose solution mimicked this effect and exhibited a substantial stabilizing effect on the viability of *Claviceps purpurea* when artificially inoculated on host rye for commercial ergot production. This empirical observation was reinforced by comparable observations made by Crowe et al.[32] who went on to suggest mechanistically that certain sugars may replace the water around polar residues in membrane phospholipids and proteins thereby maintaining their integrity in the absence of water (Figure 6). This phenomenon has been utilized by nature to protect anhydrobiotic organisms such as fungal spores, yeast cells, and cysts of brine shrimp against desiccation and other environmental stressors.[33]

The family of sugars that exhibit dehydroprotectant properties are largely mono-, di-, and oligosaccharides, and their ability to do so correlates with their ability to form glasses.[34] Working with the model disaccharide trehalose, Green and Angell concluded that the trehalose/water system passes into the glassy state and thereby arrests all long-range denaturation-type molecular motion.[35]. Working with sorbitol, Clegg *et al.* came to similar conclusions.[36] Working with cellobiose and later with maltose, sorbitol, lactitol, trehalose and sucrose, Kossovsky *et al.* have shown that "aqueous" molecular secondary structure is preserved upon surface adsorption and desiccation. The property appears to be limited to polyhydroxyloligomers because neither the monosaccharide xylitol nor the polysaccharide α-methyl cellulose

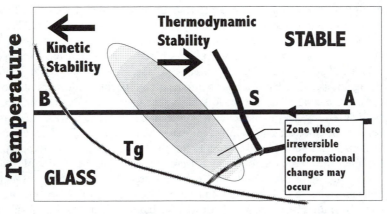

Water Content

Figure 4. The relationship between temperature, water content, and stability *(after Franks, F.)*[6] In a dilute aqueous suspension, a biochemically active molecule is structural stabile but is vulnerable to a wide range of environmental degradative forces such as hydrolysis, oxidation and racemization. In a surface immobilized or dehydrated state, a biochemically active molecule achieves greater kinetic stability at a cost of thermodynamic instability. From a dilute state (A) through supersaturation (S) with progressive water loss on the way to a solid glassy state (B), a biochemcially active molecule passes through a thermodynamically defined (entropic loss of water and enthalpy of adsorption) transition zone (stippled) where irreversible conformational changes may occur. We have observed that the disaccharides used to fabricate Aquasomes appear to stabilize biochemically active molecules in this zone during surface-induced dehydration. The dashed line represent the freeze-drying pathway between the eutectic point and T_g.

Figure 5. Molecular denaturation. The loss of the cushioning effect of water, analogous to a fly striking a windshield, induces structural changes in the three dimensional arrangement of the molecular elements. This is also known as the 'brick wall' effect.

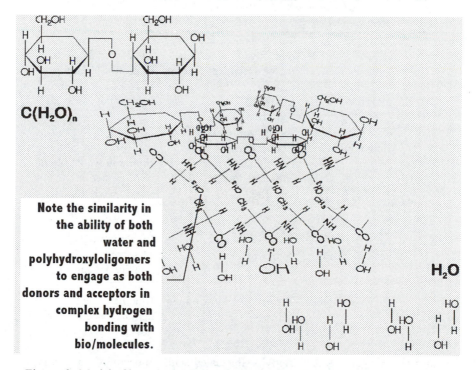

Figure 6. Model of how the hydroxyl groups of carbohydrates replace and act in a structural manner similar to water. The biophysics have been explored in detail by Crowe, Franks, Carpenter, Clegg and others. The microbiological term for molecular stability induced by polyhydroxyloligomers in the absence of water is anhydrobiosis.

exhibited the same molecular stabilization activity[37] Much of the biophysics has been subsequently explored by Franks and others.[38]

Summary

Aquasome technology is a platform system for preserving the conformational integrity and biochemical activity of macromolecules. It is opening a whole new route to vaccines, synthetic blood, drug delivery, and gene therapy (Table II). It is several years behind competing successful technologies such as liposomes, modified adenovirus delivery vehicles and genetically engineered vaccines, but the simplicity of the chemistry may allow for rapid development. Nevertheless, there is much work ahead before any applications using aquasomes are introduced commercially.

Table II: Current Status of the Aquasome-Based Molecular Delivery Technology

Use	Protein/Surface macromolecule	Rationale	Current Status
Vaccines	Antigenic envelope proteins including EBV and HIV	To be effective, protective antibodies, the objective of vaccine therapy, must be triggered by conformationally specific target molecules	Appears to be effective for a wide variety of antigens. Protective immunity not assessed. Long term toxicity not assessed.
Blood Substitutes	Hemoglobin	Cooperativity of hemoglobin is conformationally sensitive	*In vitro* properties are similar to whole blood. *In vivo* properties including safety and toxicity not fully characterized.
Pharmaceuticals	Active Drug such as insulin	Drug activity is conformationally specific	Bioactivity of drugs preserved. Long term toxicity not characterized.
Gene therapy	Genetic material	Targeted intracellular delivery	Binding and release demonstrated. Targeting not currently reliable.
Enzymes	Polypeptide such as the enzyme, DNase	Activity fluctuates with molecular conformation	Bioactivity of enzymes preserved. Long term toxicity not characterized

Acknowledgments

The author thanks the numerous undergraduate and graduate students at UCLA who have assisted in the technical understanding of aquasomes over the years. These include, in particular, A. Gelman and E. Sponsler. The author also thanks the many collaborators, especially Drs. Robin Garrrel and Rointan Bunshah of UCLA and both Alan Rudolph and Gan Moog Chow of the Naval Research Laboratories. The work has been supported by a number of sources including National Science Foundation Divisions of Chemistry, Materials Research, and Biological Sciences, the Office of Naval Research/Naval Research Laboratory core program, the Biomaterials Bioreactivity Characterization Fund of UCLA, and 923934 Ontario Inc.. The technology is the subject of patents issued and pending to the Regents of the University of California.

Literature Cited

1. Craig, E. A. *Science* **1993**, *260*, 1902
2. Agard, D.A. *Science* **1993**, *260*, 1903.
3. Jennings P.A.; Wright P.E. *Science* **1993**, *262*, 892.
4. Ptitsyn, O.B.; Semisotnov, G. V. In *Conformations and Forces in Protein Folding*; Nall, B. T.; Dill, K. A., Eds.; American Association for the Advancement of Science: Washington, DC, Chapter 10, pp 155-168.
5. Stigter, D.; Dill, K. A. In *Conformations and Forces in Protein Folding*; Nall, B. T.; Dill, K. A., Eds.; American Association for the Advancement of Science: Washington, DC, Chapter 3, pp 28-42.
6. Franks, F. *Bio/technology* **1994**, *12*, 253.
7. 100. Hnatyszyn HJ, Kossovsky N, Gelman A, Sponsler E. *J.Pharma. Sci & Tech*, **1994**, *48*, 247.
8. Kossovsky, N.; Gelman, A.; Sponsler, E.; Millett, D. *J Appl Biomat* **1991**, *2*, 251; Kossovsky, N.; Bunshah, R. F. U.S. Patent No. 5,178,882.
9. Kossovsky N, Gelman A, Sponsler E, Rajguru S, Hnatyszyn HJ, Torres M, Mena E, Ly K, Festekjian A. *J Biomed Mat Res*, **1995**; *29*, 561; and U.S. Patent No. 5,334,394
10. Kossovsky, N., Gelman, A., Hnatyszyn, H.J., Rajguru, S., Garrell, R.L., Torbati, S., Feeney, S.S., and Chow, G-M., *Biocong Chem* **1995**; *6*, 507; and Garell, R.L., *manuscript in preparation*.
11. Kossovsky, N. *Mat Technol* **1993**, *8*, 3.; Kossovsky, N.; Bunshah, R. F.; Gelman, A.; Sponsler, E.; Umarjee, D. M.; Suh, T. G.; Prakash, S.; Doer, H. J.; Deshpandey, C. V. *J Appl Biomat* **1990**, *1*, 289; Kossovsky, N.; Bunshah, R. F. U.S. Patent Nos. 5,219,577; 5,460,830; 5,462,750; 5,462,751; and 5,464,634.
12. H. James Hnatyszyn: Ph.D. Thesis, (N. Kossovsky, Thesis Advisor) University of California, Los Angeles, 1995.
13. Kossovsky N, Anderson S, Gelman A and Sponsler E. Persistent activity of DNase following its non-covalent surface immobilization on

polyhydroxyloligomeric   modified self-assembling nanostructured ceramic particles. Presented at the American Chemical Society Annual Meeting, Chicago, Aug 1995.

[14] Kossovsky, N.; Sponsler, E.; Gelman, A. U.S. Patent No. 5,306,508.

[15] Ramsey BW; Dorkin HL. *Pediatric Pulmonology* **1994** *17*, 404; Fuchs HJ; Borowitz DS; Christiansen DH; *et al. N Engl J Med*, **1994** *331*, 637.

[16] J-M. Lehn, J-M. *Science* **1993**, *260*, 1762.

[17] Whitesides, G. M.; Mathias, J.P.; Seto, C.T.; *Science* **1991**, *254*, 1312.

[18] Kossovsky, N.; Millett, D.; Gelman, A.; Sponsler, E.; Hnatyszyn, H.J.; *Bio/Technology* **1993**, *11*, 1534.

[19] Burchard, W. *TRIP* **1993**, *1*, 192.

[20] Meijer, D.K.F.; Jansen, R.W.; Molema, G. *Anti viral Research* **1992**, *18*, 215.

[21] Harding, C.V, Unanue, E.R. *Cell Regulat* **1990**, *1*, 499.

[22] Sela, M., *Science*, **1989**, *166*, 1365.

[23] Retzinger, G.S., Meredith, S.C., Takayam, K., Hunter, R.L., Kesdy, F.J., *J Biol Chem* **1981**, *256*, 8208.

[24] Atherton, E.; Sheppard, R. C.   *Solid phase peptide synthesis: A practical approach*; Practical Approach Series, IRL Press: New York, NY, 1989 pp 13-

[25] Harkins, W.D., *J Chem Phys* **1942**, *10*, 268.

[26] Kossovsky, N.; Millett, D. *Mat Res Soc Bull* **1991**, *9*, 78.

[27] Bullough PA; Hughson FM; Skehel JJ; Wiley DC. *Nature* **1994**, *371(6492)*, 37.

[28] Yang QX, Huang F, Huang T, and Gelbaum L. *Biophys. J.*, **1993**; *64*, 1361.

[29] R.W. Lewis, *J. Am. Pharm. Assoc.* **1948**, *37*, 511.

[30] RW. Lewis, *U.S. Patent* No. 2,402,902.

[31] Y.H. Loo, and R.W. Lewis, *Science* **1955**, *121*, 367.

[32] J.H. Crowe, *Am. Nat.* **1971**, *105*, 563.

[33] J.H. Crowe et al. *Biochimica et Biophysica Acta* **1988**, *947*, 367; Clegg JS; Jackson SA. *Febs Letters*, **1992**, *303*, 45.

[34] H. Levine and L. Slade, *BioPharm*. May 36-40 (1992).

[35] J.L. Green and C.A. Angell, *J. Phys. Chem.* **1989**, *93*, 2880.

[36] Clegg JS; Jackson SA; Fendl K. *J Cellular Phys.* **1990**, *142*, 386.

[37] Kossovsky N, Nguyen A, Sukiassians K, Festekjian A, Gelman A, and Sponsler E: *J Colloid Interface Sci*, **1994**, *166*, 350.

[38] Prestelski SJ, Tedeschi N, Arakawa T, and Carpenter JF. *Biophy. J.*, **1993**, *65*, 661.

RECEIVED December 27, 1995

Chapter 24

Nanoconfined Proteins and Enzymes: Sol–Gel-Based Biomolecular Materials

Bakul C. Dave[1,2], Bruce Dunn[2], Joan S. Valentine[1], and Jeffrey I. Zink[1]

[1]Department of Chemistry and Biochemistry and [2]Department of Materials Science and Engineering, University of California, Los Angeles, CA 90095

Nanostructured materials obtained by sol-gel encapsulation of biomolecules are a novel class of biomaterials. The biological macromolecules, confined within the nanometer-size pores of the matrix, show both similarities to and differences from solution characteristics. The effects of nanoconfinement on the structural and reactivity patterns of the proteins and enzymes are discussed. The applications of these nanostructured biomaterials in the area of molecular biorecognition, detection, and biosensing are also presented.

Nanostructured materials are characterized by ordered structural domains at the level of nanometers (*1*). These materials display the properties of condensed matter without a long range order. In general, four types of such materials based on the integral modulation dimensionalities of zero (nanoconfined particles), one (linear tunnel or channel structures), two (multilayers) and three (nanophases) are possible (*2*). The simplest nanostructured materials are the nanoconfined systems of zero modulation dimensionality which consist of a host matrix with a nm-size spatial cavity that can act as an enclosure for dopant molecular particles. The nanostructured materials obtained by encapsulation of biological macromolecules in sol-gel derived porous SiO_2 structures that contain a trapped bioparticle represent a particularly novel and recent example in this category (*3-4*).

Sol-gel is a low temperature solution based method for synthesis of glass (*5*). The formation of the silicate matrix is achieved by hydrolysis of an alkoxide (usually tetramethyl orthosilicate, TMOS), followed by condensation to yield a polymeric oxo-bridged SiO_2 network. Hydrolysis results in conversion of Si-OR bonds to Si-OH bonds which condense to form an oxo-bridged polymeric Si-O-Si structure. These reactions occurring in a localized region lead to formation of sol particles. The degree of cross-linking as a result of polycondensation increases, and the viscosity of the sol begins to increase. This viscous material then solidifies and leads to formation of a porous gel (*6*).

0097–6156/96/0622–0351$15.00/0

A gel is a two phase system comprised of porous solid and trapped aqueous phase. Even after the gelation point, the structure and properties of the gel continue to change. The primary cause is continuing polycondensation reactions in the solid amorphous phase that increases the cross-linking. Another process that takes place is the spontaneous shrinkage of the gel and resulting expulsion of pore liquid from the gel. This expulsion of liquid is primarily due to the formation of new Si-O-Si bonds via polycondensation of Si-OH fragments that result in contraction of the gel network. Due to these processes taking place during aging, the strength of the gel increases and pore sizes in the gel become smaller. When the pore liquid of the gel is allowed to evaporate, the gel volume decreases due to loss of liquid, and the gel shrinks. As the evaporation of the water takes place, capillary forces induce the gel network to draw together. The result is that due to drying the gel material becomes more condensed with smaller pores. The totally dried gels, termed xerogels, are approximately 1/8 the original size, show no further loss of pore liquid, and therefore, are dimensionally invariant.

It has been shown that the sol-gel materials can be used as host matrices for a variety of biological molecules (7-11). The dopant biomolecules reside in the porous network of these sol-gel composite materials as a part of nanostructured architecture. The unique nanostructured assembly of such sol-gel composites is characterized by biomolecules enclosed in the nanopores of the material. The bioparticles arranged as part of sol-gel composites are characterized by intermediate order and mobility, as opposed to the higher degrees of order available in solids or the pronounced mobilities present in solution media. In other words, the properties of both the solid and solution phases prevail in sol-gel environment. In spite of general similarity in reaction chemistry with macroscopic solution based discipline, variation in overall reaction kinetics can be observed as a direct consequence of encapsulation. Usually it is the interactions of the dopant molecules with the sol-gel matrix that determine the reaction pathways a particular system undergoes. Such a nanostructured system utilizes the properties of spatially isolated molecules in a solvent-rich environment necessary for stability of the biomolecules.

Encapsulation of organic and inorganic molecules in sol-gel glasses has been explored for almost a decade (12). The molecular sizes of such dopants are considerably smaller than average pore diameters and these low-molecular weight dopants are mobile in the solvent-filled pores. The high molecular weight biomolecules, on the other hand, are confined within the nanopores of the sol-gel matrix. This review is concerned primarily with elaborating different aspects of nanoconfined biomolecules in porous sol-gel media. A diverse range of proteins and enzymes has been stabilized in sol-gel derived glasses. A variety of proteins including globular and membrane-bound proteins, enzymes, and other biosystems are functionally active within these glasses. In almost all the cases investigated so far, the chemistry of dopant biomolecules in gel glasses is analogous to that in solution with the exception that now the system involves a porous silicate matrix. The combination of biological chemistry with that of inorganic glasses forms the central aspect this review article.

The paper is organized in three parts. First, the effects of nanoconfinement on the structure of the sol-gel trapped biomolecule are discussed. Second, from the results on apparent reactivity of these biomaterials, the effects of the matrix on the reactivity of trapped enzyme are elucidated. Finally, the interaction of the confined biomolecules with exogenous ligands/substrates and the applications of these materials in the area of molecular biorecognition are discussed. The reaction chemistries of biologically active molecules in the nanostructured materials have been crucial in establishing the role of the matrix upon the structure and reactivity of the confined proteins.

Nanoconfined Bioparticle.

Unlike isotropic solution media where the biomolecules have equal mobility and conformational flexibility in all directions, encapsulation constrains them in all three dimensions. The amorphous glasses incorporate biological molecules dispersed in an aqueous phase in the interconnected porous network. The interior of the electrostatically charged pores consists of biomacromolecule embedded into a matrix of structured aqueous phase. The nature of trapped biomolecule in such an aggregation is characterized by several important factors: 1) improved stability of the biomolecule by virtue of being confined to a pore, 2) pre-orientation, since it cannot tumble freely though it can rotate, 3) selective overpopulation in certain conformational microstates, and 4) limited excluded volume in the pore, which increases the effective concentration. These specific effects often induce selectivity for certain types of processes compared to the properties of the individual biomolecules freely moving in the solution. The partial order imposed due to presence of a rigid cage may result in unique phenomena usually not feasible in a solution medium. As a result of the confinement of the macrobiomolecules in the nanopores of the gel, significant deviations in structural dynamics and energetics can occur. It is likely that the dopant bioparticle may adopt a slightly different conformation upon encapsulation in a nanopore. Indeed, in spite of general similarity with purely isotropic solution based reactions, variations in overall stability and reaction kinetics can be observed.

Nanoconfinement of cytochrome *c*.

Cytochrome *c* (cyt *c*) is a heme-containing electron transfer protein with a molecular weight of ~12 400. The redox properties of this protein are essential for its biological function. To determine the effects of nanoconfinement on the structure and properties of cyt *c*, the protein was encapsulated in silica sol-gel. The optical absorption properties of the protein (an intense Soret band at ~400 nm and a Q band in the 550 nm region) provide useful information about the state of the protein in the confined environment (*13*). The absorption spectra of cyt *c* in aqueous buffer and the sol-gels show slight variations in both the intensity and the wavelength of the transitions. The solution spectra were obtained in 0.1 M acetate buffer (pH 4.5). The spectra in the sol-gels were obtained by equilibrating the gel samples with 0.1 M acetate buffer (pH 4.5). No leaching of the protein was observed during the equilibration experiments. The aged gels were allowed to dry and were rehydrated with buffer to obtain the absorption spectra.

The characteristic absorption spectral pattern of the cyt *c* in the solution phase and the aged silica gel sample is similar (*3*). The heme signature absorptions due to the Soret and Q bands are preserved upon nanoconfinement of the protein in the gel. Although the overall characteristics are similar, the solution spectrum shows the Soret band centered at 406-nm, but the spectrum of the protein in the aged gel sample shows a slight blue shift in the Soret band from 406- to 404-nm. Upon ambiently drying the samples (to 20% of the original weight of the aged gel), the blue shift continues during the aged gel to xerogel structural transformation. A total shift of 9-nm (404- to 395-nm) can be attributed to drying effects. These effects must arise due either to pore collapse or due to loss of solvent phase from the aged gel samples that accompanies drying. In order to determine the exact cause of the blue shifts, absorption spectra were obtained on rehydrated xerogel samples. Absorption spectra of xerogels immersed in 0.1 M acetate buffer (pH 4.5) show a distinctly red shifted Soret transition centered at 404-nm. This value is exactly the same as that observed for aged gel samples. The 9-nm shift in the Soret transition is, thus, caused by drying/rehydration of the gel samples. The reversibility of these shifts indicates absence of physical constraints as contributing effects, and establishes the cause of the absorption band shifts as arising from changes in microenvironment.

Pore shrinkage of the silica gel upon drying is an irreversible process (5-6). If pore collapse due to drying were to cause the blue shifts upon dehydration, the original spectrum of the aged gel would not be recovered upon rehydration. One, therefore, must conclude that the absorption spectroscopic changes are due to changes in microenvironment of the trapped protein caused by loss of intervening solvent phase upon drying, while the dimensions of the pore containing the protein remain constant throughout the drying process. However, the absorption spectroscopic data only pertains to the pores that contain the protein and is not applied to the free pores which undergo the usual shrinkage as evidenced by the overall volume shrinkage of the gels by ~70%.

The above results suggest that the pores that contain the protein behave differently than the free pores. In other words, the presence of the protein dopant somehow alters the behavior of the silicate porous structure. Not only is the protein conformation altered by the silica gel matrix but the structure of the silica network is also affected by the presence of the biological dopant. The biomolecule, according to the spectroscopic data resides in pore that conforms to the size of the protein. The gel network shows a wide distribution of pore sizes, but it appears that the protein is only selectively entrapped by the pores which meet its dimensional requirements. Since the protein is added to the sol and is encapsulated in the growing gel polymeric network, it can be concluded that the pore forms around the protein. The silicate fragments formed as a part of the initial hydrolysis reactions are charged as well H-bonding and the complementary properties of the protein may allow it to interact with the silica polymer. If such interactions are sufficiently attractive the protein is likely to act as a structural template around which the gel network can form.

The biomolecule dopant with its large size, comparable to the dimensions of the pore, and its large number of H-bonding groups on its surface provides an effective interacting environment that affects the spatial characteristics of the growing polymeric network and promotes growth around the dopant particle. The presence of peripheral H-bonding groups results in extensive interaction with the silicate polymer. During the initial stages of network formation, the polymeric silica containing Si-O(H)-Si, and Si-OH fragments orients around the biomolecule with a predetermined selectivity in order to minimize the energy of the ground state structure. The dopant molecule, therefore, serves as a nucleus around which further steps of polymerization can take place i.e. the biomacromolecule acts as a structural template around which the formation of a polymeric cage takes place. The biomolecule, in this way, selectively determines the dimensions of the pore in which it is going to be confined. The pore forms around the biomolecule according to the electrostatic requirements of the biomolecule. In other words, the biomolecule acts as a structural template and specifically designs its own resident pore. The sol-gel matrix thus formed acts as a specific host, in sharp contrast to its properties as host for low molecular weight dopants. Once formed, the pore in the bio-gel nanostructured material conforms to the dimensions of the dopant bioparticle, and conversely, the protein adopts to the "new" microenvironment furnished by the silicate network. These results point to the fact that the bio-gel nanostructures are, to some extent, self-organizing. The protein-gel system thus represents an example of a noncovalently interacting nanocomposite, characterized by an interdependent synergy of molecular interactions, wherein each component affects and determines the structure of the other.

Rotational Mobility of Nanoconfined Cytochrome *c*. Nanostructures are characterized by moderate order and mobility. The rate and extent of motion are important in characterizing the degree of nanoconfinement. The mobility of the nanoconfined bioparticles is an important issue to be determined. In principle, the translational motions of the particles are expected to be eliminated upon encapsulation, but rotational mobility may be preserved. If so, such site isolation may induce variations in rotational processes compared to those of the individual biomolecules freely moving in the solution.

In addition to being stiff or rigid the pore walls must be characterized as *active* or *passive* depending on whether the nanocontact interactions of the pore walls with the macromolecules are attractive, repulsive, or negligible. If sufficiently attractive or repulsive, such interactions will influence the rotational dynamics of the nanoconfined biomolecule. The pore in a gel is not a simple cage comprised only of SiO_4 tetrahedra, but also contains a great deal of adsorbed or hydrogen-bonded water. It is also electrostatically negatively charged (pI of silica gel ~2) with an overall negative charge at the pH of the buffer (~7) generally used to encapsulate biomolecules. Such a pore will show affinity towards polar and charged biomolecules. It will also readily alter peripheral hydrogen-bonding interaction of the biomolecules. In addition, due to electrostatic charges inside the porous network, the rotational ground states of hydrogen-bonding and positively charged biomolecules will be stabilized while those of anionic biomolecules will be destabilized with respect to pure solutions. As the overall energetics of the biomolecules are maintained by a large sum of relatively weaker nanocontact interactions, these effects are expected to lead to altered rotational dynamics of trapped particles.

The technique used to probe rotational dynamics is a.c. dipolar relaxation (*14*). The rotational activation energy associated with a dipolar molecule when placed in an alternating field as it tries to orient its own field is directly related to the characteristics of the immediate microenvironment. Due to a release of electrostatic energy, the impedance of the system drops at the transition frequency (~1 MHz for cytochrome *c*), which is reflected by a maximum in the plot of complex impedance vs. log of frequency (*14*). Ferricytochrome *c* is dipolar and shows an effective dipole moment of ~300 Debye units. As a result of physical caging effect and due to noncovalent interactions with the pore walls the total activation energy for rotation is expected to be raised significantly. The overall sum of the effects can be interpreted in terms of assuming that the protein inside the pore of the gel experiences an effective microenvironment with different viscosity. Techniques which are dependent upon the viscosity of the medium can be employed to derive the extent of gel-protein interactions. When applied to sol-gel glasses, the differences are quantitated in terms of *microviscosity* suggesting that the biomolecule is sensing the resistance to movement only in its immediate surroundings.

The dipolar relaxation measurements provide insight concerning the nature of the interaction between molecules and their environment. The rotational activation energy associated with an orienting dipolar molecule when placed in an ac field is directly related to the characteristics of the immediate microenvironment. The cytochrome *c* molecule has an estimated dipole moment of ~300 Debye units. The differences in dipolar relaxation energies in the sol-gel matrix as compared to aqueous media should provide useful information about the extent to which the molecule experiences an altered environment upon encapsulation in silica gel. A plot of the imaginary component of the impedance versus log frequency exhibits a maximum corresponding to the dipolar transition. In general, proteins show a dipolar transition centered at ~10^6 Hz. Using a set of interdigitated gold electrodes on a silicon substrate, complex impedance measurements were performed on the films that were immersed in buffer. Measurements were performed over the frequency range 20 Hz to 1 MHz using a two probe method. Undoped sol-gel films did not show any well-defined transition in this frequency range. Ferricytochrome *c* in 0.01 M acetate buffer (pH 4.25) undergoes relaxation at ~$10^{5.5}$ Hz whereas sol-gel thin-film encapsulated cytochrome *c* showed a similar transition centered at ~$10^{4.5}$ Hz. The activation energy (E) for dipolar relaxation process is $E = RT \ln(RT/2\pi hf)$ where R is the gas constant, h is the Planck's constant, f is frequency, and T is the absolute temperature. An additional activation barrier of ~1.1 kcal/mol can therefore be associated with matrix encapsulation. This increase suggests that interaction between the biomolecule and the matrix restricts the rotational movement of the protein to a slightly greater extent inside the sol-gel environment. The small 1.1 kcal/mol difference indicates minimal change in protein noncovalent interaction with the medium and that only a slightly perturbed microenvironment is experienced by the protein upon entrapment in the sol-gel glass film. Additionally, the data can also be used to measure the changes in microviscosity experienced by the protein in the sol-gel medium. For a given relaxation time, $t = (2\pi f)^{-1}$, the viscosity of the medium can be given as $\eta = tkT/4\pi r^3$, where η is the viscosity of the medium, k is the Boltzmann constant, T is the absolute temperature, and r is the radius of the bioparticle (14). Thus, the ratio $\eta_{solution}/\eta_{gel} \sim$ 0.1 suggests an approximately ten-fold increase in the microviscosity experienced by the bioparticle upon encapsulation in the sol-gel medium. These results provide a conclusive proof about the confinement effect and its consequences upon the rotational dynamics of cyt c biomolecule as a confined bioparticle.

Reactivity of Nanoconfined Biomolecules.

Once the proteins are confined within the porous structure of the sol-gel matrix, it becomes imperative to determine if these nanostructured materials still remain amenable to interactions with external reagents. This issue assumes central importance because the functional relevance of nanoconfined proteins and enzymes is contingent upon their reactions with suitable substrates or ligands.

A generalized reaction for catalysis by an enzyme is represented as follows

$$E + S \underset{k_{-1}}{\overset{k_1}{\rightleftharpoons}} ES \xrightarrow{k_2} E + P$$

where E is the enzyme, S is the substrate, and P is the product of the reaction (15). How does the nanoporous matrix affect the kinetics of such a reaction? For one thing, it impedes molecular motion. Ordinarily, all enzymatically catalyzed reactions involve the reactant and products separated by an activation energy barrier. The matrix can influence the kinetics of the reaction in the manner in which it may act as a source or a sink for energy. For example, how the matrix influences the ability of the reactants to reach the top of the barrier (form the activated complex), and how it enables the products to dispose of their excess energy may finally decide the kinetics of the system. In other words, the matrix determines essentially the reaction surfaces on which the reactions occur. The nanoconfinement of the dopant biomolecules is such that the matrix pore forms around the protein as template. As discussed above, substantial noncovalent nanocontact interactions exist along the periphery of the protein-gel interface. As a result of the nanoconfinement, specific enzyme-substrate or protein-ligand contact regions may differ from those in the purely isotropic solution media causing variations in probabilities for trajectories of substrate approach. Additionally, the electrostatic microenvironment provided by the matrix (in the form of altered dielectric of the medium), may determine the relative stability of reactants and products, and the rate at which charge is transferred from one place to another. Due to the overall negative charge of the matrix, the rates of the reactions that generate positively charged products can be expected to be accelerated while those of reactions generating negatively charged products will be retarded.

In addition to these chemical effects, physical effects due to nanoconfinement within a pore of finite dimension should also be taken into consideration. Upon treatment of sol-gel confined enzyme with an exogenous substrate, the substrate diffuses through the porous network and subsequently binds to the trapped enzyme. The physical aspects of nanoconfinement are expected to manifest themselves in the way the reaction commences and ensues in the nanopore. The binding of the substrate with the enzyme that results in the formation of the enzyme-substrate complex causes an increase in the volume. Within a silica gel pore, which conforms closely to the dimensions of the enzyme molecule, only a collection of bioparticles that contain critical free volume should then be able to react. The equilibrium constant of the substrate binding reaction (k_1/k_{-1}) is thus expected to be lowered in the sol-gel matrix. Additionally, due to the requirement of critical pore volume, the activation energy for the substrate binding reaction must also increase and thereby lower the rate of formation of the enzyme-substrate complex (i.e. lower k_1). Once the enzyme-substrate complex is formed in the pore of the gel, the substrate is converted to product via a transition state. The transition state formed in the pores may still retain the molecular volume of the reactants due presence of a rigid confining matrix. Consequently, local deformations (strain) must occur within the transition state to accommodate product formation. If the strain is sufficiently great, small collections of biomolecules in the transition state (in extremely constrained pores) are unstable and may tend to revert back to reactant. An overall effect of this will be an increase in activation energy for the formation of transition state and a decrease in value of k_2 in the sol-gel matrix as compared to solution media. The final step of the catalytic act involves separation of bound product from the enzyme to regenerate free enzyme. The physical caging effect due to nanoconfinement must considerably affect the dynamics of this step. For the dissociative processes, there exists a significant possibility that the matrix will hinder physical separation of the products, in which case they may recombine due to proximity effects. The cage effect will therefore result in an overall decrease in the rate of product formation (i.e. a decreased value of k_2).

Based on the preceding arguments, the reaction kinetics are expected to be altered as a direct consequence of nanoconfinement. These can be summarized as 1) an overall decrease in the rate of enzyme-substrate complex formation (lower k_1) 2) an overall decrease in the rate of product formation (lower k_2), and 3) an overall decrease in the equilibrium constant for the formation of enzyme-substrate complex (lower k_1/k_{-1}). The reactivity of nanoconfined proteins and enzymes with low molecular weight ligands and substrates is crucial to practical applications especially in the field of biosensor design and implementation. The reactions of nanoconfined proteins and enzymes are investigated in aged silica gel, and provide significant insights concerning the nature of these process.

Enzyme Catalysis. Biocatalysis is extremely sensitive to microenvironmental effects, and subtle changes in the structural, conformational, and electronic makeup of the enzyme are well known to result in extreme variations in catalysis rates. The catalysis experiments not only reveal structural integrity but more so persuasively the conformational integrity as well.

The Michaelis-Menten kinetics (*15*) provide a facile estimate of the altered reaction dynamics and the energetics of the nanoconfined enzyme systems. The well-known Michaelis constant K_m ($\sim k_{-1}/k_1$) measures the dissociation of the enzyme-substrate complex and in turn serves as an estimate of its stability. An increased value of the Michaelis constant implies that the equilibrium of the enzyme substrate complex is shifted towards the left i.e. towards the free enzyme and substrate, and suggests a relatively weaker enzyme-substrate complex. Another parameter, k_{cat} ($\sim k_2$), called the turnover number, estimates the rate of formation of the product from the enzyme-substrate complex. A facilitated product formation will result in increased turnover number (i.e. increased k_{cat}). The ratio k_{cat}/K_m, consequently, represents the *apparent* rate constant for combination of a substrate with the free enzyme. The apparent equilibrium constant for the binding of the free substrate is represented as K_m while the rate constant is given as k_{cat}/K_m. The former (K_m) is provides information about the thermodynamics while the latter (k_{cat}/K_m) is a rate constant and hence estimates the kinetics of the substrate-binding.

Glucose Oxidase. Glucose oxidase (GOx) was one of the first enzymes to be stabilized in transparent sol-gel glasses (7). An important requirement in preservation of the catalytic activity is that there must not be a substantial structural change during the process of matrix confinement. In order to quantify the enzyme activity, glucose oxidase was immobilized and the turnover number (k_{cat}) and the apparent dissociation constant (K_m) with β-D-glucose were determined. The k_{cat} for nanoconfined glucose oxidase (250 s^{-1}) is identical to that found for the enzyme in solution (251 s^{-1}). The K_m value in the sol-gel medium (0.05 M) for the glucose-glucose oxidase complex (glu-GOx) is about twice that found in the solution (0.028 M) suggesting either a decreased tendency for combination reaction or an increased aptitude for the enzyme-substrate complex to dissociate into the products. However, the unaltered k_{cat} value suggests that the rate of product formation is not changed as a result of confinement. Thus, it can be concluded that in the gel, the formation constant for the glu-GOx complex is reduced. This argument is supported by a comparison of the apparent association constant (k_{cat}/K_m) for glu-GOx system in the solution (~9000 M^{-1} s^{-1}) and in the gel (5000 M^{-1}s^{-1}). The rate of the dissociative product forming step is essentially unperturbed, but the rate of associative enzyme substrate complex formation is reduced by half in the gel matrix (see Table I).

Table I. Comparison of Reactivity for Nanoconfined and Free Enzymes[a]

Enzyme	$(K_m)_s$	$(k_{cat})_s$	$(k_{cat}/K_m)_s$	$(K_m)_g$	$(k_{cat})_g$	$(k_{cat}/K_m)_g$
Glucose oxidase	0.028 M	251 s^{-1}	8964 s^{-1} M^{-1}	0.05 M	250 s^{-1}	5000 s^{-1} M^{-1}
Oxalate oxidase	1.1 x 10^{-4} M	187 x 10^{-4} s^{-1}	170 s^{-1} M^{-1}	4.1 x 10^{-4} M	9.4 x 10^{-4} s^{-1}	2.3 s^{-1} M^{-1}
G-6-P dehydrogenase	21.7 M	0.746 min^{-1}	0.034 min^{-1} M^{-1}	78.4 M	0.271 min^{-1}	0.0034 min^{-1} M^{-1}

[a]The subscripts "s" and "g" refer to solution and aged gel, respectively.

Glucose-6-phosphate Dehydrogenase. Dehydrogenases are a broad class of enzymes. Glucose-6-phosphate dehydrogenase (G6PDH) catalyzes the oxidation of D-glucose-6-phosphate (G6P) to D-glucono-δ-lactone-6-phosphate while reducing NADP+ to NADPH. The reaction kinetics of this enzyme in solution as well as in sol-gel media were studied to elucidate differences in reactivity. The Michaelis-Menten kinetics of the enzyme reveal variation in the kinetic parameter in the sol-gel medium as compared to buffered solution. An approximately fourfold increase in the value of Michaelis constant is observed in the aged gel (78.4 M) as compared to solution (21.7 M). The increase is consistent with an overall shift in the equilibrium towards free enzyme and substrate. As may be expected, the rate constant for the formation of enzyme-substrate complex, k_{cat}/K_m, is also substantially lowered in the confined medium (0.034 min^{-1}M^{-1} in solution; 0.0034 min^{-1} M^{-1} in aged gel). The turnover number for the nanoconfined enzyme (0.271 min^{-1}) is also reduced as compared to the free enzyme (0.746 min^{-1}). It is interesting to note that upon confinement, the rate constant for the binding of the substrate is reduced by a factor of ten, while the rate constant for product formation is reduced only by a factor of three (see Table I).

Oxalate Oxidase. For the oxalate oxidase system, the value of the Michaelis constant (K_m) increased for the sol-gel nanoconfined system ($K_m /10^{-4} = 1.1$ in solution and 4.1 in aged gel). The approximately fourfold increase indicates that the binding of the oxalate with the enzyme is weaker. The apparent association constant (k_{cat}/K_m) for the oxalate-oxalate oxidase complex was substantially altered as a result of confinement of the protein the sol-gel matrix ($k_{cat}/K_m = 170$ in solution and 2.3 in aged gel). This implicates a relatively destabilized enzyme-substrate complex. Additionally, the k_{cat} parameter is also reduced in the gel ($k_{cat}/10^{-4} = 187$ in solution and 9.4 in aged gel), suggesting deactivation of the product forming step in the gel medium. In this case, while the dissociative product forming step is reduced only by a factor of 20 ($187/9.4$), the associative step is reduced considerably by a factor of 75 ($170/2.3$).

The effects of sol-gel confinement upon the structure and reactivity of trapped biomolecules are listed in Table II.

Table II. Effects of Nanoconfinement on Bioparticle Structure and Reactivity

Characteristic	Nanoconfined	Unconfined
Medium	Porous gels	Solution
Cavity size	~ 100 Å	Infinite
Degrees of freedom	0-D	3-D
Free volume	Limited	Infinite
Effective concentration	High	Low
Structure	Constrained	Unconstrained
Nanocontact interactions	Present	Absent
Conformation	Modified	Unaltered
Site symmetry	Altered in some cases	Unaltered
Mobility	Restricted	Free
Size variability	Conditional (up to 100 Å)	Unconditional
Microviscosity	Increased	Unaltered
Selectivity	Orientational	Independent
Substrate transport	Diffusion controlled	Free
Microenvironment	Solvent + matrix	Solvent
Intermolecular reactions	Possible	Ubiquitous
Unwanted reactions	Inter- and intramolecular	Chiefly intramolecular
Reactions facilitated	Intramolecular	Intermolecular
Molecular associations	Inhibited	Unchanged
Molecular dissociations	Inhibited	Unchanged

Biorecognition.

Biological macromolecules are highly specific reagents. Proteins or enzymes are loci for binding and/or catalysis of specific exogenous substrates. In general, enzyme-catalyzed reactions proceed at rates 10^8 to 10^{20} times faster than corresponding uncatalyzed reactions. The high specificity and rates of reaction make them excellent reagents for chemical analysis. This property of biorecognition towards the substrate makes them useful in analytical detection and biosensing (16). The nanoconfinement of biomolecules provides an efficient biosensing design where the motion of recognition molecule is restricted while the flow of the analytes is allowed. Based on the principles of biorecognition several systems have been designed that are capable of detection of biologically relevant entities (17).

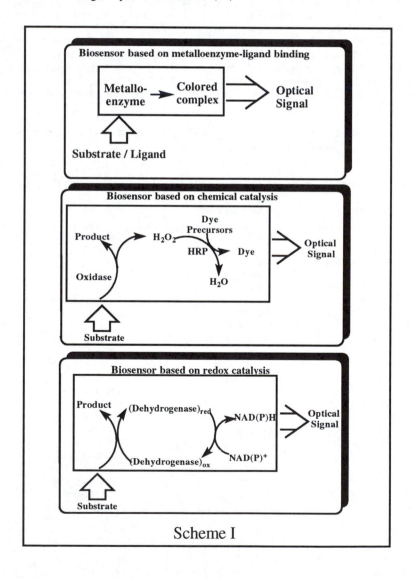

Scheme I

The detection of specific exogenous molecular entities depends on their recognition by the biomolecule to the exclusion of other concomitants present in the medium. The biorecognition properties of the encapsulated enzymes are essentially unaltered by nanoconfinement, and therefore can be effectively employed in diagnostic applications. For enzymes acting as centers of exogenous ligand-binding, the resultant changes in optical properties can be utilized for detection purposes. On the other hand, for the enzymes involved in catalysis, the changes associated with decrease of reactant or increase of product concentrations serve as useful monitors. So far, bio-gels have been successfully utilized for detection of dioxygen, glucose and oxalate. The transparent gel matrix allows optical transduction of the signal. The initiation, translation, and interpretation of optical events requires a suitable photometric light source and minimal undesirable effects from scatter, absorption and other optical side effects from the matrix. The unique feature of sol-gel derived SiO_2 glasses is their optical transparency to 300 nm coupled with the fact that the pore sizes are of the order of 10 nm so that there is minimal scattering of light. These properties make these materials suitable for sensor designs based on optical methods of transduction. Three different types of optical biosensor elements have been designed (see Scheme I). For a biosensor element based on colored metalloenzymes, the changes in absorption spectra were correlated with the concentration of the ligand, and O_2, CO, and NO sensors were based on absorption spectral changes of the heme group. However, with enzymes that do not absorb in the visible region, an alternate strategy was employed, wherein the enzyme-substrate reaction was coupled with generation of optically absorbing dye from its precursors. The glucose and oxalate sensors using immobilized oxoidase/peroxidase enzymes used this method to generate an optical signal. With dehydrogenases that use NADH (or NADPH) as cofactor, the fluorescence of the reduced form of the cofactor can be used for biorecognition purposes. A model system based on oxidation of G6P by G6PDH was investigated.

Biorecognition of Dissolved Oxygen. Hemoglobin (Hb), and myoglobin (Mb) are two heme-proteins that functionalize binding of atmospheric dioxygen. The high affinity of these proteins for O_2 coupled with the changes in visible absorption spectra provide an opportunity to develop a dioxygen sensor based on these proteins. After establishing the stability and reactivity of these proteins in biogels, it was necessary to quantify the optical changes for applications as biosensor elements. For this purpose myoglobin was chosen as a model system because of its greater sensitivity for dioxygen (K_m (in pO_2), 25 torr for Hb and 3 torr Mb). Due to the very fast reaction with gaseous oxygen, experiments have been done with dissolved oxygen (DO). The initial objective to establish proof-of-concept regarding the utility of oxygen-binding proteins as biosensors was to establish a correlation of the absorption spectral characteristics with DO concentration (*17*). As shown in Figure 1, the prominent differences in absorption maxima of the Soret transition of the deoxy and oxy form in the 420 nm region allow a convenient way to quantitate protein-dioxygen interactions. By monitoring the decrease in intensity of the 431.5 nm transition of the deoxy form, a linear relationship was established between the optical response of Mb and DO concentration (Figure 1, inset). Moreover, only one wavelength needs to be monitored to determine the DO concentration. The analytical advantages of this novel biosensing element are linear correlation, short response intervals (~1min), high specificity, reversibility of the oxy form to deoxy state and thus, cost-effectiveness in terms of reusability of the sensing bio-gel. Apart from the utility of Mb biogels as O_2 sensors, they can, also be used as a sensor for CO, by taking advantage of the binding of CO and the distinct changes in the absorption spectrum.

Biorecognition of NO. Nitric oxide, as a trace level entity, is involved in a variety of biological functions including muscle relaxation, platelet inhibition, neuro-

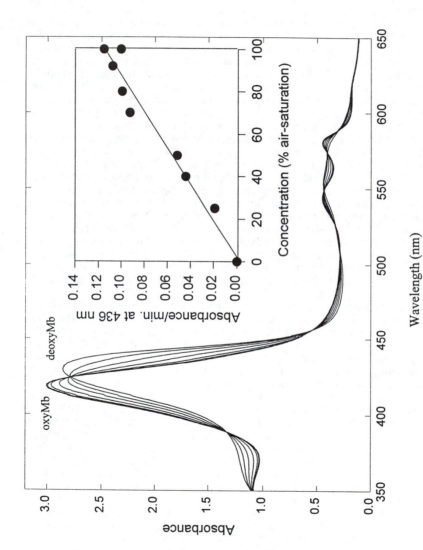

Figure 1. Optical absorption changes shown by the deoxy myoglobin containing gels upon treatment with dissolved oxygen. Inset shows the correlation between rate of absorption change at 436 nm with concentration of dissolved oxygen.

transmission and immune regulation. Because of the increasing awareness of the biological relevance of NO, there is considerable interest in developing an appropriate sensor. Although both Fe containing Hb and Mb, can bind NO, under aerobic conditions the reaction is non-specific and competitive O_2-adduct formation also takes place. In order to avoid interference from atmospheric oxygen, Mn-containing Mb was chosen as sensor element, due to its specificity for NO. The generation of NO was achieved via conversion of L-arginine to citruline catalyzed using the enzyme NO synthase. As confirmed by the characteristic absorption spectrum of MnMbNO, the biorecognition of *in situ* generated NO was demonstrated with gel entrapped MnMb. The important diagnostic features of this novel optical sensor are its ability to bind specifically to NO in an aerobic environment and to detect levels of NO generated under physiological conditions.

Biorecognition of Glucose. The monitoring of blood glucose levels has many clinical applications. The biosensors for detection of glucose utilize the enzyme glucose oxidase (GOx), which catalyzes air oxidation of β-D-glucose to give gluconic acid and hydrogen peroxide. Using the bio-gel methodology, a glucose sensor was developed using immobilized GOx in combination with horseradish peroxidase (HRP) and suitable dye precursors, to take advantage of the catalysis induced by peroxidase between the dye precursors and hydrogen peroxide. The dye precursors used were 4-aminoantipyrine and p-hydroxybenzene sulfonate which formed a quinoneimine dye with an absorption maximum at 510 nm. On exposure of gel monoliths containing the simultaneously immobilized GOx and HRP to glucose, the GOx catalyzed generation of hydrogen peroxide, in turn, triggers the second reaction i.e. the formation of dye from its precursors catalyzed by HRP. The rate of formation of colored dye is directly proportional to the concentration of the glucose present. By monitoring the absorption maximum of the dye generated, the amount of glucose in the surrounding solution can be determined. This method of biosensing takes advantage of the optical properties of the bio-gels, in contrast to the amperometric biosensors which correlate dioxygen uptake with glucose concentration.

Biorecognition of Oxalate. A similar detection approach was also used in the biorecognition of oxalate via immobilized oxalate oxidase/peroxidase and dye precursors. The peroxide generated by oxidation of oxalate to carbon dioxide reacts with the dye precursors MBTH and DMAB in a peroxidase catalyzed reaction to yield an indamine dye with an absorption maximum at 590 nm. The concentration of oxalate was then correlated with increase in absorption due to the dye. The sensitivity of the bio-gel to concentrations of the oxalate species was comparable to that in aqueous solutions as shown by the values of the Michaelis constant (K_m for the gel was 4.1×10^{-4} M as compared to 1.1×10^{-4} M for the solution). The lower detection limits that were studied were of the order of 10 μM levels.

Biorecognition of Glucose-6-phosphate. The biomaterial incorporating G6PDH along with immobilized $NADP^+$ cofactor is sensitive towards detection of G6P in solution. Upon immersion of the bio-gel monoliths in solution containing G6P, the enzyme-catalyzed dehydrogenation of the substrate generates the corresponding lactone. During this process the $NADP^+$ cofactor acts as an electron-sink and is reduced to NADPH. While the $NADP^+$ cofactor does not fluoresce in the visible, its reduced form is a strong emitter in the ~500 nm region. The increased fluorescence due to the formation of NADPH, initiated by the G6P substrate provides a very convenient and extremely sensitive detection methodology for detection of G6P. A relative preponderance of NADPH (or NADH) dependent dehydrogenases in the biological world suggests that a wide array of substrate recognition may be possible using sol-gel based biomaterials incorporating suitable enzymes.

Conclusions.

The collective results from these studies help to establish several important conclusions: 1) biological macromolecules, in general, are stable within sol-gel derived silica glasses, 2) the dopant biomolecules act as templates for the formation of self-specific pores around them, 3) due to the porous nature of the matrix, external substrates are able to interact with confined proteins, 4) optical transparency of these glasses allows interactions of photons with encapsulated biomolecules and permits optical monitoring of molecular events related to structure and reactivity, and 5) biorecognition based on proteins and enzymes encapsulated in sol-gel glasses is feasible.

The nanostructured biomaterials discussed in this paper are promising for mimicking biocatalysis, molecular recognition, energy transduction, and information storage in the solid state. Additionally, such nanostructuring of biological macromolecules allows the naturally perfected molecular architecture of biomolecules to be applied within the context of a materials science based discipline. These studies will be beneficial towards the design of new biomaterials and their applications in fields such as bioanalysis, molecular electronic devices and other related disciplines. Although presently in incipient stages, it is tempting to consider the long-range implications of the sol-gel based biomolecular devices in areas such as molecular optics, electronics, and photonics where such devices in various forms of signal storage and transmission can be utilized.

Biological cellular organization is nanostructured. The unique state of aggregation of biological cellular components is characterized by packed molecular assemblies interspersed in aqueous phase to achieve an organized nanostructured architecture. The cellular biomembranes enforce such organization. The nanostructured environment available to biomolecules in the pores of silica gels, to some extent, exemplifies the short-range structured influence of to an extrinsic organizing medium. Further studies on the effects of nanoconfinement upon the structure and reactivity of various systems should improve our understanding of different processes that are mediated by biological molecules arranged in naturally organized cellular components.

Acknowledgments. This research was supported by the National Science Foundation (DMR-9202182). We also greatly appreciate the contributions of K. E. Chung and S. Yamanaka.

Literature Cited.

1. Gleiter H. *Adv. Mater.* **1922**, *4*, 474.
2. Siegel, R. W. *Nanostruct. Mater.* **1993**, *3*, 1.
3. Ellerby, L. M.; Nishida, C. R.; Nishida, F.; Yamanaka, S. A.; Dunn, B. S.; Valentine, J. S.; Zink, J. I. *Science* **1992**, *255*, 1113.
4. Braun, S.; Rappoport, S.; Zusman, R.; Avnir, D.; Ottolenghi, M. *Mater. Lett.* **1990**, *10*, 1.
5. L. L. Hench and J. K. West, *Chem. Rev.*, **90** (1990) 33.
6. Brinker C. J.; Scherer, G. *Sol-Gel Science: The Physics and Chemistry of Sol-Gel Processing*; Academic Press, San Diego, 1989.
7. Yamanaka, S. A.; Nishida, F.; Ellerby, L. M.; Nishida, C. R.; Dunn, B.; Valentine, J. S.; Zink, J. I. *Chem. Mater.* **1992**, *4*, 495.

8. Wu, G. S.; Ellerby, L. M.; Cohan, J. S.; Dunn, B.; El Sayed, M. A.; Valentine, J. S.; Zink, J. I. *Chem. Mater.* **1993**, *5*, 115.
9. Zink, J. I.; Valentine, J. S; Dunn, B. *New J. Chem.* **1994**, 18, 1109.
10. Narang U.; Prasad, P. N.; Bright, F. V.; Kumar A. *Chem. Mater.* **1994**, *6*, 1596.
11. Avnir, D.; Braun, S.; Lev, O.; Ottolenghi, M. *Chem. Mater.* **1994**, *6*, 1605.
12. Dunn, B.; Zink, J. I. *J. Mater. Chem.* **1991**, *1*, 903.
13. Salemme, F. R. *Ann. Rev. Biochem.* **1977**, *46*, 299.
14. Grant, E. H.; Sheppard, R. J.; South G. P. *Dielectric Behaviour of Biological Molecules in Solution*; Oxford University Press, Oxford, 1978.
15. Fersht, A. *Enzyme Structure and Mechanism*; W. H. Freeman, New York, 1985.
16. Hall, E. A. H. *Biosensors*; Prentice Hall: Englewood, NJ, 1991.
17. Dave, B. C.; Dunn, B.; Valentine, J. S.; Zink, J. I. *Anal. Chem.* **1994**, *66*, 1120A.
18. Chung, K. E.; Lan, E. H.; Davidson, M. H.; Dunn, B. S.; Valentine, J. S.; Zink, J. I. *Anal. Chem.* **1995**, *67*, 1505.

RECEIVED December 27, 1995

Chapter 25

Sol–Gel Materials with Controlled Nanoheterogeneity

Ulrich Schubert, Fritz Schwertfeger, and Claus Görsmann

Institute of Inorganic Chemistry, Technical University Vienna, Getreidemarkt 9, A–1060 Vienna, Austria

Diphasic materials containing a nanometer-sized phase of one component in an oxide phase were obtained by controlled thermal treatment of xerogels in which the precursor for the nanophase is highly dispersed. The high dispersion was achieved by sol-gel processing of organically modified alkoxides. Thus, very thin carbonaceous layers partially covering the inner surface of silica aerogels developed during pyrolysis of organic groups located at the surface of the primary particles. This lead to a very efficient infrared opacification. Metal/ceramic nanocomposites were prepared by complexation of metal ions and tethering the resulting metal complexes to the oxide matrix during sol-gel processing. Subsequent thermolysis in air resulted in nano-sized metal oxide particles, which were then reduced by hydrogen to give highly dispersed metal particles.

One of the advantages of preparing oxide materials by the sol-gel method (*1*) is the possibility to control their microstructure and homogeneity. For most applications very homogeneous materials are desired. However, R. Roy already pointed out in the early eighties that the advantages of sol-gel processing can also be exploited for the preparation of di- or multiphase ceramic materials (*2*).

Our approach for getting a nanometer-sized phase **B** in the oxide phase **A** is a high, ideally molecular dispersion of the phase **B** precursor while the inorganic network of **A** is formed during sol-gel processing. This is achieved by chemically binding the phase **B** precursor to the network of **A** using organically (or organo-functionally) substituted alkoxides (*3*). The nanophase **B** is then obtained by controlled thermal treatment in a later step (Scheme 1). Two examples for this approach will be discussed: (i) Carbonaceous structures (phase **B**) in SiO_2 aerogels (phase **A**), and (ii) highly dispersed metal or alloy particles (phase **B**) in SiO_2 or TiO_2 (phase **A**).

Carbonaceous Structures in SiO_2 Aerogels

Silica aerogels (*4*) have several interesting applications, one of them being thermal insulation materials. Due to their low density and small pore radii, the heat transport via the solid aerogel skeleton and the gas phase is low. The radiative

0097–6156/96/0622–0366$15.00/0

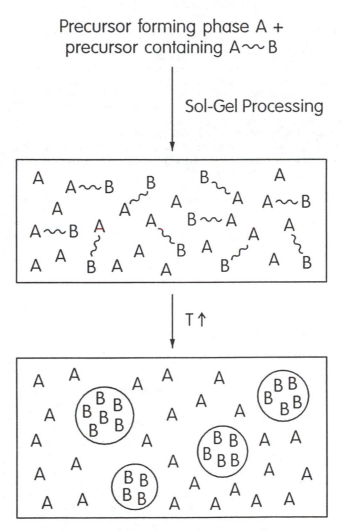

Scheme 1. Preparation of a nanometer-sized phase **B** in the oxide phase **A** by sol-gel processing, followed by controlled thermal treatment.

transport below 20°C is also low, because silica aerogels absorb heat sufficiently. However, use of silica aerogel for heat insulation at medium temperatures (50-500°C) requires reduction of the radiative heat transport. The reason is that the radiation maximum at these temperatures is at 2-8 μm, where silica has a low specific extinction (1 $m^2 kg^{-1}$). Carbon black is very well suited for infrared opacification due to its broad absorption band in the relevant range.

Mixing aerogel powders with 20% carbon soot increases the specific extinction in the range of 2-8 μm to a sufficient value of 100-200 $m^2 kg^{-1}$ (specific thermal conductivity 0.018-0.020 $Wm^{-1}K^{-1}$). However, one disadvantage of this approach is the handling of powders. Furthermore, it is not possible to further increase the specific extinction by adding a higher portion of soot. Specific extinction depends not only on the quantity of added carbon but also on its structure, size and agglomeration. For larger (agglomerated) particles, which are more easily formed with a higher soot portion, extinction decreases again (5). Another possibility to achieve infrared opacification is the addition of soot already during sol-gel processing. This approach has the same limitations concerning extinction. Additional problems arise from sedimentation of the soot particles and their influence on the formation of the gel nanostructure.

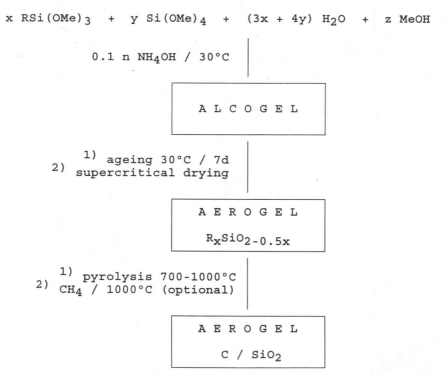

$$x \; RSi(OMe)_3 \quad + \quad y \; Si(OMe)_4 \quad + \quad (3x + 4y) \; H_2O \quad + \quad z \; MeOH$$

0.1 n NH_4OH / 30°C

ALCOGEL

1) ageing 30°C / 7d
2) supercritical drying

AEROGEL

$R_x SiO_{2-0.5x}$

1) pyrolysis 700-1000°C
2) CH_4 / 1000°C (optional)

AEROGEL

C / SiO_2

Scheme 2. Preparation of carbon structures in silica aerogels

Preparation of the Aerogel Composites. A new approach free of these problems, is to generate carbonaceous structures in silica aerogels either by pyrolysis of organically modified aerogels (Scheme 2) or by pyrolysis of gaseous organic compounds (6-8).

The structure of a silica aerogel prepared under basic conditions is shown in Figure 1 (*4*). It consists of secondary particles of about 5 nm diameter composed of smaller primary particles. Residual Si-OH and Si-OR groups are at the surface of the primary particles.

We have prepared organically modified silica aerogels by NH_4OH-catalyzed hydrolysis and condensation of $RSi(OMe)_3$ / $Si(OMe)_4$ mixtures (R = alkyl, aryl or functional organic groups), followed by supercritical drying of the alcogels with methanol or CO_2 (*9*). Incorporation of up to 20% $RSiO_{3/2}$ units into the aerogels did not disturb their micro- and nanostructure significantly, and therefore the typical aerogel properties were retained. In the organically modified aerogels the organic groups R are mainly located at the surface of the primary particles (Figure 1) and replace the Si-OH and Si-OR groups (*9*).

The inner aerogel surface should be uniformly covered by carbon for an efficient infrared opacification. The organically modified silica aerogels are ideal starting compounds for several reasons:

- The organic groups are already located at the inner surface. Therefore, there are no diffusion problems.
- The equal distribution of the organic groups on the inner surface results in a great number of well distributed nucleation centers during pyrolysis and thus leads to small carbon particles.

Optimization of the Pyrolysis Process. The pyrolysis process was carried out by heating the organically modified silica aerogels with an optimized heating rate to 700-1000°C in a standing atmosphere of argon. Passing argon over the samples during heating resulted in a considerably smaller amount of elemental carbon after pyrolysis. This shows that the gaseous compounds formed during pyrolysis play a very important role for the growth of the carbon structures.

Optimization of the pyrolysis process of organically substituted silica aerogels $R_xSiO_{2-0.5x}$ showed the following trends (*6*). Numerical values for some examples are given in Table I.

- With an increasing number of carbon atoms in the organic groups R of the starting aerogel the *carbon content* in the pyrolyzed aerogels increased. In the series of aerogels having the starting composition $R_{0.2}SiO_{1.9}$ (20% of the silicon atoms being substituted by an organic group) and densities of 230-290 kgm^{-3} the carbon contents after pyrolysis increased from 2.6% for R = CH_3 to 9.0% for R = C_3H_7. It further increased to 13.9% for R = C_6H_5, due to the larger number of carbon atoms and the aromatic nature of this group.
- The larger the organic group R and the higher the portion of organic groups in the starting aerogel (x in Scheme 2), the higher was the portion of *retained carbon*.
- Decreasing the density of the aerogels resulted in a larger loss of carbon during pyrolysis. This is understandable, since the porosity of the aerogels increases with decreasing density. The gaseous hydrocarbons formed upon hydrolysis therefore can easier diffuse out of the material.
- Once the organic groups are pyrolyzed, there was no further loss of carbon. Holding selected samples at 1000°C up to 10h did not result in a detectable decrease of the carbon content.
- The *temperature* necessary to pyrolyze all organic groups was lower with decreasing density of the aerogels. The phenyl or propyl-substituted aerogels with densities smaller than 150 kgm^{-3} were completely pyrolyzed at 700°C. The aerogels with starting densities between 200 and 300 kgm^{-3} required heating to 1000°C.

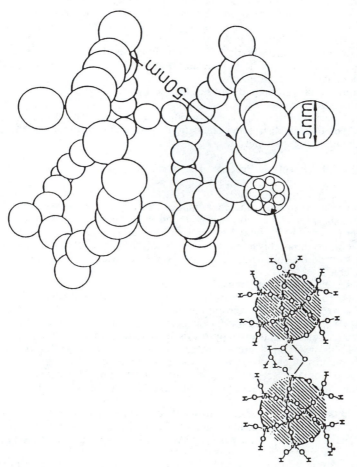

Figure 1. Structure of a silica aerogel prepared under basic conditions. In the organically modified aerogels $R_xSiO_{2-0.5x}$ ($x \leq 0.2$; R = terminal alkyl, aryl) the Si-OH groups are replaced by Si-R groups.

Table I. Carbon content (by elemental analysis) of selected C/SiO_2 aerogels, and density and surface area changes during pyrolysis.

starting composition a)	starting density [kgm^{-3}]	final density b) [kgm^{-3}]	final carbon content (wt. %)	retained carbon	final surface area c) [$m^2 g^{-1}$]
$Me_{0.2}SiO_{1.9}$	237	276 (8)	2.59 d)	33	444 (22)
$Me_{0.4}SiO_{1.8}$	208	239 (14)	6.18 d)	63	394 (35)
$Pr_{0.2}SiO_{1.9}$	230	274 (19)	18.98 d)	85	331 (35)
	134		15.09 e)	44	
$Ph_{0.2}SiO_{1.9}$	288	307 (10)	13.92 d)	72	320 (36)
	151		12.48 e)	69	
	87		11.15 e)	61	

a) Me = methyl, Vi = vinyl, Pr = n-propyl, Ph = phenyl. b) The values in parentheses are the percentual density increase relative to the starting density. c) The values in parentheses are the percent decrease of the surface area. d) Pyrolysis complete at 1000°C. e) Pyrolysis complete at 700°C.

In conclusion, to achieve a *high carbon content* in the pyrolyzed aerogels and to *retain a large percentage of carbon* during pyrolysis, aerogels substituted with large organic groups, preferably aromatic groups, should be used. The alternative possibility of increasing the portion of the organically substituted silicon atoms in the starting aerogels (x in Scheme 2) beyond 20% is less suitable, because it cannot be raised as high as necessary without destabilizing the aerogel structure.

Structure and Properties of the Composites. A density increase of 10-20% was observed during pyrolysis of the aerogels, and the specific surface area dropped by 20-35% (Table I) due to sintering processes. There was no obvious dependency of the density increase on the nature of the organic group. However, higher portions of organic groups in the starting aerogel and lower starting densities clearly promoted the densification process. When the aerogels were held at 1000°C after pyrolysis, the density continuously decreased. The starting aerogels are hydrophobic owing to the organic groups covering the inner surface. During pyrolysis the aerogels became hydrophilic (i.e. the aerogel surface became polar), because the organic groups are destroyed.

Despite these changes, the overall result of the structural investigations is that the typical aerogel nanostructure is basically retained during degradation of the organic groups and formation of the carbonaceous structures. Transmission electron micrographs showed no difference between the unpyrolyzed and the pyrolyzed aerogels. In both cases there were secondary particles in the range of 5-10 nm and pores of 20-50 nm. Since no differently shaped new particles were detected, the carbon must either have the same morphology as the silica particles or it is highly dispersed as a sheath over the silica skeleton. The latter possibility is more likely. It is supported by the small angle X-ray scattering (SAXS) data of the pyrolyzed aerogels (8). Figure 2 shows the development of the scattering intensity during heating. The slope of the curves increases upon heating from $D_s \approx 3.4$ to 3.8, indication a smoothing of the inner surface due to removal of the organic groups. The size of both the primary and secondary particles remains constant during pyrolysis or at holding the temperature at 1000°C. The development of interference structures at large scattering vectors ($q > 0.12$ Å^{-1}) starting at about 900°C may be a hint for the formation of new structures in the sub-nanometer range. However, these effects are at the lower limit of this method.

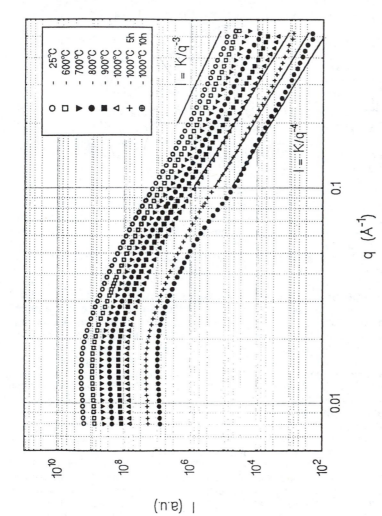

Figure 2. SAXS intensity vs. scattering vector for an aerogel with the initial composition $Ph_{0.2}SiO_{1.9}$ (starting density 288 kgm^{-3}) depending on the pyrolysis temperature and holding period at 1000°C. Reproduced by permission of Elsevier Sci. Publ. from Ref. 8.

A Raman spectrum of the same pyrolyzed aerogel (initial composition $Ph_{0.2}SiO_{1.9}$) showed two bands at 1596 and 1292 cm^{-1}. These bands are characteristic for crystalline and amorphous carbon (*10*), which are present in equal amounts. Holding the pyrolyzed aerogels at 1000°C did not change the intensity ratio and thus the carbonaceous structures. XPS only showed the lines of graphitic carbon, and the only peak in the ^{29}Si-MAS-NMR spectrum was that of SiO_2, i.e. there were no Si-C containing groups.

We conclude from these observations that the carbon structures generated during pyrolysis cover part of the silica nanospheres, from which the aerogel skeleton is composed (Figure 1). A complete and even coverage of the inner surface by carbon cannot be expected, because part of the inner surface is not accessible and preferred growth around carbon nucleation sites is probable. Furthermore, an even distribution of carbon would lead to a very thin layer (at the obtained carbon content) which could not form graphitic structures. We therefore assume that only part of the inner surface (10-15%) is covered by a carbon layer of 1-2 nm thickness.

This structural model also explains the very efficient infrared opacification (*7*). Due to the arrangement of the carbon structures generated during pyrolysis, a relatively small amount of carbon results in an efficient extinction. Thus, a specific extinction of 73 m^2kg^{-1} at 5 μm was obtained for the aerogel with the starting composition $Ph_{0.2}SiO_{1.9}$ and the starting density 288 kgm^{-3} (Table I and Figure 2) containing 13.9 wt.% carbon after pyrolysis. The extinction was even higher (180 m^2kg^{-1} at 5 μm) for the aerogel with the same starting composition but a lower starting density (87 kgm^{-3}, see Table I).

Theoretical calculations have reproduced the thermal conductivity behaviour of the C/SiO_2 aerogels by assuming a partial coverage of the silica nanospheres. Silica spheres covered with a carbon layer of a few nanometer thickness provide the same very high mass specific extinction as solid carbon spheres (*5,7*).

Pyrolysis of Hydrocarbons in the Aerogels. An even more effective infrared opacification is only possible by increasing the carbon content. However, incorporation of a significantly higher percentage of organic groups in the starting aerogels would affect their nanostructure. Therefore, the only possibility to increase the carbon content is from an external source. Our notion was that after pyrolysis of the organically substituted aerogels there would be enough carbon nuclei to facilitate pyrolysis of a hydrocarbon gas.

As a matter of fact, thermal conductivity was increased by a factor of 5-10 by treating the pyrolyzed aerogels with a CH_4/Ar mixture at 1000°C (*6*). The two aerogels with different densities discussed above, obtained by pyrolysis of $Ph_{0.2}SiO_{1.9}$, may serve as examples. The carbon content of the aerogel with the higher density was increased from 13.9 to 33.2 wt.% carbon by treatment with CH_4/Ar at 1000°C for 2h. During this treatment 61% of the CH_4 carbon was incorporated. Due to the higher carbon content, the specific extinction increased from 73 m^2kg^{-1} at 5 μm to about 1000 m^2kg^{-1}! The same treatment for the aerogel with the lower density resulted in an increase of the carbon content from 11.2 to 44.6 wt.% (75% of the CH_4 carbon was incorporated) and of the specific extinction from 180 to 1300 m^2kg^{-1}.

When the CH_4 treatment is continued, the specific extinction goes through a maximum value after 2 h, because the carbon structures get too large. Extinction is then lowered by scattering. After about 5 h carbon can no longer be integrated *into* the samples, because graphitic structures then cover and seal the outer surface of the samples.

Pyrolysis of CH_4 is obviously facilitated by the presence of carbon nuclei. We wanted to test whether the very active surface of silica aerogels alone is sufficient to generate carbonaceous structures from methane. When a carbon-free silica aerogel (i.e. an aerogel not substituted by organic groups) was heated to 1000°C and then treated with CH_4/Ar, carbon structures also developed. However, the carbon incorporation within the first 2h was rather low (about 5%). In this stage carbon nuclei are obviously formed. Then the rate of carbon formation rapidly increased. The aerogel contained about 30% carbon after 5h and 45% carbon after 10h CH_4 treatment. The specific extinction of these C/SiO_2 composites was lower (400 m^2kg^{-1}) than that of the CH_4-treated aerogels with carbon nuclei from pyrolyzed organic groups. While our work was in progress, A.J.Hunt et al. also reported the preparation of carbon particles in silica aerogels by pyrolysis of organic gases (*11*).

The main advantage of using organically substituted aerogels therefore is that due to the equal distribution of the organic groups on the inner surface many well distributed nucleation centers are formed during pyrolysis without an extensive induction period, which lead to a homogeneous distribution of carbon at the inner surface of the aerogels.

Highly Dispersed Metal Particles in SiO_2 or TiO_2

Organofunctional alkoxides of the type $(RO)_nE-X-A$ in which the two reactive centers, the functional organic group A and $E(OR)_n$, are connected by a hydrolytically stable spacer X have found wide-spread applications, mainly for $E = Si$ (*3*). One of the many possible functions of A is that of binding metal ions (M^{m+}) to give the complexes $\{[(RO)_nE-X-A]_yM\}^{m+}$. Suitable groups A are, for example, NH_2, $NHCH_2CH_2NH_2$, CN or $CH(COMe)_2$. In alkoxysilane chemistry a variety of precursors of this type, particularly $(RO)_3Si(CH_2)_nA$, are commercially available or can be easily prepared.

For $E \neq Si$ only very few derivatives of the type $(RO)_nE-X-A$ are known. This is partly because the functional group A cannot be linked to the alkoxide moiety by a hydrocarbon group, due to the hydrolytic cleavage of most relevant E-C bonds. Therefore, the grouping E-X-A must have another chemical composition than for $E = Si$. We recently prepared precursors of this type by reaction of non-silicon alkoxides with diamino acids (*12*). For example, reaction of $Ti(OEt)_4$ with lysine provided the lysinate complex $H_2N(CH_2)_4CH(NH_2)COOTi(OEt)_3$. The X-ray structure analysis of the corresponding glycine derivative (which is a dimer containing two bridging OEt groups) showed that the α-amino group is coordinated to the metal (as in Figure 3). Bonding of the carboxylate ligand is thus stengthened by the chelate effect. The second amino group in the diamino carboxylate derivatives is available for the coordination of metal ions, as schematically shown in Figure 3.

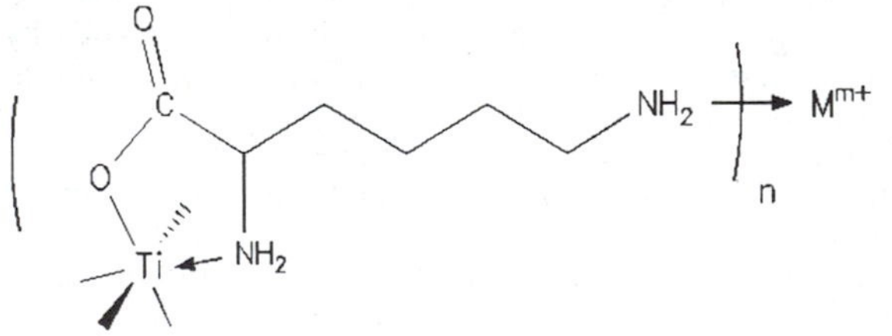

Figure 3. Coordination of metal ions by lysinate-modified titanium alkoxides.

The complexes formed by coordination of the organofunctional alkoxide derivatives $(RO)_nE-X-A$ are the same as those formed with the corresponding ligands not being substituted by a $(RO)_nE$ group. For example, when metal salts are reacted with $(RO)_3Si(CH_2)_3NHCH_2CH_2NH_2$ (DIAMO), the u.v. spectra are nearly identical with those of the corresponding ethylene diamine ($H_2NCH_2CH_2NH_2$) complexes. The same is true for $H_2N(CH_2)_4CH(NH_2)COO-Ti(OEt)_3$ (Figure 3), which results in the same type of complexes as NH_3.

The complexes formed by reaction of metal salts with the alkoxides $(RO)_nE$-X-A can be processed by the sol-gel method in the usual way. Since isolation of the complexes is either unnecessary or impractical, they are usually prepared in situ. Due to complexation of the metal ions and anchoring of the resulting metal complexes to the oxide matrix formed during sol-gel processing, aggregation of the metal ions is prevented. The metal-containing entities are homogenously distributed in the resulting gels.

Preparation of the Composites. Based on the thus achieved maximum dispersion of the metal precursor during the sol-gel step, we developed a general method of preparing nano-sized metal particles in oxide matrices (*13-15*). Contrary to conventional methods for the preparation of highly dispersed metals (*16*) on oxidic supports (impregnation, precipitation), this method allows the preparation of metal/ceramic composites with adjustable metal loadings. The inherent advantages of the sol-gel process can additionally be exploited, such as the tailoring of the microstructure of the oxide matrix, or the preparation of sols with a defined rheology suitable for coatings.

The metal/SiO_2 composites were prepared by the three-step procedure shown in Scheme 3.

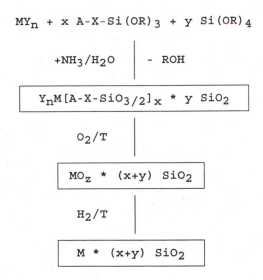

$$MY_n + x \ A\text{-}X\text{-}Si(OR)_3 + y \ Si(OR)_4$$

$$+NH_3/H_2O \quad \Big| \quad - ROH$$

$$\boxed{Y_nM[A\text{-}X\text{-}SiO_{3/2}]_x \ * \ y \ SiO_2}$$

$$O_2/T \quad \Big|$$

$$\boxed{MO_z \ * \ (x+y) \ SiO_2}$$

$$H_2/T \quad \Big|$$

$$\boxed{M \ * \ (x+y) \ SiO_2}$$

Scheme 3: Preparation of highly dispersed metals in a SiO_2 matrix by means of organofunctionally substituted alkoxysilanes (Y = counter ion).

Step 1: Sol-gel processing of a solution of the metal salt, $(RO)_3Si(CH_2)_3A$ and, optionally, $Si(OR)_4$ to facilitate formation of the oxide network. The ethylene diamine derivative $(RO)_3Si(CH_2)_3NHCH_2CH_2NH_2$ (DIAMO) proved to be very suitable, because it forms stable complexes with most transition metals. The precursor ratio and the reaction conditions predetermine important materials properties, such as composition, homogeneity of the metal distribution and porosity.

Step 2: Oxidation of the metal complex containing gels at high temperatures in air to remove all organic components. The initially employed metal salts have to contain counter-ions (e.g., carboxylates, ß-diketonates, nitrate) that can also be revoved during this step. The composites MO_z/SiO_2 are formed.

Step 3: Preparation of the metal/SiO_2 composites by reduction of the metal oxide particles.

This route provides metal/SiO_2 nanocomposite powders in which
- the metal particles have diameters of a *few nanometers*,
- the particle size distribution is very *narrow*, even when the metal loading is high,
- there is n*o agglomeration* of the metal particles,
- the metal particles are *statistically distributed* throughout the matrix,
- the *metal loading* is variable, up to 50 mol% metal.
- the *kind of metal* is easily varied.

The main advantage of this method is the narrow size distribution of the metal particles (*vide infra*). When composites are prepared by sol-gel processing without complexation of the metal ions and tethering the resulting metal complexes to the oxide matrix, the arithmetic average of the particle diameters may also be in the lower nanometer range, but the size distribution usually is very broad.

Another advantage of this approach is the possibility to impregnate porous solids with the metal complex containing sol. After drying, the oxidation and reduction step is then performed as for the preparation of composite powders. We prepared, for instance, small Pt particles on SiO_2 pellets by this method (*17*).

The procedure to prepare metal/SiO_2 nanocomposites by complexation of the metal ions and tethering the resulting metal complexes to the oxide matrix during sol-gel processing was recently extended to other systems:
- The preparation of highly dispersed *bimetallic particles* is a more complicated problem, because size distribution *and* composition distribution have to be controlled simultaneously. Preliminary studies on materials having the composition $Pd_xNi_{(1-x)} * 15 \, SiO_2$ and $Cu_{0.6}Ni_{0.4} * x \, SiO_2$ showed that the size distribution was as narrow as for monometallic particles prepared by the same method. However, the composition distribution was rather complicated (*15,18,19*). Although the two metals were intimately mixed during sol-gel processing, this was not sufficient to maintain close contact through the oxidation and reduction step. In the Cu/Ni system separate single-oxide particles were formed during the oxidation step due to their immiscibility.
- *Semiconducting Fe_2P or Ni_2P nanoclusters* in SiO_2 were prepared by using $Ph_2PCH_2CH_2Si(OR)_3$ for the initial coordination of the metals. However, the oxidation step (step 2) was omitted, and the metal complex-containing xerogels were immediately heated in the presence of H_2 (*20*).
- The method is also applicable to *other oxide matrices* than SiO_2. For example, to obtain small metal particles in TiO_2, the above discussed lysinate alkoxide derivative $H_2N(CH_2)_4CH(NH_2)COOTi(OEt)_3$ was used for coordinating the metal ions and tethering them to the titanate gel. The preparation of the metal/TiO_2 composites is then performed as that of the metal/SiO_2 composites (Scheme 3). When calcination in air is carried out at 450°C an anatase matrix is formed. Oxidation temperatures higher than 550°C result in the appearance of rutile. With copper as the transition metal, CuO particles were formed, as in the silicate system. Contrary to this, the titanates $MTiO_3$ are formed starting from cobalt or nickel acetate. Reduction with H_2 at 500°C resulted in highly dispersed metals, independent of the composition in the oxide stage.

Metal Particle Size, and Size Distribution. The average metal particle diameters in the metal/ceramic nanocomposites depend on
- the *kind of metal*. For example, very small particles (< 5 nm) were obtained for Cu $*$ SiO_2 or Pd * 2 SiO_2, medium sized (10 - 20 nm) for Ag * 4 SiO_2, Co * 6 TiO_2 or Ni $*$ 6 TiO_2, and larger ones (20 - 50 nm) for Ni * SiO_2, Cu * 4 TiO_2 or $Cu_{0.6}Ni_{0.4}$ * 5 SiO_2.

- the *metal loading*, in some cases. For example, while the average metal diameter in Pd * x SiO_2 hardly depends on x, it increases from 5.9 nm in Ni * 33 SiO_2 to 50.3 nm in Ni * SiO_2 (when the composites are prepared with DIAMO).
- the kind and molar ratio of the *complexing group* used in the first step of the synthesis.
- the *oxidation conditions* (step 2).

The latter two parameters have the biggest influence on the particle size and the size distribution in the final composites. If one strives for carbon-free composites (necessary for some, but not each application), the oxidation temperature has to be high enough to ensure complete oxidation of all organic components, but should not be higher than necessary to avoid excessive growth of the metal particles and broadening of the particle size distribution. The particle size does not change very much during reduction (step 3), because in most cases $T_{red} < T_{ox}$.

Depending on the metal (which in some cases catalyzes the combustion of the organic groups) and on the organic groups, typical optimized oxidation conditions are between 400 - 550°C for 1h. Figure 4 shows how the metal particle diameters, obtained from TEM measurements, in the composite PtO * 62 SiO_2 (prepared from $Pt(acac)_2$ + 2 DIAMO + 60 $Si(OEt)_4$) varied, if the parameters of step 1 (Scheme 3) were kept constant, and the oxidation conditions (step 2) were modified (*17*). The average size was calculated either by arithmetic averaging ($D_N = \Sigma n_i D_i / \Sigma n_i$; D_i = particle diameter, n_i = numbers of particles) or by weighting relative to the surface ($D_A = \Sigma 4\pi n_i (D_i/2)^2 D_i / \Sigma 4\pi n_i (D_i/2)^2$) or the volume ($D_V = \Sigma^4/_3 \pi n_i (D_i/2)^3 D_i / \Sigma^4/_3 \pi n_i (D_i/2)^3$) of the particles. If there are a few big particles, they may make up a high percentage of the overall mass of the dispersed metal. Therefore, a few big metal particles affect D_A and D_V more than D_N. The closer the value of D_N, D_A and D_V for a particular sample, the smaller is its particle size distribution.

The average particle diameters were in the range of 3.5 nm, and the size distribution is very narrow (D_{min} = 1.7 nm, D_{max} = 11.6 nm) if oxidation was carried out at 450 - 550°C. There was only little influence of the oxidation period. It should be noted that in this particular composite oxidation of the organic groups was only complete at 550°C, i.e. if the oxidation was carried out at lower temperatures, the resulting composites contained residual carbon-containing groups (less than 0.5%). An oxidation temperature of 750°C for 1 h resulted in both an increase of the particle size (D_N = 6.2 nm) and a broadening of the particle size distribution (D_{min} = 3.6 nm, D_{max} = 52.0 nm). These effects were even more pronounced, if the composite is held at 750°C for 10 h. Oxidation at 950°C resulted in a broad particle size distribution, the largest particles being nearly 2 μm in diameter. The extensive particle growth at high temperatures was not unexpected and reflects the usual sintering behaviour.

The kind of the complexing group used in the first step of the synthesis also has a pronounced influence on the resulting metal particle diameters. The mean particle diameter in Ni * 15 SiO_2 decreases from 22.9 nm to 2.6 nm (same reaction conditions during step 2 and 3), if the acetylacetone derivative $[CH_3(O)C]_2CH-(CH_2)_3Si(OR)_3$ is used for the complexation of the nickel ions during sol-gel processing instead of the ethylene diamine derivative DIAMO.

This shows that the particle growth during calcination is a rather complex process, in which not only the usual growth mechanisms (Ostwald ripening or cluster-cluster growth) are involved, but also mechanisms in which the organic pyrolysis products participate.

There should be no excess of the complexing group beyond the stoichiometric amount necessary for the coordination of the metal ions. The effect of an excess of the complexing group was shown for nanocomposites of the composition

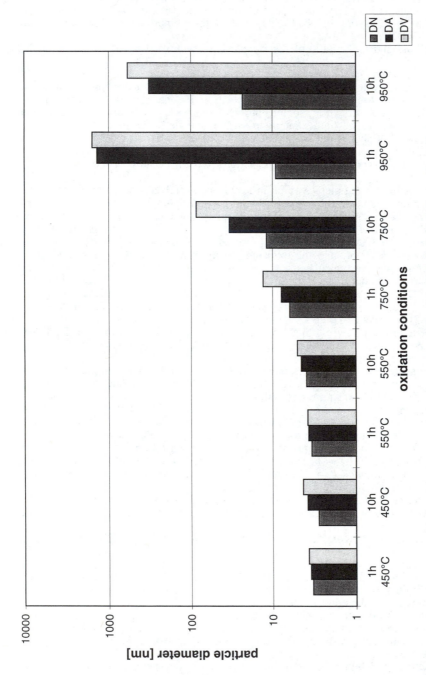

Figure 4. Development of the metal particle size of $PtO * 62\ SiO_2$, depending on the oxidation conditions (step 2 in Scheme 3). See text for the definition of D_N, D_A, and D_V (Note that the diagram is on a logarithmic scale).

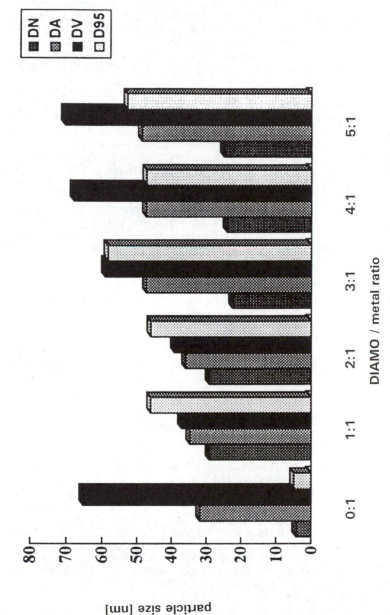

Figure 5. Size distributions of composites having the composition $Cu_{0.6}Ni_{0.4}$ *
5 SiO_2. See text for the definition of D_N, D_A, D_V, and D_{95}.

$Cu_{0.6}Ni_{0.4}$ * 5 SiO_2 prepared from nickel and copper acetate, DIAMO, and $Si(OEt)_4$ (*19*). The molar amount of (DIAMO + $Si(OEt)_4$) was kept constant to obtain a series of composites having the same composition. An increasing portion of DIAMO in the starting mixture resulted in larger oxide particles after the oxidation step. This may be due either to the different network structure caused by the lower degree of crosslinking or to the higher local temperatures caused by the exothermic oxidation of a greater amount of organic groups. The particle size was retained during reduction. Therefore, an increasing portion of DIAMO also resulted in larger alloy particles.

However, the complexing of the metal ions, has a beneficial effect on the particle size *distribution*, which was determined from TEM measurements. As seen from Figure 5, complexation by one or two DIAMO moieties per metal atom leads to particularly narrow size distributions. Without any complexation (DIAMO/ metal ratio 0:1), the majority of the metal particles are the smallest among all samples (95% of all particles have diameters below 5 nm [= D_{95} value]), and therefore D_N is very small. However, there are some very big particles ($D_{max} = 132$ nm) which leads to a high D_V. Therefore, the particle size distribution in the 0:1 sample is much broader than that of the 1:1 and 2:1 samples. Raising the DIAMO/metal ratio beyond 2, broader metal particle size distributions are again observed (increasing values of D_A and D_V). In conclusion, the narrowest size distributions (but not necessarily the smallest average particles sizes) are obtained, if only the amount of DIAMO needed for complexation of all metal ions (1 - 2 molar equivalents) is added to the starting solution.

Literature Cited
1.   Brinker, C. J.; Scherer, G. *Sol-Gel-Science, the Physics and Chemistry of Sol-Gel Processing*, Academic Press: London, 1990.
2.   Roy, R.A.; Roy, R. *Mat. Res. Bull.* **1984**, *19*, 169.
3.   Review article: Schubert, U.; Hüsing, N.; Lorenz, A. *Chem. Mater.* in press.
4.   Fricke J. *Aerogels* (Springer Proc.Phys. Vol.6, Heidelberg, 1986). Gesser, H. D.; Goswami, P. C. *Chem. Rev.* **1989**, *89*, 765; Fricke, J.; Emmerling A. *Struct. Bonding* **1992**, *77*, 37; Hench,L. L.; West J. K. *Chem. Rev.* **1990**, *90*, 33.
5.   Kuhn, J. *Physik in unserer Zeit* **1992**, *23*, 84.
6.   Schwertfeger, F.; Schubert, U. *Chem. Mater.* in press.
7.   Schwertfeger, F.; Kuhn, J.; Bock, V.; Arduini-Schuster, M.C.; Seyfried, E.; Schubert, U.; Fricke, J. in *Thermal Conductivity* **1994**, *22*, 589, Tong, T.W. (Ed.), Technomic Publ., Lancester.
8.   Kuhn, J.; Schwertfeger, F.; Arduini-Schuster, M.C.; Fricke, J.; Schubert, U. *J.Non-Cryst. Solids* **1995**, *186*, 184.
9.   Schwertfeger, F.; Glaubitt, W.; Schubert, U. *J. Non-Cryst. Solids* **1992**, *145*, 85. Schwertfeger, F. Hüsing, N.; Schubert, U. *J. Sol-Gel Sci. Technol.* **1994**, *2*, 103. Schwertfeger, F.; Emmerling, A.; Gross, J.; Schubert, U.; Fricke, J. in *Sol-Gel Processing and Applications*, Attia Y.A., Ed., Plenum Press, New York 1994, p.351. Hüsing, N.; Schwertfeger, F.; Tappert, W.; Schubert, U. *J.Non-Cryst.Solids*, **1995**, *186*, 37. Schubert, U.; Schwertfeger, F.; Hüsing, N.; Seyfried, E. *Mat. Res. Soc. Symp. Proc.* **1994**, *346*, 151. Hüsing, N.; Schubert, U. *J. Sol-Gel Sci. Technol.* in press.
10.   Tuinstra, F.; Koenig, J.I. *Chem. Phys.* **1970**, *53*, 1126.
11.   Hunt, A.J.; Cao, W. *Mat. Res. Soc. Symp. Proc.* **1994**, *346*, 451. Lee, D.; Stevens, P. C.; Zeng, S. Q.; Hunt, A. J. *J. Non-Cryst. Solids* 1995, *186, 285*.
12.   Schubert, U.; Tewinkel, S.; Möller, F. *Inorg. Chem.* **1995**, *34*, 995.
13.   Breitscheidel, B.; Zieder, J.; Schubert, U. *Chem. Mater.* **1991**, *3*, 559.

14. Schubert, U.; Breitscheidel, B.; Buhler, H.; Egger, C.; Urbaniak, W. *Mat. Res. Soc. Symp. Proc.* **1992**, *271*, 621.
15. Schubert, U.; Görsmann, C.; Tewinkel, S.; Kaiser, A.; Heinrich, T. *Mat. Res. Soc. Symp. Proc.* **1994**, *351*, 141.
16. Romanowski, W. *Highly Dispersed Metals*, Ellis Horwood Publ.: Chichester, 1987.
17. Görsmann, C. *Ph.D. Thesis*, University of Würzburg, 1995.
18. Mörke, W.; Lamber, R.; Schubert, U.; Breitscheidel, B. *Chem. Mater.* **1994**, *6*, 1659.
19. Kaiser, A.; Görsmann, C.; Schubert, U. *J. Sol-Gel Sci. Technol.* in press.
20. Burnam, K., J.; Carpenter, J. P.; Lukehart, C. M.; Milne, S. B.; Stock, S. R.; Glosser, R.; Jones, B. *Mat. Res. Soc. Symp. Proc.* **1994**, *351*, 21.

RECEIVED November 9, 1995

Chapter 26

Sol–Gel Processing of Silica–Poly(vinyl acetate) Nanocomposites

C. L. Beaudry and L. C. Klein[1]

Ceramics Department, Rutgers University, P.O. Box 909, Brett and Bowser Roads, Piscataway, NJ 08855–0909

Sol-gel processing of organic/inorganic composite materials has attracted much study in recent years. The low temperature sol-gel process and its inherent versatility allow microstructural control on the nanoscale, leading to many interesting materials. In this preliminary study, we have synthesized silica/poly(vinyl acetate) (SiO_2/PVAc) composite materials and investigated the effect of formamide substitutions for ethanol as the solvent. The interaction between SiO_2 and PVAc is primarily hydrogen bonding between silanols on the silica surface and the carbonyl groups of the PVAc. The time to gel, skeletal density, and the linear shrinkage were significantly dependent on the presence and quantity of formamide. Nitrogen sorption was used to evaluate the xerogels structures. The thermal behavior was studied with thermogravimetry/Fourier transform infrared spectroscopy (TG/FTIR).

Sol-gel processing has been used to prepare inorganic oxides by hydrolyzing alkoxides. In addition, sol-gel processing lends itself quite well to the synthesis of inorganic-organic hybrid materials. First, the low temperature aspect of sol-gel processing can be exploited to create many inorganic-organic composite materials unattainable with typical high temperature ceramic processing (*1*). Second, sol-gel processing can provide intimate mixing of chemically dissimilar materials at room temperature.

There are several ways in which organic materials may be incorporated into inorganic matrices. The strength of the interactions between the phases can divide these types of hybrid materials into two classes (*1*). Class I hybrid materials correspond to weak phase interactions such as van der Waals interactions, hydrogen bonding (*2*), or simple mechanical blending of the inorganic and organic phases (*3*). Class II hybrid materials possess strong covalent or iono-covalent bonds, usually via -Si-C- linkages (*1,4*), between the inorganic and organic phases. Class II hybrids are discussed first because they represent the larger group. Both of these groups contain some subdivisions, usually differing by way of preparation or specific function (*5*).

[1]Corresponding author

0097–6156/96/0622–0382$15.00/0

Covalently Bonded Organic/Inorganic Gels

In the absence of an organic constituent, a conventional silicon alkoxide sol-gel process results in a three dimensional network of inorganic polymer consisting of Si-O-Si units. In a linear organic polymer, the forces acting between repeating units in a polymer backbone are C-C covalent bonds, whereas those acting in bulk between polymer chains are weaker Van der Waals bonds. However, in a sol-gel process carried out with a precursor containing a direct inorganic/organic link, which is not subject to hydrolysis, it is possible to form homogeneous composite materials with the organic/inorganic link in tact. Undoubtedly, the largest group of hybrid silica gels are linked via a Si-C bond that is stable and does not undergo hydrolysis under the sol-gel processing conditions.

Inorganic/Organic Interpenetrating Networks (Class I Hybrids)

Interpenetrating networks are another way to combine organic and inorganic constituents. Interpenetrating networks are classified by a time element, i.e. sequential or simultaneous. The ability of silica sols based on tetraethylorthosilicate and tetramethylorthosilicate (TEOS,TMOS) to solvate some organic molecules enables their polymerization in situ, in the environment of organic polymer solutions. Of course the number of polymers which can form solutions with sol-gel formulations is limited, but the simplicity of the method encouraged several authors to use it as an efficient way to get interesting organic/inorganic composites.

Polymerization of silica precursor in the highly viscous environment of organic polymer solution forces silica particles to form around the polymer chains. From the point of view of chemical structure, these materials are semi-sequential inorganic/organic interpenetrating networks, by analogy to interpenetrating polymer networks (IPN's) (6). The term semi reflects the linear structure of the solvated polymer in which the crosslinked inorganic network is formed. As a result highly transparent, homogeneous, lightweight materials can be obtained. Of course, the ultimate properties are complex functions of the TEOS-polymer concentration, sol formulation chemistry, and length of the polymeric chain, as well as other parameters of the process (7,8).

As a general rule, regardless of the Tg of the starting polymer, the stiffening of the system is expected, resulting in the increase of Tg or its disappearance. Also, density does not follow the inorganic and organic increments and remains low even with high inorganic contents, showing the dominance of polymer-like behavior. What is interesting is that a porous structure (open or closed pores) is easy to design by the polymer concentration, as well as water to TEOS and to acid catalyst ratios.

It is assumed that the polymers which give transparent sol-gel hybrids when mixed with polymerizing TEOS are those capable of hydrogen bond formation with hydroxylated SiO_2 (2). Thus, strong interactions between the silanols, known to exhibit the character of Bronsted acids, and some specific groups of the polymer being hydrogen acceptors, are responsible for homogeniety and the high degree of two phase mixing. The hydrogen bonding is evident through the IR spectra, where the hydrogen acceptor group band is found to be broad and of low intensity, similarly to Si-OH absorption band.

Several linear polymers bearing hydrogen acceptor groups like amide, carbonyl, carbinol, were successfully used to produce transparent semi-sequential inorganic/ organic IPN's in the form of films or monoliths. These polymers were poly(N,N-dimethacrylamide), poly(vinyl pyrrolidone), poly(acrylic acid) and its copolymers, and poly(vinyl acetate) (2).

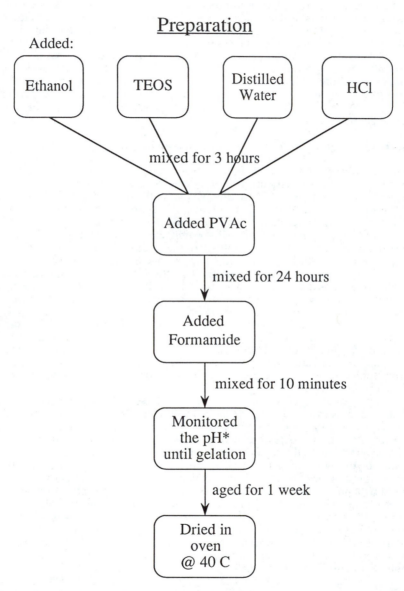

Figure 1. Flowchart for the preparation of SiO_2/PVAc composites by the sol-gel process.

Poly(vinyl acetate)

Poly(vinyl-acetate), the polymer chosen for this study, is functionally incapable of covalently bonding with the polymerizing silicon alkoxide, TEOS. Therefore, the SiO_2/PVAc composites may be considered class I hybrid materials. Reasons for interest in mixing PVAc with silica include its excellent optical transparency, a refractive index close to that of silica, and its solubility in ethanol-water mixtures. The solubility of PVAc in ethanol-water mixtures allows the polymer to be added directly to a typical alkoxide sol-gel formulation of alkoxide, alcohol, water, and catalyst, followed by the polymerization of the inorganic network around the organic polymer domains.
     Polyvinyl acetate,

$$-(CH_2-CH)_n-$$
$$|$$
$$O$$
$$|$$
$$C=O$$
$$|$$
$$CH_3$$

is known as a thermoplastic that is used in a variety of coating and adhesive applications (*9*). A high degree of mixing is expected from strong interactions through hydrogen bonding between the silanol groups (Si-OH) of the hydrolyzed alkoxide/silica surface and the carbonyl groups (C=O) of the PVAc.

Silica/PVAc Crack-free Monoliths

Two difficulties associated with sol-gel processing of large monolithic (crack-free) solids are [1] fracture during drying, caused by capillary stresses, and [2] long processing times. The addition of a more compliant phase, such as an organic polymer, may help to reduce stress on the oxide network during drying. Alternatively, chemical additives have been shown to reduce processing times and prevent drying fracture (*10*). Formamide, $HCONH_2$, one of the most studied so-called drying control chemical additives (DCCA's), has been chosen for this study.
     In this work, we prepared SiO_2/PVAc composite materials and investigated the effects of formamide substitutions for ethanol. The time to gel, skeletal density, and the linear shrinkage are reported and are strongly dependent on the presence of formamide. TG/FTIR was used to study the thermal behavior (*11*). The composite materials xerogel structure were evaluated with nitrogen sorption.

Experimental

All samples were prepared with the TEOS:H_2O:HCl molar ratio equal to 1:8:0.0897. The solvent medium (80 ml) was varied from pure ethanol to 50/50 volume% ethanol and formamide. Typically, 60 ml of TEOS, 38.6 ml of distilled water, and 2 ml of concentrated HCl were mixed with the ethanol fraction of the solvent for 3 hours. The ethanol was dehydrated, 200 proof, from Pharmco Products Inc. The TEOS was obtained from Dynasil A, Dynamit-Nobel and the HCl was reagent grade from Fisher Scientific. After mixing for three hours, PVAc (M.W. 83,000 from Aldrich) was added (15% by weight based on SiO_2 content of the TEOS) to the appropriate compositions and mixed for 24 hours. The compositions without PVAc were also mixed for 24 hours. The rest of the solvent, formamide, was added and mixed for ten minutes. A processing flowchart is shown in figure 1.
     After the final mixing, the sols were poured into polypropylene test tubes or polystyrene petri dishes and sealed to prevent solvent evaporation. The time from

Table I. Properties of Silica Gels and Silica/PVAc Composites

Sample	Solvent (vol%) Etoh/Form	t_{gel}	Skeletal Density (g/cm^3)	Linear Shrinkage (%)	Residual Weight @ 900°C (%)
OFRM	100/0	40 days†	2.00*	58.1	81
12.5FRM	87.5/12.5	50.5 hours††	1.65	54.4	59
25FRM	75/25	35.9 hours	1.46	46.2	52
37.5FRM	62.5/37.5	31.3 hours	1.38	40.6	34
50FRM	50/50	27.6 hours	1.33	35.2	26
OFRM-PVAc	100/0	40 days	1.60*	57.4	72
12.5FRM-PVAc	87.5/12.5	68.0 hours	1.60	52.0	53
25FRM-PVAc	75/25	51.9 hours	1.44	44.1	37.5
37.5FRM-PVAc	62.5/37.5	42.7 hours	1.36	38.5	30
50FRM-PVAc	50/50	38.0 hours	1.31	33.2	23.5

† +/- 0.5 days
†† +/- 0.5 hours
* Dried @ 150°C

sealing the test tubes to the time when, tilting the test tube, no fluidity was observed, was defined as the gelation time (t_{gel}). During gelation at room temperature ($\approx$21-22ºC) the solution pH* was monitored. pH* is used to indicate that the pH is measured in nonaqueous media. After gelation , the gels were aged at RT for one week, opened, and dried at 40ºC until constant mass was achieved.

The skeletal density was measured with helium pycnometry (Micromeritic Accupyc 1330) with 10 purges and the run precision software (with at least a 0.05% volume tolerance level).

Nitrogen isotherms were obtained with a Coulter Omnisorp 360 system. All samples were ground into a powder and degassed at 150ºC overnight. A fixed dosing (75 torr for each dose) method was used to acquire a full isotherms. The specific surface area was calculated by the BET method.

Thermogravimetry (TG) was performed on a Perkin-Elmer TGA 7 system. Infrared spectra of the evolved gases during the TG heating, were obtained with a Perkin-Elmer TG/FTIR system consisting of a TGA 7 analyzer coupled to an external optical bench of a Perkin-Elmer 1700-X FTIR interface. The samples were analyzed in bulk form, directly from the drying oven.

For all of the TG runs the experimental conditions were the same - a temperature range of 35 to 900ºC at a heating rate of 10ºC/min in air. The sample mass was kept between 24-27 mg and the balance and sample purges were set at 40 and 20 cm³/min. The experimental conditions used for the TG/FTIR analysis were: sample mass, 25-30 mg; heating rate, 40ºC/min; and purge gas flow, 40 cm³/min of air. The silica content of the composite material was assumed to be the mass of the sample at 900ºC, where the organic phase has been burned out. In the present study, samples will be designated by XFRM-Y; where X = volume% of formamide as the solvent and the presence of PVAc is indicated if substituted for Y; e.g. 12.5FRM-PVAc refers to a composition synthesized with 12.5 volume% formamide as the solvent and PVAc.

Results

After the drying step, most of the compositions were monolithic, transparent to translucent materials. Compositions containing PVAc, appeared increasingly translucent as the percentage of formamide was increased. Formamide has been shown to increase the mean pore size of silica gels (*12*), indicating that the increased translucency arises from the increase in the mean pore size, that contributes to scattering. It is also possible that the residual formamide, after drying, within the pores contributes to scattering.

In Table I, the gel times, the skeletal densities, the linear shrinkage, and the residual weight are listed for all compositions. The substitution of formamide for a fraction of the ethanol clearly reduces the time to gel. A substitution of 12.5 volume% of formamide for ethanol reduced the gelation time from 40 days to 1-2 days. Increasing the amount of formamide from 12.5 to 50 volume% further cut the gel time in half for compositions with and without PVAc. These results correlate to the observation that the pH* of the formamide containing sols increases with time to approximately 4.0 (Figure 2). The rise in pH* occurs faster in the compositions containing more formamide. Rosenberger et al. (*13*) explained this effect by concluding that the formamide undergoes hydrolysis producing formic acid and NH₃, leading to stepwise increases of the sol pH. It is not suprising to observe a reduced gelation time in this pH* range. If time to gel is plotted versus the starting sol pH*, there is a minimum around 4.0 (*14,15*), corresponding to maximum in the condensation rate. Essentially, formamide added to a an acid catalyzed sol (based on TEOS) allows efficient hydrolyis (initial acidic conditions) followed by efficient condensation (pH$\approx$4). Not only does the gel time decrease, reducing the processing

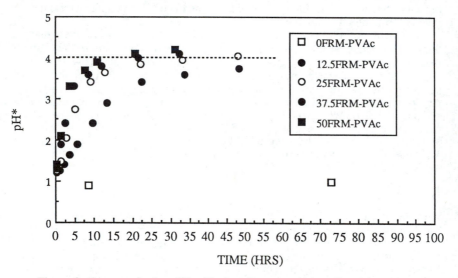

Figure 2. Increase in the pH* with time for compositions listed in table I.

time, the gels dry as crack-free monoliths (disks of diameter $\geq$5 cm and 4-5 mm thick). Compositions of similar dimensions without formamide fractured into many pieces. Smaller cylindrical samples of this composition did survive the drying stage, but the processing time from mixing the precursors to a dried gel was over two months. Thus, the addition of formamide faclitated the production of monolithic materials in a timely manner.

Table I also lists the skeletal densities of the dried gels measured by helium pycnometry. The results do not follow a simple weighted average of the respective density of amorphous SiO_2 (2.20 g/cm^3) and PVAc ($\approx$1.2 g/cm^3). The continous decrease in density with incremental additions of formamide suggests that formamide interferes with skeletal condensation. There is considerable formamide present in the dry gels as indicated by the thermogravimetry results. The rather low skeletal density values suggest interaction such as hydrogen bonding, between the phases. This supports the existence of hydrogen bonding between silanol groups on the silica gel surface and the carbonyl groups of formamide (*16*) and PVAc (*2*).

The dimensional changes during aging and drying were also dependent on the formamide content (table I) The linear shrinkage was calculated (for the cylindrical samples) by measuring the length of gel initially and the final length after drying. The decrease in the linear shrinkage was approximately a linear function of the formamide content. Compositions without formamide decreased in length by almost 60%, while compositions with a substitution of 50 volume% of formamide as the solvent only decreased in length by $\approx$35%. The addition of 15 weight% of PVAc contributed to a linear shrinkage reduction of $\approx$2% for all compositions. By increasing the PVAc content to 30% by weight for the 12.5 volume% formamide level, the linear shrinkage was reduced an additional 2%.

In order to relate the effect of formamide on the gelation process and the subsequent structure of the xerogels, nitrogen sorption was performed. The results are provided in table II. The addition of formamide to acid catalyzed silica and silica/PVAc gels result in mesoporous, rather than microporous solids. Increasing the substitution of formamide for ethanol increases the BET surface area and pore volume. On the other hand, the addition of PVAc resulted in decrease of the BET surface area and pore volume for equal levels of formamide.

Thermal analysis of the SiO_2/PVAc materials show the effect of the addition of PVAc (*11*). The TG curve (figure 3) reveals a three stage weight loss consisting of the lower temperature double reaction and a minor reaction at higher temperatures. The double reaction is better resolved by the corresponding derivative plot as shown. The higher temperature reaction, labeled B, is the decomposition of PVAc, while the lower temperature reaction is believed to be the removal/decomposition of the formamide with small amounts of water and ethanol (reaction A). In support of these conclusions are: (1) in all samples, the magnitude of the weight loss for reaction A increases as the amount of formamide increases, and (2) the weight loss for reaction B is not observed in the samples without PVAc. Furthermore, a 12.5FRM gel with double the amount of PVAc exhibited a significant increase in the weight loss for reaction B.

Discussion

Thermogravimetry was used to study the thermal behavior (see reference 11 for more details) of the gels. In all samples without PVAc, a distinct double stage weight loss was observed with a more gradual weight loss occurred at higher temperatures. The lower temperature weight losses were attributed to solvent removal, and the minor second stage, at higher temperatures, was due to oxidation of the residual alkyl groups at approximately 225 to 350°C (*17*). The small weight losses due to organic removal observed in the silica gels were not apparent in the composite gels containing PVAc. The overlap of reaction temperatures for the respective decompositions, alkyl

Table II. Nitrogen sorption data

Sample	Isotherm Type	BET S.A.[†] (m^2/g)	C	Vol_{ads}[††] (cc/g STP)	Pore Volume (cm^3/g)
0FRM	I (IV)	N/A	N/A	226.2	0.295 (0.346)
25FRM	IV	652	93	521.1	0.806
50FRM	IV	711	92	621.6	0.962
0FRM-PVac	I	N/A	N/A	154.9	0.231
25FRM-PVAc	IV	479	50	440.1	0.681
50FRM-PVAc	IV	555	50	536.6	0.830

† P/P_o 0.05 to 0.25 and correlation of 0.9999 or better
†† Volume adsorbed @ P/P_o = 0.95

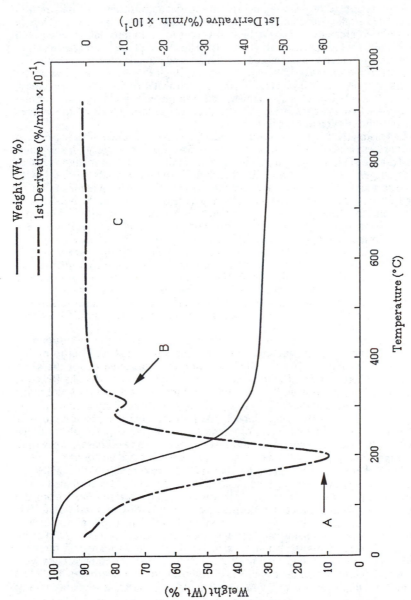

Figure 3. Thermogravimetric weight loss for 37.5FRM-PVAc with 3 regions highlighted: A- adsorbed, physically trapped species; B- PVAc decomposition, removal; C dehydroxylation of silica surface.

groups and PVAc, and the fact the PVAc is a major component of the composite, may have overshadowed the small weight losses due to decomposition and oxidation of unreacted alkyl groups.

The magnitude of the weight loss due to decomposition and oxidation is directly related to the concentration of unreacted alkyl groups in the gel. Thus, a fully hydrolyzed alkoxide will not exhibit a weight loss. In most alkoxide derived gels there will be some unreacted alkyl groups. The extent of the hydrolysis reaction is influenced by experimental conditions such as water to alkoxide molar ratio (R) (18) and pH of the sol (19). It is well known that high R values (R≥4) and acidic conditions promote fast hydrolysis. For the experimental conditions in this study, R=8 and sol pH≈1, hydrolysis is considered complete, explaining the minimal weight loss due to organic oxidation. The gradual weight loss at higher temperatures is due to dehydroxylation of the silica surface.

Increasing the amount of formamide as the solvent resulted in an increase in total weight loss (table I). When formamide was first used it was found that samples fractured during sintering, leading to the conclusion that there is formamide present in the dried gels (10). Additionally, residual formamide tends to react with atmospheric water causing the "dried" gels to "sweat" (20). This phenomenon was observed for all compositions containing formamide depending on the humidity of the particular day. The question is, how much formamide is actually present in a dried gel? Using TG/FTIR to analyze the evolved gases during heating, it was clear that the majority of the solvent weight loss was formamide. Futhermore, the magnitude of the weight loss in range of reaction A dramatically increased with increasing formamide. For a substitution of 12.5 volume% formamide approximately 35% of the initial mass has evolved while for a 50 volume% substitution more than 70% of the initial mass has been lost in this range. With respect to the skeletal density and linear shrinkage decreasing and residual formamide with increasing substitutions of formamide, it is obvious that formamide interferes with the skeletal condesation.

The addition of PVAc, has some interesting effects on the corresponding weight loss curve. The temperature range for reaction A is shifted out to higher temperatures. As more PVAc is added, the weight loss decreases for the same temperature indicating that the PVAc makes solvent removal more difficult. Reaction B is significantly changed with PVAc content. In fact, over a relatively small range of reaction temperature (≈290-375°C) it is possible to quantitatively account for the PVAc added to the system. This supports the fact that hydrogen bonding and/or mechanical interlocking are the forces of interaction acting between the silica surface and PVAc. If covalent bonds were to exist between the two phases, we would expect a higher decomposition temperature for the stronger covalent bonds.

The higher temperature reaction (labeled C), approximately ≈400 to 850°C, is more pronounced with increasing amounts of PVAc. Reaction C is attributed to the dehydroxylation of the silica surface. Hydroxyl groups are removed from the surface by a condensation reaction to form siloxane bonds (Si-O-Si) and water. It is possible that the amount of PVAc in the system will affect this phenomenon. Initially, the polymer domains may hinder residual silanol groups (SiOH) from condensing further. This is also suggested by the linear shrinkage differences between compositions with and without PVAc. PVAc containing compositions were consistently 2% longer than their non-PVAc counterparts. Furthermore, TG/FTIR results indicate that pyrolysis of PVAc is complete at ≈550°C, at a heating rate of 10°C/min., which may allow the condensation reaction to proceed, thus deferring the initial dehydroxylation reaction to higher temperatures.

Although formamide reduced processing time of monolithic solids, its use is far from ideal for synthesis of inorganic/organic material. Residual formamide is troublesome and difficult to remove, and may even cause fracture during heating (10). Futher complicating the removal of formamide is the high boiling point of formamide (210°C) with respect to the stability of the organic polymer. Obviously, we would like to synthesize crack-free inorganic/organic composites, in a timely manner, without the use of formamide.

Conclusions

The addition of formamide to acid catalyzed SiO_2 and SiO_2/PVAc gels resulted in optically transparent to translucent crack-free monoliths with relatively short processing times. Increasing the amount of formamide, decreased the gel time, reducing the processing time. The substitution of formamide for a fraction of ethanol had drastic effects not only on the gel time, but the linear shrinkage, skeletal densities, as well as their appearance. The interactions between the phases are strong enough that a range of composite materials, containing SiO_2, PVAc, and pores can be produced.

Acknowledgments The financial support of the Center for Ceramic Research, a New Jersey Commission on Science and Technology Center, is greatly appreciated. The authors would also like to thank Anna Wojcik and Seiji Yamazaki for their useful discussions.

Literature Cited:

1.   Sanchez, C.; Ribot, F. *New J. Chem.* **1994**, *18*, 1007.
2.   Landry, C. J.; Coltrain, B. K.; Wesson, J. A.; Zumbuladis, N.; Lippert, J. L. *Polymer* **1992**, *33*, 1496.
3.   Lin, H.; Day, D. E.; Stoffer, J. O. *J. Mater. Res.* **1993**, *8*, 363.
4.   Huang, H.; Orler, B.; Wilkes, G. L. *Polymer Bull.* **1985**, *14*, 557.
5.   Wojcik, A. B.; Klein, L. C. *Applied Organometallic Chemistry* **1995**, to appear.
6.   Sperling, L. H. *Interpenetrating Polymer Networks and Related Materials* Plenum Press: NY, **1981**.
7.   Wojcik, A. B.; Klein, L. C. *SPIE* **1993**, *2018,* 160-166.
8.   Yamazaki, S.; Klein, L. C. *J. of Non-Cryst. Solids* **1995**, submitted.
9.   "Polyvinyl Acetate", Union Carbide Chemicals and Plastics Co., Inc., Product Bulletin SC-1127, **1989**.
10.   Wallace, S.; Hench, L. L. In *Better Ceramics Through Chemistry*; Brinker, C. J.; Clark, D. E.; Ulrich, D. R., Eds.; Materials Reseach Society Symposia Proceedings Vol. 32; North Holland: NY, NY, 1984; 47-52.
11.   Beaudry, C.; Klein, L. C. *J. of Thermal Analysis* **1995**, to appear.
12.   Boonstra , A. H.; Bernards, T. N. M.;Smits, J. J. T. *J. of Non-Cryst. Solids* **1989**, *109*, 141.
13.   Rosenberger, H.; Bürger, H.; Schütz, H.; Scheler, G.; Maenz, G. *Zeitschrift für Physikalische Chemie Neue Folge* **1987**, *153*, 27.
14.   Klein, L. C.,; Woodman, R. In *Int. Symp. on Sol-Gel Science and Technology* , Pope, E. J. A.; Sakka, S.; Klein, L. C., Eds.; Ceramic Transactions vol. 55, American Ceramic Society, Westerville, OH, 1995, 105-116.
15.   Coltrain, B. K.; Melpolder, S. M.; Salva, J. M. *In Ultrastructure Processing of Advanced Materials*; Ulmann, D. R.; Ulrich, D. R., Eds.; Wiley & Sons, Inc., NY, 1992; pp. 69-76.
16.   Artaki, I.; Zerda, T. W.; Jonas, J. *J. of Non-Cryst. Solids* **1986**, *81*, 381.
17.   Gadalla, A. M.; Yun, S. J. *J. of Non-Cryst. Solids* **1992**, *143*, 121.
18.   Gottardi, V.; Guglielmi, M.; Bertoluzza, A.; Fagnano, C.; Morelli, M. A. *J. of Non-Cryst. Solids* **1984**, *63*, 71.
19.   Aelion, R.; Loebel, A.; Eirich, F. *J. Am. Chem. Soc.* **1950**, *72*, 5705.

20.   Hench, L. L. In *Better Ceramics Through Chemistry*; Brinker, C. J.; Clark,
       D. E.; Ulrich, D. R., Eds.; Materials Reseach Society Symposia Proceedings
       Vol. 32; North Holland: NY, NY, 1984; 101-110.

RECEIVED December 11, 1995

INDEXES

Author Index

Affiliation Index

Subject Index

X

Y

Z

Bestsellers from ACS Books

The ACS Style Guide: A Manual for Authors and Editors
Edited by Janet S. Dodd
264 pp; clothbound ISBN 0–8412–0917–0; paperback ISBN 0–8412–0943–X

Understanding Chemical Patents: A Guide for the Inventor
By John T. Maynard and Howard M. Peters
184 pp; clothbound ISBN 0–8412–1997–4; paperback ISBN 0–8412–1998–2

Chemical Activities (student and teacher editions)
By Christie L. Borgford and Lee R. Summerlin
330 pp; spiralbound ISBN 0–8412–1417–4; teacher ed. ISBN 0–8412–1416–6

Chemical Demonstrations: A Sourcebook for Teachers,
Volumes 1 and 2, Second Edition
Volume 1 by Lee R. Summerlin and James L. Ealy, Jr.;
Vol. 1, 198 pp; spiralbound ISBN 0–8412–1481–6;
Volume 2 by Lee R. Summerlin, Christie L. Borgford, and Julie B. Ealy
Vol. 2, 234 pp; spiralbound ISBN 0–8412–1535–9

Chemistry and Crime: From Sherlock Holmes to Today's Courtroom
Edited by Samuel M. Gerber
135 pp; clothbound ISBN 0–8412–0784–4; paperback ISBN 0–8412–0785–2

Writing the Laboratory Notebook
By Howard M. Kanare
145 pp; clothbound ISBN 0–8412–0906–5; paperback ISBN 0–8412–0933–2

Developing a Chemical Hygiene Plan
By Jay A. Young, Warren K. Kingsley, and George H. Wahl, Jr.
paperback ISBN 0–8412–1876–5

Introduction to Microwave Sample Preparation: Theory and Practice
Edited by H. M. Kingston and Lois B. Jassie
263 pp; clothbound ISBN 0–8412–1450–6

Principles of Environmental Sampling
Edited by Lawrence H. Keith
ACS Professional Reference Book; 458 pp;
clothbound ISBN 0–8412–1173–6; paperback ISBN 0–8412–1437–9

Biotechnology and Materials Science: Chemistry for the Future
Edited by Mary L. Good (Jacqueline K. Barton, Associate Editor)
135 pp; clothbound ISBN 0–8412–1472–7; paperback ISBN 0–8412–1473–5

For further information and a free catalog of ACS books, contact:
American Chemical Society
Customer Service & Sales
1155 16th Street, NW, Washington, DC 20036
Telephone 800–227–5558